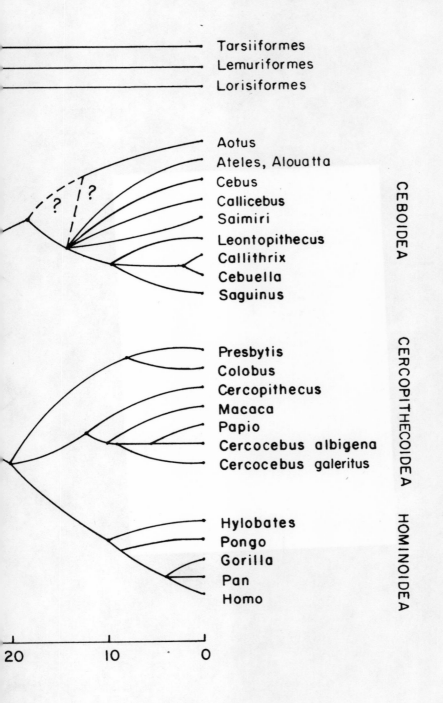

Tarsiiformes
Lemuriformes
Lorisiformes

Aotus
Ateles, Alouatta
Cebus
Callicebus
Saimiri
Leontopithecus
Callithrix
Cebuella
Saguinus

? ?

CEBOIDEA

Presbytis
Colobus
Cercopithecus
Macaca
Papio
Cercocebus albigena
Cercocebus galeritus

CERCOPITHECOIDEA

Hylobates
Pongo
Gorilla
Pan
Homo

HOMINOIDEA

20   10   0

PRIMATE COMMUNICATION

PRIMATE COMMUNICATION

EDITED BY
CHARLES T. SNOWDON, CHARLES H. BROWN,
AND
MICHAEL R. PETERSEN

CAMBRIDGE UNIVERSITY PRESS
Cambridge
London   New York   New Rochelle
Melbourne   Sydney

Published by the Press Syndicate of the University of Cambridge
The Pitt Building, Trumpington Street, Cambridge CB2 1RP
32 East 57th Street, New York, NY 10022, USA
296 Beaconsfield Parade, Middle Park, Melbourne 3206, Australia

First published 1982

Printed in the United States of America

Library of Congress Cataloging in Publication Data
Main entry under title:
Primate communication.
Papers presented at a symposium held July 5–6,
1980 in Parma, Italy.
Includes indexes.
1. Primates – Congresses.   2. Animal communication
– Congresses.   I. Snowdon, Charles T.   II. Brown,
Charles H., 1947–   .   III. Petersen, Michael R.
QL737.P9P6722   599.8'0459   82-1219
ISBN 0 521 24690 3   AACR2

The diagram reproduced as endpapers in this book is adapted
from V. M. Sarich and J. E. Cronin (1976) "Molecular systematics of the primates," which appeared in *Molecular Anthropology* (M. Goodman et al., eds.) published by Plenum Press, New
York.

Monkeys . . . very sensibly refrain from speech, lest they should be set
   to earn their livings.

Kenneth Grahame
The Golden Age: Lusisti Satis (1895)

Contents

Contributors

MARY CATHERINE ALVEARIO *Monell Chemical Senses Center, Philadelphia, Pennsylvania 19104*

CRAIG BIELERT *Primate Behavior Research Group, University of Witwatersrand, Johannesberg 2001, South Africa*

CHARLES H. BROWN *Psychology Department, University of Missouri, Columbia, Missouri 65211*

DOROTHY L. CHENEY *Department of Anthropology, University of California, Los Angeles, California 90024*

SUZANNE CHEVALIER-SKOLNIKOFF *Human Interaction Laboratory, Department of Psychiatry, University of California, San Francisco, California 94143*

BERTRAND L. DEPUTTE *Station Biologique de Paimpont, Université de Rennes, Plelan le Grand 35380, France*

GISELA EPPLE *Monell Chemical Senses Center, Philadelphia, Pennsylvania 19104*

JEAN-PIERRE GAUTIER *Station Biologique de Paimpont, Université de Rennes, Plelan le Grand 35380, France*

ANNIE GAUTIER-HION *Station Biologique de Paimpont, Université de Rennes, Plelan le Grand 35380, France*

DAVID A. GOLDFOOT *Wisconsin Regional Primate Research Center, Madison, Wisconsin 53706*

UWE JÜRGENS *Max-Planck-Institut für Psychiatrie, 8 Munich 40, West Germany*

YAIR KATZ *Monell Chemical Senses Center, Philadelphia, Pennsylvania 19104*

E. B. KEVERNE *Department of Anatomy. University of Cambridge, Cambridge, CB2 3D4, England*

JOHN D. NEWMAN *Laboratory of Developmental Neurobiology, National Institute of Child Health and Human Development, Bethesda, Maryland 20205*

MICHAEL R. PETERSON *Department of Psychology, Indiana University, Bloomington, Indiana 47401*

JOHN G. ROBINSON *Department of Zoology and Florida State Museum, University of Florida, Gainesville, Florida 32611*

ROBERT M. SEYFARTH *Department of Anthropology, University of California, Los Angeles, California 90024*

HARRIET J. SMITH *Laboratory of Developmental Neurobiology, National Institute of Child Health and Human Development, Bethesda, Maryland 20205*

CHARLES T. SNOWDON *Department of Psychology, University of Wisconsin, Madison, Wisconsin 53706*

DAVID SYMMES *Laboratory of Developmental Neurobiology, National Institute of Child Health and Human Development, Bethesda, Maryland 20205*

PETER M. WASER *Department of Biological Sciences, Purdue University, West Lafayette, Indiana 47907*

Preface

In the summer of 1979 we received a call for suggested symposia for the VIIIth Congress of the International Primatological Society to be held in Italy in 1980. We decided that it would be timely and appropriate to hold a symposium on primate communication. Several years had passed since the last major symposium on primate communication (at the meetings of the American Association for the Advancement of Science in Montreal in 1964), and each of us was aware that several major developments in the field had occurred since then.

We decided to organize a symposium that would reflect these new developments and outline new directions for future research. We selected participants to reflect a diversity of approaches (field, captive social groups, traditional laboratory studies), a diversity of species (from callitrichids through hominoids), and a diversity of modalities (vision, audition, olfaction, gustation). The symposium was held in Parma on July 5 and 6, 1980. Of the contributors to the present volume, Chevalier-Skolnikoff, Goldfoot, Newman and Symmes, Robinson, Seyfarth and Cheney, and Waser were unable to attend. However, draft papers from these contributors were included in discussions at the symposium.

Early drafts of all papers were circulated to those contributors whose work was related. Thus, the final chapters presented here reflect their influences and critiques.

We are grateful to the organizer of the VIIIth Congress of the International Primatological Society, Professor B. Chiarelli, for the opportunity to organize the symposium. We are especially grateful to our hosts at Parma, Professors Danilo and Marisa Mainardi, for providing us excellent meeting facilities and superb hospitality.

We thank Alexandra Hodun for her editorial assistance and Marie Kestol for her excellent services in typing. We also thank Ann Ivins, Susan Milmoe, Beverly Smith, and Toby Muller of Cambridge University Press. Finally, we thank all the contributors, who were cheerful in accepting our editorial comments and generally prompt in returning revised drafts. It is the totality of work presented here that makes primate communication an exciting and satisfying research field for all of us.

C. T. S.
C. H. B.
M. R. P.

General introduction

CHARLES T. SNOWDON, CHARLES H. BROWN,
and MICHAEL R. PETERSEN

More than 15 years have passed since the appearance of the last major review of research on primate communication – Stuart Altmann's *Social Communication Among Primates* (1967). At that time, the study of primate communication was in its infancy; techniques for the analysis of communication were just developing and were not widely applied to primate species. For example, the first studies using spectrographic analyses of primate vocalizations had appeared only five years earlier (Rowell and Hinde 1962; Andrew 1963). Of the 17 papers in the Altmann volume only five dealt with the actual structure and content of communication signals. The remaining papers dealt with reproductive behavior, agonistic behavior, and social dynamics without reference to the specific signals mediating these behaviors.

In the ensuing 15 years techniques for analyzing communication signals have developed rapidly. We can now provide physical descriptions of most auditory signals, some olfactory and visual signals, and a few tactile signals. With the physical description of signals now reasonably well refined and accessible, students of primate communication can attend to other problems relating to communication.

Our primary reason for organizing this volume was to focus on the new techniques that have been developed for the analysis of signals and the study of their relationship to communicator and recipient and to explore the relationship, if any, that exists between primate and human communication. A recent book by Steklis and Raleigh (1979) approached this last point by examining the neurobiological relationships between nonhuman primate communication and human speech and language. The present volume complements that of Steklis and Raleigh by focusing on studies devoted to behavioral, as opposed to neurobiological, concerns.

Our second reason for organizing this volume was to provide a much needed forum for the presentation of the excellent research on primate communication that has addressed the question of natural communication systems in primates. For the past decade research on natural primate languages has been overshadowed by studies of ape–human communication using American sign language or an alternative artificial language analog. The ape–human communication studies have been valuable in elucidating the cognitive abilities of great apes to acquire some degree of mastery of a synthetic language that is biologically irrelevant to the natural history of the ape. Such an endeavor, although inherently interesting, is akin to assessing the cognitive capacity of human beings by training them to imitate the vocalizations made by chimpanzees in the wild. That few humans can successfully imitate a chimpanzee is not evidence of man's cognitive deficiencies; it does reflect the different role that chimp vocalizations have played in the chimpanzee's natural history compared with that of man. From this perspective, the accomplishments of the tutored apes are truly remarkable, yet they provide no information about the communicative behavior of primates in the wild nor about the organization and evolution of natural languages.

The studies in this volume were selected because they deal in some way with the spontaneous communication of primates. The chapters include studies from the field, studies using animals in large captive enclosures, and studies using controlled laboratory conditions. We tried to include representatives of all sensory modalities of communication and representatives of evolutionary, ecological, ontogenetic, psychophysical, perceptual, cognitive, and social approaches to communication. The final result succeeds in meeting our initial goals, with the exception of equivalence among modalities. More chapters focus on auditory communication than on any other modality. This emphasis does not simply reflect the biases of the editors, each of whom studies auditory communication. Rather, it reflects the current technical state of the field, that is, auditory signals can be studied much more easily, with much better quantification and description, than signals in other modalities. We would hope that future reports on the state of the art of primate communication would be able to display as much variety and detail of studies of visual, olfactory, and tactile communication as there is here of auditory communication.

HOW ARE STUDIES OF PRIMATE COMMUNICATION RELEVANT TO UNDERSTANDING THE EVOLUTION OF LANGUAGE?

Recent research on teaching great apes to communicate with human beings using sign language or an artificial language analog (Gardner and Gardner 1969, 1978; Premack 1971; Rumbaugh 1977; Patterson 1978) has argued that chimpanzees and gorillas are able to behave linguistically in an essentially humanlike fashion (but see Terrace, Petito, Sanders, and Bever 1979). Recent work on the neurobiology of primate communication and language has indicated that similar brain structures are involved in both apes and human beings and that the neurological differences between species are quantitative not qualitative (see Steklis and Raleigh 1979). As great apes are our closest biological relatives, advocates of the ape–human communication studies have argued that their demonstrations indicate an evolutionary continuity of linguistic function from primates to human beings.

On the other hand, several linguists have criticized the ape–human studies by arguing that human language is fundamentally, not trivially, different from all forms of animal communication (Sebeok and Umiker-Sebeok 1980). To these linguists, no study on animals could ever be informative about human language origins. The stress on the uniqueness of human language is an argument not much different from the creationist arguments of Darwin's time and the "scientific creationism" of today. The separation of human beings as entities fundamentally different from animals is inconsistent with modern evolutionary thought, as is the argument for the fundamental uniqueness of human language.

There is a third way to approach the question of the evolution of language. No modern biologist or psychologist would deny that some attributes are uniquely human; this would include certain aspects of speech and language. Neither would a modern biologist or psychologist deny the existence of continuities between animals and human beings in communication as well as in other traits. Biologists and psychologists can accept that in detail each species has a unique communication system; they would no more expect to find an isomorphism between human language and chimpanzee communication than they would expect to find one between, say, redwing blackbirds and yellow-headed blackbirds. However, just as one would expect to find some similarities in certain call structures, communicative function, and the complexity of the communication system between closely related species like the redwing and yellow-headed blackbirds, so one would expect to find some similarities in structure, function, production, and perception between human and animal communication. A biologist or psychologist would not expect to find homologies on each of these points. In many of these cases the similarities would be analogies.

We now view evolution as a branching process rather than a linear process (see endpapers). Human beings represent a branching from other hominid apes that occurred approximately 4 million years ago (Sarich and Cronin 1976). Since that time, human beings, chimpanzees, and gorillas have developed different adaptations. We also know that many traits have evolved independently in different evolutionary lines (Mayr 1958). Thus certain traits appear to have converged. This notion of evolutionary convergence is of the utmost importance to those interested in the relationship between animal communication and human language. We need not restrict ourselves to study only those species (chimpanzees and gorillas) most closely related to human beings. We may also study species that are similar to human beings in other ways. For example, Marler (1970) has argued convincingly that songbirds are useful models for understanding certain speech phenomena. Many species discussed in this volume, although not as closely related phylogenetically to human beings as are gorillas and chimpanzees, do display interesting parallels in aspects of speech and language.

One further point on which biologists and psychologists would probably disagree with the human–ape communication research concerns the use of artificial communication systems and special training techniques.

Such extensive training and the use of artificial communication systems do not tell us how chimpanzees or gorillas communicate spontaneously to their conspecifics. As human beings appear to acquire language in the normal course of development with little explicit training (Lenneberg 1967), it seems reasonable to consider as worthy of study for parallels to human language only those phenomena exhibited by animals in their natural course of development. To understand the evolution of language, we must understand those communication systems of other species that have been subjected to the selective pressures of evolution. Any other type of communication will produce false data concerning the evolution of language.

There are two summary points to be made. First, no one species will be an adequate model for the evolution of human language, as no nonhuman animal is going to display a communication system isomorphic with human language. Second, one may find through the study of several primate species with different environmental adaptations some analogs to phenomena of human language. There is an evolutionary continuity to be found, but it is a yarn made of several threads, each coming from a different source.

WHY STUDY THE NATURAL COMMUNICATION OF PRIMATES?

It may appear from the preceding discussion that our main purpose in this volume is to demonstrate the existence of parallels between primate communication and human language. However, that is not the major goal of most contributors, who see an important value in studying primate communication for its own sake. The preceding section presented an anthropocentric argument for the study of primate communication. There are also several simiocentric arguments that can be developed. As the Altmann (1967) book made clear, communication is important in almost every form of social behavior; hence, an understanding of communication in other species is a necessary prerequisite to the understanding of their social behavior. Because scientists do tend to approach their subjects with an anthropocentric bias, it is as instructive to learn about the differences between nonhuman and human primates as it is to learn of their similarities. If we find that the perception of sounds or odors by monkeys is different from ours, it makes us more aware of the potential great diversity of perceptual mechanisms. The understanding of signal structure and signal usage in animals living in different habitats enables us to understand more about the coevolution of signals, environments, and animals.

Another reason for studying primate communication is that the order Primates is perhaps the order most susceptible to massive extinction. At present it is estimated that 52 of the 120 anthropoid species are either endangered or threatened.[1] It has been estimated that by the end of this cen-

---

[1] Determined from the U.S. List of Endangered and Threatened Wildlife (revised 1-17-79) and the Appendices of the Convention on the International Trade of Endangered Species of Wild Fauna and Flora (4-13-80).

tury much of the tropical rain forest, which is the habitat for most primate species, will be completely destroyed except for a few islands or reserves. It is critical that we develop an understanding of communication in many primate species both to obtain a record of the biological variability represented by primate species that may not be with us in another few decades and to be able to make intelligent decisions about the size and location of islands or reserves. Knowing how far "long calls" travel and how they function will help us determine how large home ranges must be for different species and thus how many groups could be kept in a given reserve. Understanding sexual signaling can be of value in attempting to preserve at least some endangered species in captivity for future generations to study. Understanding the role of communication in developing and maintaining group cohesion and relationships between individual animals can help us evaluate whether a given aggregation of animals is going to function successfully as a breeding population. Understanding the normal course of the development of communication may help us to predict at an early age which animals will be good breeders and group leaders and which will not.

The studies in the present volume include representatives of all 5 of the nonhuman families of the Anthropoid suborder of Primates. Data on 4 species of Callitrichids, representing 2 of the 5 genera, are presented by Epple, Alveario, and Katz, and by Snowdon. Jürgens, Smith, Newman, and Symmes, Newman and Symmes, and Robinson cover 2 species of Cebids from 2 of the 11 genera. Gautier and Gautier-Hion, Waser, Brown, Seyfarth and Cheney, Petersen, Bielert, Goldfoot, Keverne, and Chevalier-Skolnikoff present data on 16 species of Cercopithecids, representing 6 of the 14 genera. Deputte presents data on 1 Hylobatid species, and Chevalier-Skolnikoff presents data from the 3 genera of Pongids. (See endpapers for the phylogenetic relationship between these species based on serum-protein differences.)

Somewhere in the examination of our primate origins lie the clues to our linguistic roots – yet we are on the verge of eliminating many of those species that might reveal the linguistic nature of humankind. We stand to lose much of the evidence that derives from examining primate diversity within the next two or three decades. We must learn as much as we can as soon as possible.

ADVANCES IN THE DISCIPLINE

The chapters that follow present several exciting advances in the study of animal communication. First, there is a greater sophistication in the techniques used to study animals in the field. Gautier and Gautier-Hion describe a telemetry system that allows them to record the vocalizations of each individual in a free-ranging group on a separate channel. No longer is there a problem in trying to record sounds that are too faint or in confusing the call of one animal with another. The precise timing of relationships between animals calling within a social group can be maintained. Seyfarth and Cheney describe the first successful use of auditory playbacks in a field situation to reach an understanding of the semantic referents of the various alarm calls of vervet monkeys. Robinson brings to the analysis of similar sounding vocalizations a precise correlation of spacing activities between animals within a group with the various forms of calls given by monkeys while moving. With these techniques, field researchers in the future can gain much more information than the simple descriptions of repertoires that have usually been obtained.

Another advance is in the analysis of signals from the animal's point of view rather than from that of the observer. Both Petersen and Snowdon describe techniques for studying how monkeys perceive their own vocalizations as well as those of other species. A second type of signal analysis relates to the design features of signals, that is, the ideal characteristics of signals in a particular environment. Although much has been written on theoretically ideal design features, little is known about whether various aspects of ideal features are in fact ideal to the animals perceiving them. Brown's chapter shows that psychophysical studies of monkeys can help us to understand the effective design features of signals.

There is also an increasing interplay between laboratory and field studies. Brown's psychophysical results lead to the prediction of differences in the structures of calls used by terrestrial versus arboreal primates. Results reported by Waser and by Snowdon indicate how these predictions are confirmed by field studies. Gautier and Gautier-Hion use telemetry to study the relationship of communication to social behavior in monkeys, and Smith, Newman, and Symmes obtain similar results by using discriminant analyses to identify individual vocalizers and thus study the communication of individual social relationships.

Several chapters deal with cognitive issues in primate communication, a topic rarely considered in previous years. Smith, Newman, and Symmes demonstrate a sort of metacommunicative function of the "chuck" call of squirrel monkeys. It is given only between females who are close friends. Petersen demonstrates brain lateralization of perception of species-specific calls similar to the lateralization of language functions in human beings. Snowdon presents data on complexities of signal structure and how components of these complexities can be determined. He further discusses simple grammars governing the sequencing of calls. Seyfarth and Cheney show that the signals of monkeys have external referents. Thus, monkeys communicate not only about their internal state but also about external events. Chevalier-Skolnikoff demonstrates that Piaget's theory of intellectual development is a useful model for understanding the ontogeny and end point of communication abilities in monkeys and great apes.

Several new experimental techniques for the study of communication are presented here. To determine the hedonic valence associated with each type of vocalization, Jürgens made an innovative use of brain self-stimulation through electrodes that elicit different vocalizations in squirrel monkeys. Epple, Alveario, and Katz show how scent marks can be analyzed and how hormonal and social manipulations can be used to determine whether a scent-marking behavior is biologically or experientially determined. Bielert presents a model paradigm for determining whether one putative sexual signal is, in fact, the primary signal involved in sexual arousal. Goldfoot and Keverne describe several experimental techniques for the investigation of possible pheromonal communication of sexual state in monkeys.

Finally, the ontogeny of communication, which has been severely neglected in the past, has emerged as a vigorous area of study. Three different approaches are presented. Newman and Symmes discuss techniques modeled after those used successfully with bird song, which eliminate or restrict early experience to determine the relative roles of genetics and experience in the development of vocal communication. Epple, Alveario, and Katz provide a model system for separating hormonal from social influences in the expression of aggressive communication, and Chevalier-Skolnikoff presents a Piagetian cognitive approach to the ontogeny of communication.

Because several chapters discuss a variety of themes, we have grouped them with those chapters with which they show the greatest similarities. The book is divided into five sections: (I) Affective and social aspects of primate communication; (II) Social and environmental determinants of primate vocalizations; (III) Perceptual and psycholinguistic approaches to primate communication; (IV) Ontogeny and primate communication; and (V) Single- versus multiple-channel communication: Communication of reproductive state.

AREAS FOR FUTURE RESEARCH

Compiling a state-of-the-art volume allows the identification of those research problems nearing solution and those that remain for future work. There are several areas that need further research. We have already mentioned the need for more sophisticated studies of visual, olfactory, and tactile communication. The study of ontogeny has just begun and we need to evaluate the utility of the Piagetian approach for a variety of species. We need a demonstration species like the white-crowned sparrow to determine whether vocalizations are mainly innate, as the current evidence indicates, or whether experience plays a role in vocal development in primates. Much more work needs to be done using cognitive and psycholinguistic approaches to the analysis of a species' repertoire. Does the cognitive ability of a species bear any relationship to its communicative complexity? Can one be predicted from the other? We need studies of several more species to see how each perceives its species-specific calls. We also need to collaborate more closely with neurophysiologists. It is ironic that the communication of the species about which we know the most neurologically – the squirrel monkey – has never been the subject of a field study. Conversely, animals whose vocal repertoire and perceptual abilities we know quite well from field studies have never been studied neurologically. We need to make special efforts to study communication in threatened and endangered species before we lose forever the opportunity to study them. And we need to learn more about how communication is altered by environmental changes and disruptions and how these changes in turn affect the social behavior and viability of primate populations.

The chapters that follow indicate a substantial advance over the past 15 years and hold considerable promise for the next decades of study of primate communication. The field has progressed from simple, descriptive studies cataloging differences in primate signals to the use of sophisticated experimental paradigms for use with both captive and wild populations. The more complex models and paradigms for the analysis of communication that have been developed reflect the general acceptance of primate communication as a highly complex, nonreflexive activity. There has been an active focus on determining the signal characteristics that are important to primates rather than on interpreting signals from an anthropocentric view. There is a greater sophistication in determining the contextual situations in which calls are given and in developing predictive models that lead to the determination of how signals are used. Finally, the study of ontogeny is progressing from infancy to adolescence.

REFERENCES

Altmann, S. A. 1967. *Social communication among primates*. Chicago: University of Chicago Press.

Andrew, R. J. 1963. The origins and evolution of the calls and facial expressions of the primates. *Behaviour* 20:1–109.

Gardner, R. A., and Gardner, B. T. 1969. Teaching sign language to a chimpanzee. *Science* 165:664–72.

    1978. Comparative psychology and language acquisition. *Annals of the New York Academy of Sciences* 309:37–76.

Lenneberg, E. H. 1967. *Biological foundations of language*. New York: Wiley.

Marler, P. 1970. Birdsong and speech development: Could there be parallels? *American Scientist* 58:669–73.

Mayr, E. 1958. Evolution and systematics. In *Behavior and evolution,* ed. A. Roe and G. G. Simpson, pp. 341–62. New Haven: Yale University Press.

Patterson, F. G. 1978. The gestures of a gorilla: language acquisition in another pongid. *Brain and Language* 5:56–71.

Premack, D. 1971. Language in chimpanzee? *Science* 172:808–22.

Rowell, T. E., and Hinde, R. A. 1962. Vocal communication in the rhesus monkey (*Macaca mulatta*). *Proceedings of the Zoological Society of London* 138:279–94.

Rumbaugh, D. M. 1977. *Language learning by a chimpanzee*. New York: Academic Press.

Sarich, V. M., and Cronin, J. E. 1976. Molecular systematics of the primates. In *Molecular anthropology,* ed. M. Goodman, R. E. Tashian, and J. H. Tashian, pp. 141–70. New York: Plenum.

Sebeok, T. A., and Umiker-Sebeok, J. 1980. *Speaking of apes: a critical anthology of two-way communication with man*. New York: Plenum.

Steklis, H. D., and Raleigh, M. J. 1979. *Neurobiology of social communication in primates*. New York: Academic Press.

Terrace, H. S., Petito, L. A., Sanders, R. J., and Bever, T. G. 1979. Can an ape create a sentence? *Science* 206:891–902.

PART I

Affective and social aspects of primate communication

Until recently, research in primate communication was conducted by two rather separate schools of investigators – those working in the laboratory and those studying animals in the field. Over the years it has become evident that these schools should represent mutually complementary, not separate, approaches. Often technical advances first made possible in laboratory situations have had far-reaching effects in field studies, and, conversely, field studies have tempered the speculations of laboratory investigators with an ecological validity. Some of the most promising studies of today are concerned with animals living in normal social groups in large captive housing, where one can combine the control of the laboratory with some reality of the field.

A development of this sort has occurred with respect to animal vocalizations. Marler (1956), Thorpe (1958), Rowell and Hinde (1962), and Andrew (1963) were among the first to recognize the general utility of the sound spectrograph, a tool first developed for applied engineering, then adapted for human speech analysis, and finally used in the analysis of animal signals. Prior to this, animal signals were best transcribed by phonetic approximations of the sound – a scheme clearly inadequate for its task, as rather few animal sounds resemble human phonemes. The sound spectrograph was soon used to catalog the vocal repertoires of a variety of primates (Winter, Ploog, and Latta 1966; Grimm 1967, Struhsaker 1967; Marler 1970, 1972, 1973; Gautier 1974). Once it became possible to visually represent and categorize primate vocalizations, more sophisticated questions were addressed: (1) Under what circumstances are different vocalizations emitted and what is the relative status of the vocalizer? (2) Does a structure or syntax exist that governs the sequencing and exchange of vocal signals? (3) How do vocal signals represent the affective state and intentionality of the vocalizer? Although these questions received some discussion by early investigators, only recently have methodological and technological developments permitted a sophisticated analysis.

The chapters presented in Part I are methodologically innovative in their approach to these questions. In keeping with the focus of this volume, the authors have directed their inquiry toward "natural" communication systems, yet they employ new strategies that represent a conceptual "hybridization" of the manipulative orientation of the comparative psychologist and the descriptive orientation of the field biologist.

Gautier and Gautier-Hion employ biotelemetry to describe both quantitatively and qualitatively the vocal signals in monkeys. Relatively few studies have gone beyond a simple descriptive catalog of the vocal repertoire of nonhuman primates, a consequence of two classes of difficulties: those associated with identifying the vocalizer and those representing sampling problems. Under field conditions it is often impossible to identify the source of many vocal signals. Hence, the Gautiers have attached laryngeal microphones with miniature radio transmitters to each member of the social unit. The vocal behavior of each group member is identified with a different transmitter frequency; thus the identification of the emitter of each signal is possible in semi-free-ranging groups. Though the biotelemetry technique has been used only with a captive colony, it has potential for use in a field setting. In both captive and field situations, a variety of sampling problems arise. Short-range soft calls are usually underrepresented in most samples, and the magnitude of the skewedness of these samples is heightened for social units that are incompletely habituated to human observers. In addition, because vocal profiles are likely an expression of the size and membership of the various age and sex classes of the social unit, the demographic structure of the group is a factor in the validity of the sample. Finally, field researchers typically concentrate their recording efforts in social contexts during which animals are more easily approached, less disturbed, or possibly more vocal. Yet this strategy also is likely to result in sampling biases. The biotelemetry approach advanced by the Gautiers solves these difficulties.

The problem of identifying individual vocalizers was approached quite differently by Smith, Newman, and Symmes, who take advantage of individual differences in the acoustic attributes of the call, or its acoustic "voice print," to determine the identity of the vocalizer. Through the employment of discriminant-function analysis, Smith and her associates were able to identify accurately the vocalizer of chuck calls in a captive squirrel monkey colony. They were then able to use this information to study the role of chuck vocalizations in the social bonding of squirrel monkeys. Their results show that the chuck has a metacommunicative function, that is, its use communicates the nature of social relationships. Furthermore, this approach permits the recognition and identification of the participants of vocal exchanges, or "conversations," between group members. Both discriminant-function analysis and biotelemetry hold considerable promise for studying the relationship of social status, biological context, and the semantic and syntactic structure of vocal exchanges.

A cumulating body of contextual evidence suggests that the vocalizations of some primates are semantically active and may communicate specific objects, ideas, or intentions (see Seyfarth and Cheney, chap. 10, this volume), yet the idea of vocal signals revealing affective state has received considerable long-standing support (Rowell and Hinde 1962; Green 1975). These two views are not incompatible of course, as human speech is quite effective at revealing both affective and semantic information. Jürgens has developed an innovative and elegant scheme for assessing the affective valence of squirrel monkey calls. The squirrel monkey vocal repertoire, like those of most primates, is comprised of a limited number of standard call types with potentially hundreds of variants or intermediate call types. Jürgens has addressed the problem of acoustic diversity by attempting to organize vocal signals according to a few key parameters, paralleling the periodic organization of chemical elements. Jürgens's scheme organizes calls according to their acoustic similarity, transitional frequencies (a measurement of the association of different calls in similar contexts), and the emotional quality, or hedonic state, of the vocalizer.

Jürgens operationally measures the hedonic state associated with different vocalizations through the use of electrical brain stimulation. Many calls in the squirrel monkey's vocal repertoire can be electrically evoked through the stimulation of appropriate brain structures. The hedonic valence, or emotional state, associated with any call may be assessed by the animal's behavior of inducing or terminating brain stimulation in a self-stimulation cage. The methodological and technological elegance of Jürgens's research may contribute significantly to our understanding of the motivation of various primate vocalizations. Parenthetically, it should be added that the neural stimulation approach adopted by Jürgens is limited to the squirrel monkey, because only in this species has the neuroanatomy of vocal production been studied adequately. It is ironic to note that this best-studied primate has never been the subject of a published field study.

The three chapters included in this section reveal the methodological ingenuity of many investigators of primate communication. The most exciting and promising approaches represent a conceptual blend of the traditional orientations of both the comparative psychologist and the field ethologist. The synthesis of these two approaches is likely to play a major role in future advances.

REFERENCES

Andrew, R. J. 1963. The origin and evolution of the calls and facial expressions of the Primates. *Behavior* 20:1–109.
Gautier, J.-P. 1974. Field and laboratory studies of the vocalizations of talapoin monkeys (*Miopithecus talapoin*) structure, function, ontogenesis. *Behavior* 49:1–64.
Green, S. 1975. Variation of vocal pattern with social situation in the Japanese monkey (*Macaca fuscata*): a field study. In *Primate behavior: Vol. 4, Developments in field and laboratory research,* ed. L. A. Rosenblum, pp. 1–102. New York: Academic Press.
Grimm, R. J. 1967. Catalogue of sounds of the pigtail macaque (*Macaca nemestrina*). *Journal of Zoology* 152:361–73.
Marler, P. 1956. The voice of the chaffinch and its function as a language. *Ibis* 98:231–61.

1970. Vocalizations of East African monkeys: I. Red colubus. *Folia Primatologica* 13:81–91.

1972. Vocalizations of East African monkeys: II. Black and white colubus. *Behaviour* 42:175–97.

1973. A comparison of vocalizations of red-tailed monkeys and blue monkeys, *Cercopithecus ascanius* and *C. mitis*, in Uganda. *Zeitschrift für Tierpsychologie* 33:223–47.

Rowell, T. E., and Hinde, R. A. 1962. Vocal communication by the rhesus monkey, (*Macaca mulatta*). *Proceedings of the Zoological Society of London* 138:279–94.

Struhsaker, T. T. 1967. Auditory communication among vervet monkeys (*Cercopithecus aethiops*). In *Social communication among primates,* ed. S. A. Altmann, pp. 238–324. Chicago: University of Chicago Press.

Thorpe, W. H. 1958. The learning of song patterns by birds with especial reference to the song of the chaffinch, *Fringella coelebs. Ibis* 100:535–70.

Winter, P., Ploog, D., and Latta, J. 1966. Vocal repertoire of the squirrel monkey (*Saimiri sciureus*): its analysis and significance. *Experimental Brain Research* 1:359–84.

1 · Vocal communication within a group of monkeys: An analysis by biotelemetry

JEAN-PIERRE GAUTIER and ANNIE
GAUTIER-HION

METHODOLOGICAL DIFFICULTIES IN THE STUDY OF PRIMATE COMMUNICATION

The majority of studies of primate vocal communication conducted over the past two decades have been concerned with cataloging and describing vocal repertoires. Hence, the acoustical structures of various vocalizations have been described for a number of species, and hypotheses have been developed about the functional significance of classes of different vocalizations. Though these studies have varied in their sophistication and detail, they have still permitted interspecific comparisons of overall repertoires and these comparisons have yielded generalizations regarding the nature, pattern, and complexity of primate signaling (for a review, see Gautier and Gautier 1977). To date, relatively few studies have advanced beyond a descriptive catalog to more detailed analytical inquiries. Few investigations have addressed either the ontogeny of vocal behavior (see Newman and Symmes, chap. 11, this volume) or the status and social context of the participants of vocal exchanges (see Smith, Newman, and Symmes, chap. 2, this volume; Green 1975). An advanced quantitative analysis would record each vocal signal emitted in balanced social groups, identify the age, sex, and status of each vocalizer, and evaluate vocal displays in respect to their social context, antecedents, and consequences. The serious methodological obstacles that must be overcome to achieve this desired level of analysis may be divided into two classes: those associated with identifying the vocalizer and those that result from sampling problems.

The problem of individual identification

Visual observation allows identification of the vocalizer for only small groups at very close range in open habitats. As the group size increases, the number of vocalizations and the swiftness of their exchange increases

We particularly wish to thank Patrice Quinton (Institut de Recherche en Informatique et Systèmes Aléatoires, Université de Rennes) for his valuable assistance in computer analysis.

greatly, and it becomes impossible to visually identify the emitter of each signal. For example, paroxysms of vocal activity occur during the choruses of mangabeys, chimpanzees, and talapoin monkeys. During these episodes the din is astounding, as most group members vocally contribute to the display (Marler and Lawick-Goodall 1971; Gautier 1974; Gautier and Deputte 1975). The problem of vocalizer identification is further heightened in forest habitats in which only a small portion of the social unit may be visible at the same time. In the forest-dwelling cercopithecine monkeys this problem is compounded because vocal production is essentially nasal with little facial movement (Gautier 1975) to visually confirm the identity of the emitter. Thus the problem of vocalizer identification becomes one of extraordinary magnitude for research on many primate species in natural groups in their natural habitats.

Sampling difficulties

Serious sampling problems arise when investigators collect data that do not represent the natural exchange of vocal signals. The difficulties in achieving representative samples fall into four categories: those resulting from differences in vocal amplitude; those resulting from differences in the habituation of the vocalizers to the presence of human observers; those resulting from differences in the composition of the social units; and those resulting from differences in the logistics of recording and observation.

Sampling and vocal amplitude. In most studies the catalog of vocalizations has been biased toward long-range, loud, easily recorded vocalizations, whereas low-amplitude signals often have been overlooked or ignored (however, see Smith, Newman, and Symmes, chap. 2, this volume). Yet in some *Cercopithecus* groups, for example, loud calls comprise less than 8% of the vocalizations emitted (see our later discussion).

Sampling and vocalizer habituation. Vocal exchanges from individuals and social units habituated to the presence of human observers differ markedly from the vocal behavior of unhabituated primates (Marler 1976). A dramatic example is given by a number of Gabonese primates subjected to heavy hunting pressure. Recordings from these populations exhibit an increase in alarm-chirp vocalizations relative to the records of habituated groups. Furthermore, habituation, or the tolerance of human approach, is relative and difficult to measure or assess. Thus samples obtained at the beginning of a study may not be comparable with those taken at the end of the study, as the study group becomes progressively less sensitive to the presence of the observer. The problem of habituation is particularly acute regarding the recording of soft, short-range vocalizations requiring a reduction in the distance between observer and vocalizer.

Sampling and the composition of the social unit. The demographic structure of a group influences both the overall rates of vocalizations and the frequency of usage of each vocal type. Thus the vocal profile of a troop including four infants may be significantly different from that of a troop of the same species without any infants. Consequently, comparisons of the vocal systems of two or more troops should consider differences in the size and membership of the various age and sex classes.

Sampling and the logistics of recording. The human observer is often inclined to develop logistic strategies that optimize the acquisition of high-fidelity tape recordings. Indeed, the inexperienced field observer is often shocked by the disparity between listening to live displays and listening to playbacks of the faint renditions captured on tape. Thus observers learn to select periods and contexts during which the animals are more easily approached, less disturbed, or even more vocal. But these logistic strategies may lead to sampling biases. For example, Marler (1976) has observed that chimpanzee vocalizations recorded at provisioned feeding sites exhibit an atypically high proportion of aggressive calls.

Singly and collectively, these four sampling difficulties in the data pools of individual studies may lead to biases that prohibit meaningful comparisons between species. Moreover, even within a study, these difficulties may lead to inaccuracies in recording the daily distribution of vocal behavior and consequently result in failures to discern the vocal expression and mediation of social relationships.

If ethologists are to discover the function and organization of communication systems, it is essential that these sampling difficulties be overcome and that techniques that reveal the identity of the vocalizer be developed. The study of communication has been impeded by these problems in a manner closely paralleled in the neurosciences, for example, when the origin of electrical potentials has not been specified or when the electrodes used in a brain study are sensitive to some neural events but insensitive to others.

A promising solution to these problems in primate communication may emerge from the application of biotelemetry (Gautier 1979). By fitting a larnygophone and a miniature radio transmitter to all the members in a social group, it may be possible to determine the identity of the emitter of each vocalization. Thus it becomes possible to determine quantitative and qualitative differences in the use of the vocal repertoire according to the age, sex, and status of the vocalizer. Furthermore, the social context of each vocalization and the vocal response of the recipients of the signal may be monitored more fully.

BIOTELEMETRY IN PRIMATE GROUPS

We have recently initiated an exhaustive study through biotelemetry of vocal exchanges in a captive cercopithecine group. The monkeys are equipped with laryngeal microphones connected to miniature high-frequency, frequency modulation (FM) radio transmitters. The radio transmitter attached to the harness (Figure 1.1) of each monkey is tuned to a different carrier frequency to allow the identification of the individual

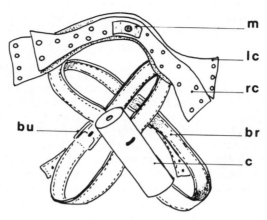

Figure 1.1. Diagram of a harness: (m) microphone, (lc) leather collar, (rc) rubber collar, (br) braces, (bu) buckles, (c) container for the oscillator and the battery, which is placed on the back of the animal.

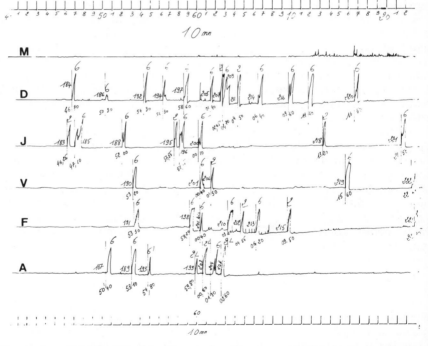

Figure 1.2. Graphic recording of a vocal exchange among the six group members. Each track corresponds to one animal: (M) adult, (D) infant female, (J) adult female, (V) subadult female, (F) subadult male, (A) juvenile male. Lateral tracks give the time in seconds.

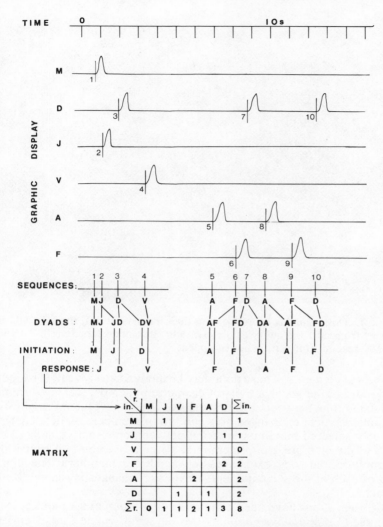

Figure 1.3. Schematic representation of the matrix, building from a graphic display of vocalizations with the determinations of sequences and the initiation/response dyads.

emitting each signal. Each frequency channel is monitored by a separate FM receiver, which, in turn, is recorded on a separate track of a multi-track tape recorder. In this manner sequences of the vocal displays among all group members are accessible to study. The chronology of vocal sequences is readily displayed by connecting each track of the magnetic recorder to a separate channel of a graphic recorder (electroencephalograph), as shown in Figure 1.2. The following data may then be entered in a computer: the identity of the vocalizer, the call type, and the delay be-

Figure 1.4. Three animals equipped with their harnesses: left, the subadult male (F); third from left, the adult female (J); right, the infant female (D). The adult male (M), second from left, is not equipped.

tween vocal elements. These data may be analyzed to reveal: (1) A quantified record of the vocal activity of each group member; (2) measurement of the delay time between different calls of one or several animals; (3) descriptions of vocal exchanges between group members as defined by an arbitrarily selected maximum interval between consecutive calls (2 sec) given by one or more group members; (4) measurements of transitional frequencies of the vocal exchanges between group members. The identification of each of these components of vocal exchanges is schematized in Figure 1.3.

Our observations have been conducted on a group of six monkeys – an adult male *Cercopithecus pogonias* (M), an adult female *C. ascanius* (J), and their four hybrid offspring, a 74-month-old subadult female (V), a 54-month-old subadult male (F), a 40-month-old juvenile male (A), and a 14-month-old infant female (D) – who live in an 11 × 4 × 4 m wire-mesh enclosure. Three members of this social group are shown in Figure 1.4.

For illustrative purposes we have elected to describe a sample of six sequences (totaling 129 min) of the usual activities of the monkeys: (1) three feeding periods, recorded at 8 A.M., corresponding to the first meal of the day; (2) a resting period, recorded at 2 P.M., dominated by resting and grooming; (3) a roosting period, recorded at 6 P.M., characterized by a regrouping of the monkeys prior to nightfall; (4) a warning sequence provoked by the sound of trees falling in the neighboring forest.

Table 1.1. *Characteristics of the repertoire*

Vocal types	Description	Context–function	Subclasses
0	Boom (low pitched)	Intragroup: rallying	0
		Intergroup: spacing	
1M	Bark (loud; 1–3 units)	Intragroup: rallying	1M
	Stereotyped	Intergroup: spacing	
	Unstereotyped	Warning	1M
1	Chirp (high pitched)		
	Standard	Warning	1
	Light	Light warning	1'
2	Grunt (low pitched)		
	Standard	Cohesion	2
	False	Crouching	2L
3	Trill (medium pitched)	Isolation	3
4	Staccato grunt		
	Standard	Aggression	4
	Light	Warning	4L
5	Gecker (atonal; rhythmic)		
	Standard	Discomfort	5
	Light	Discomfort	5'
	Chirp	Discomfort + warning	5–1
6	Trill (high pitched)		
	Standard	Contact	6
	Light	Contact	6'
	Chirp	Contact + warning	6–1
7	Scream (noisy; whistled)	Distress	7
2–6	Grunt-trill (low + high pitched)	Cohesion-contact	2–6

Categories of the vocal repertoire

Adult male and female repertoires are easily categorized and have been described previously (Marler 1973; Gautier 1975). Hybrid offspring create a problem of vocal categorization, however, for whereas adult male and female calls are acoustically different, the calls of hybrid young possess a hybrid structure. Taking into account all the degrees of hybridization in the calls of the young, 100 different vocalizations can be recognized for the total group (Gautier, in preparation a). For the purpose of this chapter they have been merged into 18 functional subclasses, which have homologous equivalents in both species even though their fine acoustical structure may be different. They are derived from the 9 primary vocal types presented in Table 1.1.

Vocal activity according to age and sex

The observed group had a high level of vocal activity, with 3,855 calls given during 129 min. The mean call frequency for the group was nearly

Table 1.2. *Mean call frequency by individual*

Individuals	Number of calls	Call frequency (calls/min)
A♂ (M)	308	2.37
A♀ (J)	467	3.60
Sa♀ (V)	399	3.08
Sa♂ (F)	608	4.69
J♂ (A)	745	5.75
I♀ (D)	1,328	10.25
Σ Individuals	3,855	29.75

Note: All the interindividual differences are significant (χ^2: $p < .05$) except those between the adult and subadult females.

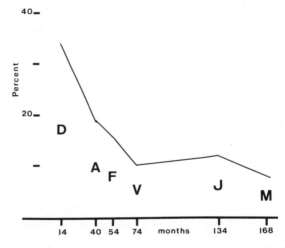

Figure 1.5. Age-related changes in the vocal activity expressed by the relative percentage of calls given by each animal among the total sample (N = 3,855).

30 calls/min, as given in Table 1.2. Differences among individuals are obvious. The frequency of vocal behavior decreased regularly with increasing age for the four young, from a maximum of 10.25 calls/min for the infant. However, the oldest young did not differ significantly from the adult female. In contrast, the adult male's vocal activity was significantly less than that of the adult female.

Figure 1.5 shows a relative decrease in vocal output occurred with age. When the calls given during the warning context were eliminated from analysis, the adult male's activity was still the lowest and constituted only 3% of the overall vocal output of the group. In females, this decrease with

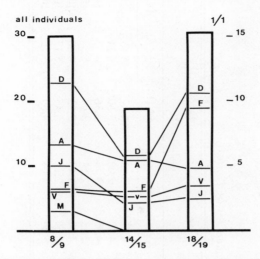

Figure 1.6. Changes in the mean call frequency (calls/min) during the three daily periods for the group (histogram) and for each individual (lines M to D).

age stopped at sexual maturity; vocal activity is probably maintained at a constant level after this period, irrespective of the age of the subject.

Vocal activity and time of day

The pattern of vocal activity changed throughout the day, showing a bimodal profile with two similar peaks in the morning and evening and a significant decrease in the middle of the day (Figure 1.6). These changes could have resulted from the dominant activity at different daily periods and/or an internal rhythm. However, individual animals showed obvious differences: The adult male vocalized only in the morning; the adult female was also more vocal in the morning but maintained her vocal activity, throughout the day; the subadult female and the juvenile male showed no rhythm at all, and the subadult male called more frequently in the evening. Individually, bimodality was found only in the infant. However, if the contribution of this animal is eliminated, the bimodality of the group as a whole is still maintained.

Vocal activity and context

The amount of vocal activity emitted during the six different sequences is given in Table 1.3. The mean call frequency was highest in the warning episode and lowest in the resting period. It had an intermediate value for the three feeding sequences and the roosting sequence. The vocal output of each group member varied according to the context (Figure 1.7).

Table 1.3. *Mean group and individual call frequencies by context*

Sequences	Duration (min)	Number of calls	Mean group call frequency (calls/min)	A♂ (M)	A♀ (J)	Sa♀ (V)	Sa♂ (F)	J♂ (A)	I♀ (D)
						Mean individual call frequencies (calls/min)			
1. Feeding	19.466	601	30.873	3.13	6.57	3.03	2.67	5.55	9.91
2. Feeding	29.666	884	29.798	0.30	4.00	3.00	3.50	6.74	12.27
3. Feeding	19.923	565	28.359	0.05	4.00	3.76	2.36	6.57	11.64
4. Resting	19.716	372	18.867	0.00	2.13	2.54	3.04	5.38	5.78
5. Roosting	29.999	921	30.700	0.00	2.40	3.40	9.53	4.80	10.57
6. Warning	10.816	512	47.334	21.91	2.60	2.22	5.45	5.17	10.00
All sequences	129.589	3,855	29.747	2.37	3.60	3.08	4.69	5.75	10.25

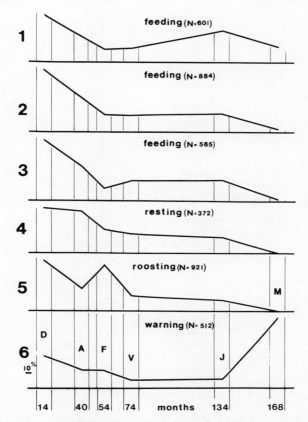

Figure 1.7. Relative vocal activity according to the context for each group member (N = total number of calls).

Whereas the adult male's vocalizations were almost completely restricted to the warning and morning feeding sequences, the other group members were much less vocal during the warning sequence, although their vocal patterns in all other contexts were similar.

Frequency of usage of the repertoire

Differences in individual behavior suggest a differential use of the repertoire. Figure 1.8 shows that the trill was the most frequently given call, followed by the grunts, the barks, and the chirps. The gecker, grunt-trill, isolation trill, and boom were far less frequent events. The use of each vocal type changed with age and sex, as shown in Table 1.4. The booms and barks were uttered only by the adult male, whereas the alarm chirps were mainly characteristic of the adult female. The production of trills de-

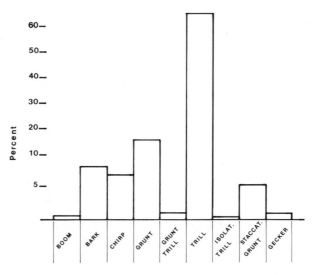

Figure 1.8. Relative percentage of use of the nine vocal types by the group.

Table 1.4. *Use of repertoire by individual*

Vocal types	A♂ (M)	A♀ (J)	Sa♀ (V)	Sa♂ (F)	J♂ (A)	I♀ (D)	Σ individuals
Boom	4	0	0	0	0	0	4
Bark	303	0	0	0	0	0	303
Chirp	0	144	48	38	41	16	257
Grunt	0	174	82	71	81	171	579
Grunt-trill	1	0	9	2	12	5	29
Trill	0	137	232	456	549	1,073	2,447
Isolation-trill	0	1	0	3	0	2	6
Staccato grunt	0	35	26	31	56	52	200
Gecker	0	6	2	7	6	9	30
All vocal types	308	467	399	608	745	1,328	3,855

creased regularly and rapidly with age, being 4.6 times more numerous in the infant compared with the oldest subadult. The grunts were characteristically given by both the youngest infant and her mother. In contrast, some calls, such as the staccato grunt, were given in more comparable proportions by all group members except the adult male.

Despite the small sample in each individual class, one can try to describe ontogenetic changes in the usage of the vocal repertoire. Figure 1.9A, which reproduces the data in Figure 1.5, shows changes in the rates of vocalizations with age, and B and C show the distribution of use of

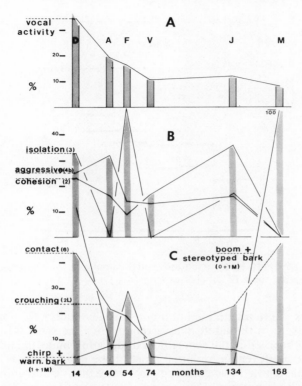

Figure 1.9. (A) Change in the vocal activity with age (see Figure 1.5) expressed by the relative percentage of all calls given by each individual; (B) changes in isolation (type 3), aggressive (4B), and cohesion (2) calls as a function of age; and (C) changes in contact (6) and crouching (2L) calls and in chirp + warning bark (1 + 1M) and boom + stereotyped bark (0 + 1M) as a function of age.

each call type with age. In this figure, calls have been grouped into seven major functional categories.

The results suggest that calls initially linked with physical contact or close interindividual distance (contact trill + crouching) showed a decrease in frequency with age, that is, they either disappeared in adults or were maintained at a low level. Calls given in alarm situations showed an inverse trend: The alarm chirps of the young male were progressively replaced by barks in the sexually and socially mature male (Gautier 1975). The emission of booms and barks, accompanied by stereotyped displays, which serve to regulate the distribution of the population (Gautier 1975; Gautier and Gautier 1977), appears suddenly in the vocal repertoire of the adult male at sexual maturity. Calls associated with the general cohesion of the social unit showed more complex patterns but appeared to be mainly characteristic of the adult female. Aggressive calls, typically given

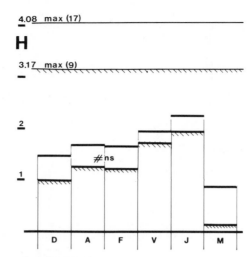

Figure 1.10. Diversity H_i for each individual calculated for the nine vocal types (bottom ᴡᴡᴡ) and their 17 subclasses (top ——).

by subadults, were maintained at a constant level after 54 months of age but were absent in the adult male's repertoire. The most striking differences were observed in the repertoires of the adult male and adult female. Whereas the adult female retained most of the calls given by the young, the adult male emitted two call types that were unique and constituted the main part of his repertoire. One animal, the subadult male, deviated from the general pattern by retaining some calls usually given by infants and decreasing the frequency of some calls normally uttered by animals of his age. These idiosyncratic variations will be discussed later.

By measuring the index of diversity (H_i) according to Shannon's formula (Hutcheson 1970; Shannon and Weaver 1949; Baylis 1976; Marler and Tenaza 1977), it is possible to compare individual differences in the pattern of usage for each of the 9 vocal types and the 17 vocal subclasses. H_i reaches a maximum value when all the vocal classes are equally used (H_i max for 9 calls: 3.17; for 17 subtypes: 4.08). From Figure 1.10, one can note a progressive increase in diversity with age from infancy to adulthood; all the interindividual differences were significant except those between the juvenile and the subadult male. The greatest difference was found between the adult male and female, with the former showing the lowest diversity. The low diversity of the adult male was due to the use of only three call types, that of the infant to the use of only one.

ANALYSIS OF THE VOCAL EXCHANGES

In this preliminary account, only dyadic exchanges are analyzed. This analysis assumed that a call is given in response to another if it occurs

Table 1.5. *Number (and percentage) of calls during routine activities and in warning context in dyads and while solitary*

Individuals	Routine activities		Warning context	
	Calls in dyads	Solitary calls	Calls in dyads	Solitary calls
A♂ (M)	49 (80.0)	12 (20.0)	228 (96.0)	9 (4.0)
A♀ (J)	160 (91.0)	15 (9.0)	57 (100.0)	0 (0.0)
Sa♀ (V)	154 (96.0)	6 (4.0)	39 (93.0)	3 (7.0)
Sa♂ (F)	270 (89.0)	33 (11.0)	67 (97.0)	2 (3.0)
J♂ (A)	268 (88.0)	36 (12.0)	70 (91.0)	7 (9.0)
I♀ (D)	441 (80.5)	107 (19.5)	119 (91.0)	12 (9.0)
Total	1,342 (86.5)	209 (13.5)	580 (95.0)	33 (5.0)

within 2 sec. More elaborate exchanges could be described by analyzing triads and so on. Two matrices have been constructed: One includes the vocal exchanges given during routine activities (feeding, resting, roosting); the other concerns a warning episode. Chi-square values were calculated between the observed frequency of the responses given by each group member to the initiator of the exchange and its expected value. The latter was measured taking into account all the responses given by each animal during vocal interactions.

Routine activities

Fifty-four minutes of recording have been analyzed to construct the matrix dealing with routine activity. Among the 1,551 calls given during this time period, 86.5% (N = 1,342) were included in vocal exchanges. However, some interindividual differences occurred when the number of calls included in vocal interactions was compared with the number of calls given alone (see Table 1.5). For example, in addition to showing the lowest level of vocal behavior, the adult male was significantly less likely to engage in vocal interactions. Although the overall vocal output of the infant female was the highest, her calls were also less likely to occur in vocal exchanges than were those of every other group member except the adult male. Participation in vocal exchanges did not differ significantly among the adult female and her oldest offspring, although the subadult female contributed more to vocal exchanges than did her two younger brothers. Thus, during routine activities the communicative value of the signals given by each member may differ. The adult female and her oldest offspring were vocally the most responsive animals or their vocalizations were the most suitable signals for inducing responses. The inverse was found for the adult male and the youngest infant. Whereas the low participation of the adult male may be related to the infrequency of his interactions within the group, the low participation of the infant could be due to her immaturity, that is, her vocal signals were not yet functionally ad-

Table 1.6. *Matrix of dyads for routine activities*

	Answering animal						
Initiating animal	A♂ (M)	A♀ (J)	Sa♀ (V)	Sa♂ (F)	J♂ (A)	I♀ (D)	Total
A♂ (M)							
Observed value	0	15	4	5	3	9	36
Expected value	1.24	4.24	4.17	7.41	7.79	11.14	
χ^2 value	.41	29.13	.03	.64	3.13	.36	
Significance	ns	+++	ns	ns	ns	ns	
A♀ (J)							
Observed value	11	26	20	15	19	41	132
Expected value	4.52	15.55	15.30	27.18	28.57	40.84	
χ^2 value	9.22	8.26	1.50	7.24	4.21	.005	
Significance	++	++	ns	++	ns	ns	
Sa♀ (V)							
Observed value	4	13	15	30	31	43	136
Expected value	4.69	16.02	15.76	28.01	29.44	42.08	
χ^2 value	.009	.52	.01	.11	.05	.01	
Significance	ns	ns	ns	ns	ns	ns	
Sa♂ (F)							
Observed value	4	14	26	59	40	67	210
Expected value	7.24	24.74	24.34	43.25	45.46	64.97	
χ^2 value	1.34	6.01	.08	8.50	.90	.06	
Significance	ns	+	ns	++	ns	ns	
J♂ (A)							
Observed value	8	18	24	34	41	81	206
Expected value	7.10	24.27	23.88	42.42	44.59	63.73	
χ^2 value	.03	1.94	.01	2.32	.34	8.00	
Significance	ns	ns	ns	ns	ns	++	
I♀ (D)							
Observed value	9	37	32	72	92	82	324
Expected value	11.17	38.17	37.55	66.72	70.14	100.24	
χ^2 value	.38	.02	1.11	.62	12.04	6.60	
Significance	ns	ns	ns	ns	+++	++	
Total	36	123	121	215	226	323	1,044

Note: Level of significance: $df = 1$; $+++ = p < .001$; $++ = p < .01$; $+ = p < .05$; $ns = p > .05$.

justed to the appropriate context, a deficiency that should reverse during ontogeny.

Even though each individual responded to every group member, vocal exchanges were not equally distributed within the group (Table 1.6). The adult male mainly responded to the adult female; his responses to others were distributed randomly as expected from his vocal output. The adult female appeared more responsive to the adult male and to herself and less responsive than expected to the subadult male; she responded as expected to her other offspring. Responses given by the subadult female were equally distributed to every group member. Responses of the subadult male were preferentially given to himself, and he interacted less than expected with his mother. Finally, the two youngest offspring exchanged more calls than expected with each other. Two positive mutual relationships can be deduced from these results: One was between the adult male and the adult female, and the other was between the infant and the juvenile. By contrast, a negative reciprocal relationship was found between the subadult male and his mother.

Warning context

Of the 580 calls recorded during the warning episode, 95% were included in vocal exchanges and few interindividual differences were found. However, vocal exchanges between the adults and between the adults and the subadults were significantly higher than were those between the two youngest offspring in this context ($\chi^2 = 5.98, p < .05; \chi^2 = 8.7, p < .01$). The matrix of dyads given in Table 1.7 shows that both the adult male and female preferentially responded to their own calls. All subjects responded less than expected to the adult male; and he, in turn, responded less than expected to every group member. All but two nonadult individuals responded to themselves with expected frequencies. The reciprocal and univocal positive affinities shown during routine activities were not preserved during the warning episode.

Two main vocal categories were emitted during the warning episodes: barkings of the adult male and chirps of other group members. When all the vocal responses of the group members except those of the adult male are combined, we find that although group members responded less than expected to the adult male only one positive and reciprocal response appears between two of them, the juvenile male and the subadult female. The others are given without individual affinities. Consequently, chirps appeared highly contagious, a characteristic well adapted to the spreading of the alarm. By contrast, the male's barks, which were both loud and numerous, appeared to inhibit the vocal behavior of other individuals. The fact that neither chirps nor barks induced preferential vocal relationships seems to indicate that in terms of social organization these calls differ from those given in routine situations. These calls are adapted to signaling alarm but appear unsuited to the creation and reinforcement of affinitive bonds between group members.

An evaluation of the biotelemetry technique

The preceding data indicate that some primate groups are remarkably vocal. The quantity of observations collected in a relatively short study

Table 1.7. *Matrix of dyads in warning context*

Initiating animal	Answering animal						
	A♂ (M)	A♀ (J)	Sa♀ (V)	Sa♂ (F)	J♂ (A)	I♀ (D)	Total
A ♂ (M)							
Observed value	147	11	4	17	8	30	217
Expected value	90.62	20.37	14.13	25.77	22.86	43.23	
χ^2 value	101.24	7.29	12.02	5.16	17.26	8.01	
Significance	+++	++	+++	ns	+++	++	
A ♀ (J)							
Observed value	9	13	6	7	8	8	51
Expected value	21.30	4.79	3.32	6.06	5.37	10.16	
χ^2 value	12.44	15.97	1.69	.04	1.04	.37	
Significance	+++	+++	ns	ns	ns	ns	
Sa ♀ (V)							
Observed value	1	4	5	7	10	9	36
Expected value	15.03	3.38	2.34	4.28	3.79	7.17	
χ^2 value	22.47	.005	2.27	1.41	10.31	.29	
Significance	+++	ns	ns	ns	++	ns	
Sa ♂ (F)							
Observed value	21	7	3	7	4	18	60
Expected value	25.06	5.63	3.91	7.13	6.32	11.95	
χ^2 value	.98	.17	.05	.02	.66	3.63	
Significance	ns	ns	ns	ns	ns	ns	
J ♂ (A)							
Observed value	10	3	9	10	10	18	60
Expected value	25.06	5.63	3.91	7.13	6.32	11.95	
χ^2 value	16.40	1.00	6.52	1.01	2.02	3.63	
Significance	+++	ns	+	ns	ns	ns	
I ♀ (D)							
Observed value	30	11	7	14	15	21	98
Expected value	40.93	9.20	6.38	11.64	10.33	19.52	
χ^2 value	5.67	.25	.003	.41	2.32	.07	
Significance	+	ns	ns	ns	ns	ns	
Total	218	49	34	62	55	104	522

Note: Level of significance: $df = 1$; $+++ = p < .001$; $++ = p < .01$; $+ = p < .05$; $ns = p > .05$.

may require an extended interval for data analysis. At present we have no substitute for a trained human listener to audit and classify each recorded vocalization. Nevertheless, the biotelemetry technique offers some distinct advantages (Gautier, in preparation, b). Figure 1.11 compares two

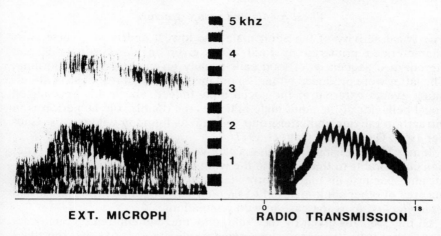

EXT. MICROPH   RADIO TRANSMISSION

Figure 1.11. Comparison of two spectrographic analyses of a cohesion–contact call, as recorded by an external microphone (left) and a laryngeal microphone (right).

spectrographic analyses of cohesion–contact calls as recorded from an external microphone and a laryngeal microphone. As may be seen in this figure, both background noises and echoes are greatly reduced by the laryngeal microphone, and the resulting superior signal-to-noise ratio is a considerable aid in acoustical analysis. Furthermore, the approach yields a virtually perfect sample of the vocal behavior of the social unit for any selected interval. At present our miniature transmitters have a maximum range of 250 m, yet more powerful transmitters may be developed. Thus the softest vocalization may be recorded at considerable distances from human observers. Because the biotelemetry technique is very flexible and adaptable, observations may be conducted under highly reproducible conditions and thus permit exhaustive ontogenetic studies. Finally, the technique is adaptable to the field, permitting the comprehensive study of natural groups in their natural habitat.

VOCAL BEHAVIOR OF PRIMATE GROUPS

The observation of our captive group shows clear differences in the vocal profile of different group members of a monkey family, based upon the total frequency of calls given, their daily rhythm of production, the context in which they are produced, the quantitative and qualitative differential usage of the repertoire, and finally the different frequencies of vocal interactions among group members. These differences may be related to many factors, including ontogeny, individual roles within the social unit, interindividual affinitive relationships, and the resulting social organization. Observations of populations in the wild bear on some of these possibilities.

Vocal profile and group structure

The vocal activity of the adult male, the lowest and least diverse, was characterized primarily by loud barks given in alarm situations and stereotyped sequences of loud calls ritually given early in the morning. The latter were produced either "spontaneously" or in response to stimulating events external to the social unit. Indeed, the highly specialized vocal behavior of the adult male indicates the double role he performs in intragroup rallying and intergroup spacing, as found in wild troops (Gautier 1975; Gautier and Gautier 1977).

It may be surprising to find that a captive male maintained a social function comparable to that found in the wild; however, the emission of rallying and spacing calls may be correlated with the presence in the surrounding area of other adult males of *C. nictitans, C. pogonias, C. cephus,* and *C. neglectus,* as is the case in wild populations.

Of the 308 calls given by the male, only one call emitted on one occasion (grunt-trill) concerned close-range pacific interactions, yet this class of vocalization comprised 79% of those emitted by the entire group. In fact, the male has a relatively peripheral position with regard to intragroup relationships; most of the interactions in which he was involved, such as grooming and, more rarely, playing, depended on the initiative of other members, especially that of the female. However, most pacific activities occur at close range and may be mediated by visual exchanges not requiring acoustic displays. Thus the male appears socially linked to his group mainly through his affinitive relationships with the adult female. A similar peripheral position has been noted for the adult males of harems of other *Cercopithecus* species (e.g., *C. ascanius,* Struhsaker and Leyland 1979; *C. cephus,* Quris, Gautier, Gautier, and Gautier-Hion 1981).

The adult female possessed the most diversified repertoire, a relatively high level of vocal activity, and she was involved in many vocal interactions. Her participation in vocal exchanges with the adult male, especially when he uttered stereotyped sequences of loud calls, could have a value in the maintenance of the pair bond. In the wild, the loud calls similarly attract adult females (personal observation).

The low frequency of vocal exchanges between the mother and her subadult son seems to result from the son's peculiar behavior during the roosting period. The son attempted to join the sleeping subgroup that included his mother, but she repelled him. His vocal rate was high (10.33 calls/min), and his calls, mainly crouching calls, were more numerous than those of other animals (33.33%) with the exception of the infant female (39%). During this period the mother was unusually silent (6%; 1.77 calls/min). As crouching calls normally characterize young animals, this may be a case of regressive behavior.

The vocal activity of the subadult female qualitatively and quantitatively resembled that of her mother, and her vocal diversity was only slightly lower. The main difference lay in the absence of preferential responses to the adult male's loud calls, which may be related to the absence of a sexual bond between them.

Although they contributed to the overall activity of the group, the infant female and the juvenile male had a minimally diversified vocal repertoire because of the prevalence of contact and cohesion calls. These vocalizations emphasized their social interdependency and close affinitive relationship. The decrease in the frequency of these calls in the older siblings corresponded to their progressive independence, which was characterized also by the absence of any preferential relationships with either adults or young. The same trend was found for alarm chirps and seems to reflect the same maturational process.

The confirmation of these observations requires repeated longitudinal studies conducted at regular time periods on the same subjects and comparative work on groups with different compositions of the same or different species.

Comparisons among different types of social structure

Among the five sympatric species of forest cercopithecines we have studied, three fundamental types of social structures have been described: (1) the family group as seen in *C. neglectus,* including one adult male, one adult female, and one or two offspring (more rarely two adult females, Gautier-Hion and Gautier 1978); (2) the harem as displayed in *C. nictitans, C. pogonias,* and *C. cephus,* including from 10 to 25 individuals (Gautier-Hion and Gautier 1974); and (3) the multimale troop as characterized by *Miopithecus talapoin,* including up to 100 individuals (Gautier-Hion 1971).

These species show quite different degrees of sexual dimorphism in their body weight. Sexual dimorphism is greatest in *C. neglectus,* followed by *C. nictitans,* lowest in *M. talapoin,* and has an intermediate value in *C. pogonias* and *C. cephus* (Gautier-Hion 1975). The results of detailed analyses of the vocal repertoires of these species (Gautier 1975) show that sex differences in vocal profiles are lowest in *M. talapoin,* highest in *C. neglectus* and *C. nictitans,* and have an intermediate value in *C. pogonias* and *C. cephus.* Thus the emergence of gender-specific vocal profiles is broadly correlated with increasing sexual dimorphism.

In species in which sexual dimorphism favors males, the main responsibilities of the adult male of the troop may be predator warning and by long-distance vocal signaling to assume territorial defense and intragroup rallying. By contrast, intragroup social activities mainly, if not exclusively, concern adult females and their young. These activities are mediated by close-range vocalizations.

Two extremes can be chosen to illustrate the relationship between social structure and the communication system. Within the small social unit of *C. neglectus,* it has been found that interindividual distances are surprisingly great, being up to 10 m in more than 80% of the cases and above 30 m in 34%. Close interactions between group members are infrequent; the adult male almost never interacts with his young infant, he rarely plays with his oldest offspring, and he is groomed only by the adult female. The latter individual is the most socially active animal: She nurses her young, initiates grooming with every group member, and even appears (at least in captive animals) as the main initiator of sexual relationships (personal observation). The vocal repertoire of this species is quite meager, including only six vocal types, all of discrete physical structure. As this species does not produce alarm chirps, the signaling of predation is primarily assumed by long-range calls of the male, which elicit silent freezing behavior in group members. Most calls in the repertoire are given by unweaned infants, and their vocal output rapidly decreases after weaning, especially in males. The vocal activity of the adult female is limited primarily to the use of cohesion grunts, which are exchanged with offspring and occur before troop movement, during troop encounters, and after alarm situations. Moreover, cohesion grunts frequently precede and appear to "stimulate" the emission of loud calls by the adult male.

A contrast is provided by talapoin monkeys (*M. talapoin*) that live in habitats similar to that of *C. neglectus*. In talapoin troops, social exchanges are numerous and complex, generally including several members. Troop movements are always accompanied by vocalizations, the frequency of which increases with the rate of movement and the density of the forest. Interindividual distances are very close, most commonly less than 10 m. The detection and signaling of predation concerns all troop members and leads to mobbing behaviors. Moreover, some situations, including aggressive interactions, regrouping after an alarm, and entering a large feeding site, occasion chorusing (all the calls of the repertoire are then uttered by more or less all troop members). No territorial behavior is observed in talapoins, and adult males do not play a specific role in intertroop spacing.

The vocal repertoire of the species is large (up to 17 call types) and the signals are highly graded. Although the vocal behavior of each age and sex class possesses some quantitative and qualitative limitations, the usage of the repertoire is generally equally distributed among all group members including adult males.

An additional example contrasts *C. neglectus* and *C. pogonias*. This latter species lives in harems consisting of about 15 monkeys. In these harems the adult male is frequently "seconded" by a second male, who accompanies him during warning sequences. Intragroup social exchanges are quite frequent, although the adult male maintains a relatively peripheral position; cohesion grunt-trills are frequently heard and may be uttered in long series of exchanges. Clearly, among the four *Cercopithecus* species described, *C. pogonias* is the most "talkative."

As previously described, the social structure of *C. neglectus* is apparently very simple, and the vocal activity quite reduced. The analyses of the repertoires of *C. neglectus* and *C. pogonias* have shown close similarities in the acoustic structure of their calls: in fact, the same basic physical structures are found. However, whereas in *C. neglectus* the six fundamental vocalizations are of discrete types, in *C. pogonias* they have given rise to 17 more or less graded signals. Thus in the latter species signals are more numerous and more graded; animals are more talkative, and

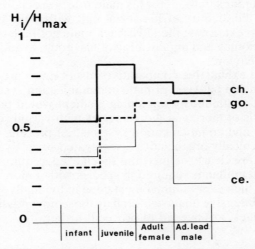

Figure 1.12. Homogeneity or heterogeneity of use of a vocal repertoire: comparison of relative diversity H_i/H_{max} of (Ce) our ceropithecine group (17 vocalizations), (Ch) chimpanzees (14 vocalizations), and (Go) gorillas (15 vocalizations) (chimpanzee and gorilla data from Marler and Tenaza 1977).

except for the adult male, they retain the ability to emit most calls of the repertoire.

It is tempting to correlate these differences in the complexity of vocal systems with parallel differences in the complexity of social organizations. Further studies are necessary to prove the existence of a possible causal relationship. Similar correlations have been explored by Marler (1976) for the chimpanzee and the gorilla. We have computed the relative diversity of usage of the overall repertoire for the *Cercopithecus* group analyzed here and for the two apes described by Marler. Four individual classes are considered for each group. Some common trends and specific differences are shown in Figure 1.12. This figure shows that age and sex class inequality in the use of the repertoire is minimal in the chimpanzee and maximal in the *Cercopithecus*. Yet, broadly speaking, in all three groups the diversity is lower in infants than in juveniles, and the diversities found in the adult females' repertoires are relatively comparable. Among these primates the main differences lie in the adult males' contribution: The greatest specialization is found in *Cercopithecus*, whereas in chimpanzees and gorillas, both sexes share most of the vocal categories.

Marler relates the high diversity index found for chimpanzee vocal usage to the high instability in chimpanzee society. All the animals in the population must face the same environmental and social constraints on an individual basis. In fact, not many responsibilities appear to be assigned to a particular social class. The male *Cercopithecus* is confronted with a completely different situation, for his main role is specialized and reduced to maintaining the identity of his social unit within the population. Between these two extremes, the silverback male gorilla has a more active role in the functioning and organization of his group, in addition to his role in territorial behavior.

Clearly, more exhaustive comparative observations are required to reveal the relationship between primate communication systems and social organization. A continuously tightening gradient would probably emerge from the analysis of more and more species with various complexities of social structure and communication systems. At present, it appears that the most hierarchically organized and segregated social systems, in which individual roles are clearly defined and specialized, require a simpler communication system than is required in species with a more fluid social organization, in which every animal must cope individually with all of life's contingencies. Intensive analyses, such as those now possible with biotelemetry techniques, are needed to proceed further.

REFERENCES

Baylis, J. R. 1976. A quantitative study of long-term courtship: II. A comparative study of the dynamics of courtship in two New World cichlid fishes. *Behaviour* 59:117–161.

Gautier, J.-P. 1974. Field and laboratory studies of the vocalizations of talapoin monkeys (*Miopithecus talapoin*): structure, function, ontogenesis. *Behaviour* 49:1–64.

    1975. Etude comparée des systèmes d'intercommunication sonore chez quelques cercopithécinés forestiers africains. Mise en évidence de corrélations phylogénétiques et socio-écologiques. Doctoral thesis, Université de Rennes.

    1979. Biotelemetry of the vocalizations of a group of monkeys. In *A handbook on biotelemetry and radio-tracking,* ed. C. J. Amlaner, Jr., and D. W. Macdonald, pp. 535–44. Oxford: Pergamon Press.

    In preparation (a). Structure of the vocalizations of hybrid monkeys of *Cercopithecus pogonias grayi* and *C. ascanius katangue.*

    In preparation (b). Radiotransmission of the vocalizations of monkeys: Defects, qualities, and prospects.

Gautier, J.-P., and Deputte, B. 1975. Mise au point d'une méthode télémétrique d'enregistrement individuel des vocalisations: application à *Cercocebus albigena. Terre el la vie* 29:298–306.

Gautier, J.-P., and Gautier, A. 1977. Communication in Old World monkeys. In *How animals communicate,* ed. T. E. Sebeok, pp. 890–964. Bloomington: Indiana University Press.

Gautier-Hion, A. 1971. L'écologie du talapoin du Gabon. *Terre et la vie* 25:427–90.

    1975. Dimorphisme sexuel et organisation sociale chez les cercopithecines forestiers. *Mammalia* 39:365–74.

Gautier-Hion, A., and Gautier J.-P. 1974. Les associations polyspécifiques de cercopithèques du plateau de M'passa, Gabon. *Folia Primatologica* 22:134–77.

1978. Le singe de Brazza: une stratégie originale. *Zeitschrift für Tierpsychologie* 46:84–104.

Green, S. 1975. Variations of vocal pattern with social situation in the Japanese macaque *Macaca fuscata:* a field study. In *Primate behavior:* Vol. 4, developments in field and laboratory research. ed. L. A. Rosenblum, pp. 1–102. New York: Academic Press.

Hutcheson, K. 1970. A test for comparing diversities based on the Shannon formula. *Journal of Theoretical Biology* 29:151–4.

Marler, P. 1973. A comparison of vocalizations of red-tailed monkeys and blue monkeys, *Cercopithecus ascanius* and *C. mitis,* in Uganda. *Zeitschrift für Tierpsychologie* 33:223–47.

1976. Social organization, communication and graded signals: the chimpanzee and the gorilla. In *Growing points in ethology,* ed. P. P. G. Bateson and R. A. Hinde, pp. 239–79. Cambridge University Press.

Marler, P., and van Lawick-Goodall, J. 1971. Vocalizations of wild chimpanzees (sound film). New York: Rockefeller University Film Service.

Marler, P., and Tenaza, R. 1977. Signaling behaviour of apes with special reference to vocalizations. In *How animals communicate,* ed. T. E. Sebeok, pp. 965–1033. Bloomington: Indiana University Press.

Quris, R., Gautier, J.-P., Gautier, J. Y., and Gautier-Hion, A. 1981. Organisation spatio-temporelle des activités individuelles et sociales dans une troupe de *Cercopithecus cephus. Terre et la vie* 35:37–53.

Shannon, C. E., and Weaver, W. 1949. *The mathematical theory of communication.* Urbana: University of Illinois Press.

Struhsaker, T. T., and Leyland, L. 1979. Socioecology of five sympatric monkey species in the Kibale forest, Uganda. *Advances in the Study of Behavior* 9: 159–228.

2 · Vocal concomitants of affiliative behavior in squirrel monkeys

HARRIET J. SMITH, JOHN D. NEWMAN,
and DAVID SYMMES

PRIMATE VOCAL COMMUNICATION AS A MODEL SYSTEM

The development of the specialized behaviors involved in sending and receiving information from conspecific animals has paralleled the development of complex sociality. Fruitful models for the study of social communication have been identified at several phylogenetic levels (Sebeok 1977), and it is clear that visual, tactile, olfactory, and acoustic channels have been exploited. In forest-dwelling primates, acoustic signaling is undoubtedly the most efficient mode of exchanging information with distant conspecific animals when speed and accuracy are important for survival. Ample testimony exists as to the range and complexity of vocal communication in these animals, and the possibility that some roots of man's remarkable spoken language are hidden in the vocal exchanges of monkeys and apes has intrigued many investigators.

The concept of a vocal lexicon in nonhuman primates (a set of discrete phonological entities employed by most adult members of the species) has gained acceptance as field studies describing the vocal signals used by the species of interest have been published. These studies have progressed from listing "call types" recorded over relatively prolonged periods of observation to grouping according to more narrowly defined functional contexts in which certain calls predominate. Examples of these classes are calls used primarily in alarm, long-distance contact, intraspecific aggression, territoriality, and so on.

To progress further in comprehending primitive language systems it is necessary to look closely at the structural variability of particular phonological entities and to test hypotheses about the kinds of information that may be transmitted. The study of that variability has revealed in some cases continuous and graded transitions among structural forms, an observation that substantiates the idea dating from Darwin's time that a continuously varying internal state can be externalized in graded communica-

We would like to acknowledge with thanks the contribution of Deborah Bernhards in collecting behavioral data in this study.

tive acts. Many early workers initially concluded that vocal signals reflect only "mood," or level of arousal, of the speaker (Marler 1965). In our current view, this approach is too restrictive and, in fact, may be a hindrance to progress (see also Seyfarth and Cheney, chap. 10, this volume). Investigation of the calls of the squirrel monkey has shown that some vocal forms such as the isolation peep are both discrete and stable (Symmes, Newman, Talmage-Riggs, and Lieblich 1979), whereas others such as the cackle appear to occur with a wide variety of structural variants (Schott, 1975). Most calls probably contain both discrete and graded components, and overemphasis on the latter may discourage the experimental testing of the information content of the former.

Another challenge to understanding primate vocal communication is the absence of clear behavioral responses to sounds in many social contexts. Efforts to apply the concept of the "sign stimulus" to vocal behavior (Mattingly 1972), using as a basis for such speculation those calls that *do* produce reliable listener responses, have diverted attention from an area of great promise: those vocalizations directed to a few chosen listeners at close range that do not appear to be associated with altered arousal or emotional tone. These "quiet" calls occur during play and affiliation, and the idea that the roots we seek may be found in the soft, semiprivate exchanges among monkeys whose survival depends on maintaining stable, peaceful social contacts is relatively new and quite promising.

THE STUDY OF AFFILIATIVE BEHAVIOR

Many investigators working with a wide variety of species have described behavior characterized by quiet, nonaggressive contact. The most prominent behavior of this kind is that directed toward infants, usually, but by no means invariably, by mothers. Among adults, quiet behavior includes resting in proximity, sharing sleeping or shaded areas, grooming, quiet truncal contact (such as huddling in squirrel monkeys), and less common forms such as mouth-to-mouth contact and tail entwining. Although play behavior is of interest as well, it is infrequently observed in adults. Consequently, the focus here is on affiliation.

Both common and relatively easy to recognize, affiliative interactions and their vocal concomitants have been described in a few reports. Among Old World primates, vocal signals apparently associated with affiliative behavior have been described in the Japanese macaque (Green 1975), the talapoin monkey (Gautier 1974), the chimpanzee (van Lawick-Goodall 1968), and the mountain gorilla (Fossey 1972), and several authors have mentioned such signals in New World primates (Winter, Ploog, and Latta 1966; Dumond 1968; Eisenberg 1976; Moynihan 1976). But no systematic descriptions appear to exist. These accounts are sparse compared with the extensive discussion of alarm signals and territorial defense calls, a testimony perhaps to the problems associated with studying affiliative behavior in the field (see also Gautier and Gautier-Hion, chap. 1, this volume). Many questions remain unanswered. Do the vocal behaviors constitute an invitation, or acceptance of an invitation, to engage in more direct forms of affiliation? Are they repeated affirmations of a "mood" or emotional state compatible with direct affiliation? Do they predict other behavior the speaker plans to undertake? Do they contain information about addressees or social relationships, that is, are they "diectic" in the sense Green and Marler (1979:125) have used the term? Our experimental studies have been directed toward answering some of these questions.

In considering our approaches to these tasks, we began with rather general expectations about the vocal concomitants of affiliation in squirrel monkeys. The chuck call is a commonly given call with several elements or structural components, the sequencing of which seems to be governed by a few rules. In their description of the vocal repertoire of *Saimiri*, Winter, Ploog, and Latta (1966) included the chuck call but suggested that it functioned primarily as a spacing mechanism. Because our observations in pilot studies suggested that chucks were related to affiliative behavior, we decided to examine the relationship more closely under relatively controlled laboratory conditions. A brief discussion of the several methodological problems that must be considered in attempting to correlate short-range vocal behavior with complex social interactions follows.

Habituating the study group

Success in studying affiliative behavior in a primate group depends on reducing the fear of the study group toward the human observers. Therefore, a process of habituation must take place, during which the group gradually becomes accustomed to the presence of the observers. An illustration of the significance of this habituation process in the field on the expression of affiliative behavior in mountain gorillas has been noted (Marler 1976). An additional problem arises in studies of captive individuals where the group is formed not by the monkeys themselves but by the investigators. We have witnessed sufficient cases of social incompatibility in artificially assembled groups to count this as a potential impediment to studies of affiliative behavior. Our study group of two male and four female Gothic-arch squirrel monkeys, captured as part of a naturally formed troop in the Peruvian Amazon (see Newman and Symmes, chap. 11, this volume), were kept together in captivity for approximately one year prior to initiation of formal data collection. During this time the study group was frequently exposed to the observers. As the habituating period progressed, the tendency for the entire group to clump together as a response to stress gradually decreased, and the affiliative preferences characteristic of socially integrated squirrel monkey groups (Hopf 1978) became evident.

Measuring behaviors of interest

Prior studies have made available schema for categorizing behavior that cover virtually the complete behavioral repertoire. In the squirrel monkey, Hopf, Hartmann-Weisner, Kuhlmorgen, and Mayer (1974) provided

Table 2.1. *Behavior categories*

Behavior	Definition
Affiliation	Sitting within 10 cm of another monkey, with or without body contact
Play	Wrestling, chasing, tail grabbing, or inhibited biting
Mount	Grasping flanks of another monkey from the rear, with or without intromission (originator or target thereof)
Aggression	Pushing, pulling, grabbing, biting (originator or target thereof)
Food steal	Grabbing food being held or eaten by another monkey
Genital display	Abducting thigh and directing exposed genitalia toward a particular monkey
Urine mark	Rubbing urine into a perch with the anogenital region
Urine wash	Depositing urine on one hand and smearing it on the sole of the foot on the ipsilateral side; often repeated with opposite hand and foot
Self-direct	Self-play, grooming or exploring
Self-huddle	Sitting quietly in a resting posture, tail draped over one shoulder, more than 10 cm from the nearest monkey
Ingestion	Eating or drinking
Environmental explore	Touching, manipulating, or sniffing any inanimate object
Residual	Sitting, standing, or locomoting more than 10 cm from the nearest monkey

exclusive descriptions for more than 100 behavioral units observed in both wild and captive squirrel monkey groups. However, many of these behavioral units are rarely observed (at least in captive groups), and several (e.g., grasping various parts of a partner's body) may be combined into a single category without significant loss of information. We have therefore eliminated some of Hopf's categories and combined others, arriving at a list of 13 categories for scoring ongoing behavior (Table 2.1). Although such a scheme is likely to miss some of the finest grain of behavioral detail, its simplicity is a virtue in the context of studying concurrent social and vocal behavior.

With respect to the measurement of affiliative behavior, we have followed Strayer, Bovenkirk, and Koopman (1975) in adopting a single affiliation score based on two observational variables. These authors found through the application of factor-analysis techniques that two measures (huddling and quiet sitting within 10 cm) formed a factor they called social attraction. We also found a significant correlation between these measures and combined them for our study of affiliation.

During formal observation periods, two observers collected behavioral data from the same subject during 5-min focal periods. The first observer recorded the behavior of the focal animal, the direction of social interactions, and the position of group members with respect to the focal animal. All vocalizations during the focal periods were tape-recorded for subsequent spectrographic analysis. By a series of soft coded taps on the microphone, the second observer marked those vocalizations for which a definite source could be identified visually. A continuous time signal recorded with the vocalizations also formed the time base for behavioral observations, permitting subsequent accurate recovery of temporal order in the two classes of data.

Quantifying variability in primate vocalizations

A major goal in the analysis of the vocal concomitants of affiliative behavior should be an assessment of the structural variability of vocal forms employed by the subjects during social interactions. The first step in this process involves collecting high-quality sound recordings for spectrographic analysis of the vocalizations, followed by identification of consistently occurring structural attributes. We accomplished this by mounting several condenser microphones in protective housings near the perches and mixing their outputs onto a single channel of a JVC KD-II stereo cassette tape recorder. The resulting tapes have a frequency response of ±3 dB from 0.1 to 15 kHz. After some screening by ear, vocalizations were analyzed by a VII 700 sound spectrograph at half original speed.

With this system we were able to classify by call type more than 95% of the sound spectrograms. Our study recorded 20 call types previously described by several authors as within the vocal repertoire of this species (Table 2.2). Since we were particularly interested in the chuck call, we examined the acoustic structure of this call in great detail (more than 2,000 individual chucks). Our analysis (Newman, Smith, and Talmage-Riggs, unpublished) identified the steeply descending frequency component noted by Winter, Ploog, and Latta (1966) as universally present and diagnostic for chucks. This structural element, to which we have given the name "mast," may be repeated and/or combined with other elements in a range of contexts. It was readily apparent in our spectrographic sample, and the identification of chucks was achieved without difficulty. Three common variants are shown in Figure 2.1.

At this stage of our analysis, we have not attempted to elucidate the possible significance in affiliative behavior of differences in number of masts or other properties of chucks. However, some details of chuck morphology, such as the interval between successive masts and the peak frequency of a mast, are sufficiently characteristic of individual monkeys to provide a basis for post hoc statistical recognition of each individual's calls, as described later.

Table 2.2. *Call-type categories*

Vocalization	Characteristics
Single chuck[a,f]	Single rapidly descending FM element with constant slope (mast)
Double chuck[a,f]	Sequence of two masts
Traller[c,f]	Sequence of more than two masts
Err single[f]	Err chuck variant; err preceding single chuck
Err double[f]	Err chuck variant; err preceding double chuck
Err traller[f]	Err chuck variant; err preceding traller
Yap[a,f]	Single mast starting above 10 kHz followed by loud cackle
Kecker[a,f]	Sequence of two or more masts, each preceded by a loud cackle and an ascending FM element (flag)
Vit[d]	Single rapidly ascending FM element with constant slope
Vit twitter[d]	Sequence of two or more vits
Chirp[a]	Combination of a peep and 1 short twitter syllable
Twitter[a,d]	Sequence of two or more twitter syllables; no cackles
Isolation peep[a,e]	Long, tonal call without major slope reversals
Peep[a]	Abbreviated version of the isolation peep, usually lower in frequency
Play peep[a]	Like vit but higher in frequency, uttered during play
Squeal[a]	Similar to peep but frequency trends upward and slope is not constant
Err[a]	Series of soft clicklike pulses (similar to purring)
Cackle[a]	Low fundamental frequency (below 1 kHz) with numerous overtones; may be noisy (harmonics obscured)
Shriek[a]	Broadband noise
Oink[b]	Cackle followed, without interruption, by a vit or squeal

Source note: [a] Winter, Ploog, and Latta (1966); [b] Winter (1968); [c] Talmage-Riggs, Winter, Ploog, and Mayer (1972); [d] Newman, Lieblich, Talmage-Riggs, and Symmes (1978); [e] Symmes, Newman, Talmage-Riggs, and Lieblich (1979); [f] Newman, Smith, and Talmage-Riggs, unpublished.

Indentifying the sources of affiliative vocalizations

Since vocalizations uttered during affiliative behavior are likely to involve several individuals calling softly and in rapid sequence, attributing a series of vocalizations to their correct sources presents an obvious problem. Several possible solutions are available. One would be to attach microphone and radio-transmitter units to each animal and to use telemetry to transmit the signals to an FM receiver and recorder, as described by Gau-

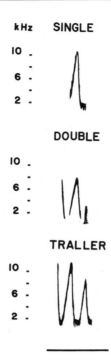

Figure 2.1. Sound spectrograms (retouched) of three typical chuck variants, all of which were recorded in this study: single chuck, double chuck, and traller. Time calibration = 0.2 sec.

tier and Gautier-Hion (chap. 1, this volume). This method may be the only practical one for studying nocturnal species. However, there are also disadvantages with this method. The subjects of interest may show poor tolerance to wearing transmitters and thereby exhibit unnatural behavior, and capturing the animals to attach the devices or to replace batteries in long term studies is both difficult and disruptive. The use of conventional microphones and recorders, and a large observer team to provide simultaneous coverage of a greater number of animals is an alternative method, but it is likely to be impractical in many research settings and increases the difficulty of habituating the animals to the observers. Furthermore, the problem of identifying obscured subjects remains. A third method, the one we adopted, is based on the growing literature demonstrating individual idiosyncracies ("vocal signature") in the calls of primates and other animals (Marler and Hobbett 1975; Waser 1977; Lillehei and Snowdon 1978), for individual differences in the structure of affiliative sounds could be used to identify retrospectively the sources of unattributed vocalizations. In a test of this approach, using only visually attributed chucks from a pilot group and the present study group, we were able to classify correctly 95% of chucks from both groups through the use of discriminant-function analysis. The analysis was based on several measurements of each chuck spectrogram (Smith, Newman, Hoffman, and Fetterly, in press). Using the profiles of the six members of the present study group, we assigned each chuck of unknown source to the most probable individual. Thus, all chucks for which good quality spectrograms could be produced were included in the data analysis, with confidence that the error rate for attributing chucks to individuals was negligible. Of 892 chucks used as primary vocal data in this study, 707 were attributed retrospectively by this method.

EXPERIMENTAL OBSERVATIONS

We collected vocal and other behavioral data from our six-member social group (four females, two males) for a total of 40 hours, utilizing the data acquisition techniques summarized above. Observations were generally made early in the morning (the animals were housed indoors and a 12/12 light-dark cycle was maintained by timers). These observations were made during the season of the year when the monkeys were reproductively inactive. Approximately 1,200 vocalizations were recorded during this period, from which 892 clear chuck spectrograms were obtained. (This number does not include the pilot work leading to development of statistical profiles.)

Vocalizations associated with affiliative behavior

The distributions over behavior of all vocalizations that occurred more than twice during this study are shown in Figure 2.2. Chuck calls were the most commonly uttered vocalization during both affiliative and nonaffiliative behavior, although the rate of chucks during affiliative behavior was significantly higher than the rate during nonaffiliative behavior (χ^2 (1) = 148.41, $p < .001$).[1] Expected frequencies were calculated by multiplication of the total frequency of chucks by the percentages of time spent in affiliative and nonaffiliative behavior. None of the other vocalizations recorded revealed a significant relationship in terms of rate with affiliative behavior.

Finding that rate of chuck production was much higher during affiliative than nonaffiliative behavior, we looked for a simple linear correlation between the amount of time spent in affiliative behavior and the frequency of chucks produced by individual monkeys. A negligible positive correlation was found ($r = .02, p > .05$). As the number of chucks uttered did not increase as a function of the total time spent in affiliative contact, we may conclude that chucks were not totally affiliation-dependent; additional sources of variability in the contexts associated with chuck production were implied.

[1] All χ^2 statistics in which $df = 1$ were computed with Yates's correction.

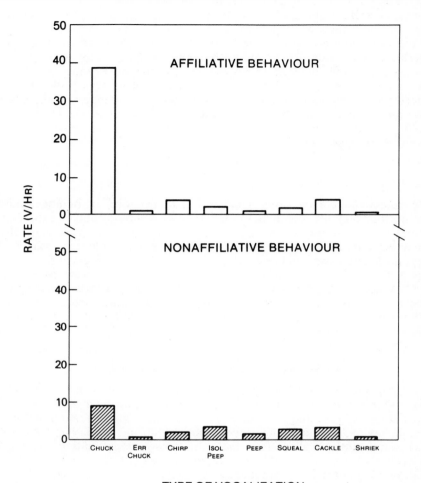

TYPE OF VOCALIZATION

Figure 2.2. Distribution of identified vocalizations during affiliative and nonaffilia-tive behavior, as rates of vocalizing per hour.

Sex differences in chuck usage

A striking difference related to sexual differentiation of vocal behavior was observed during affiliation. Females vocalized significantly more fre-quently than males ($\chi^2(1) = 40.772$, $p < .001$) and used a wider range of vocalizations. Peeps, squeals, and cackles were the only vocalizations re-corded from males in this context. No chucks were attributed to males, so the rate of chuck production given in Figure 2.2 is the female rate. All further discussion of chucks refers only to female monkeys.

Table 2.3. *Affiliative preferences*

Group member	Affiliative score[a]					
	Male 1	Male 2	Female 1	Female 2	Female 3	Female 4
Male 1	—	.23(3)	.39(1)	.03(5)	.09(4)	.26(2)
Male 2	.26(2)	—	.35(1)	.06(5)	.11(4)	.22(3)
Female 1	.29(2)	.23(3)	—	.30(1)	.02(5)	.16(4)
Female 2	.04(5)	.08(4)	.60(1)	—	.11(3)	.16(2)
Female 3	.16(3)	.18(2)	.06(5)	.14(4)	—	.47(1)
Female 4	.25(1)	.19(4)	.22(3)	.10(5)	.24(2)	—

Note: Figures in parentheses denote rank.
[a] Time spent affiliating with specified partner divided by total time in affiliation with all group members.

The role of preferred partners

To determine whether an association existed between strength of affiliative relationship and chuck production during affiliative behavior, the relative preferences of each monkey for its five potential partners were ranked from one to five on the basis of total time spent with each individual in affiliative contact. Where no preferences for particular individuals were expressed, each animal could be expected to spend about 20% of its total time in affiliation with each of its potential partners. Strong mutual affiliative preferences were observed in each animal toward one or two individuals within the group as shown in Table 2.3, but other associations between individuals hovered around, or slightly below, chance levels, indicating no preference for, or even slight aversion to, these individuals.

The rate of chuck production was compared under the following conditions: (1) Source monkey was not engaged in affiliative behavior, (2) source monkey was engaged in affiliative behavior with a nonpreferred partner (third-ranked individual), and (3) source monkey was engaged in affiliative behavior with a preferred partner (first-ranked individual). A repeated-measures analysis of variance revealed a significant difference in the rate of chucks uttered under these three conditions ($F (2/6) = 6.595$, $p < .05$). Figure 2.3 shows the chuck rate pattern for individual females.

The relationship between strength of affiliative preference and rate of chuck production was manifested by all four females, regardless of individual differences in overall rate. About 20% of affiliation time was spent with third-ranked partners, so the extremely low chuck rate seen during that type of affiliation was not the result of lack of opportunity. Rather, the nature of chuck production appeared to be discontinuous; preferred partners uttered chucks when engaged in affiliative behavior, whereas nonpreferred partners did not.

This finding generalized to the pattern of chucks produced during other

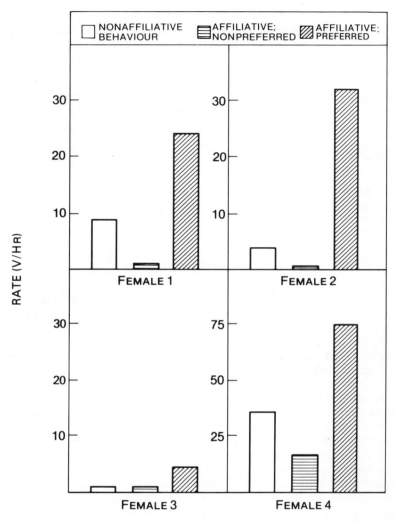

Figure 2.3. Pattern of chuck production by individual females during affiliative and nonaffiliative behavior.

social behavior. Although all females were mounted by males, only one female gave chucks during mounting (when she was mounted by her preferred partner). Only a few instances of chuck usage during aggressive interactions were recorded, but in five of the six cases the vocalizer was either the preferred partner of the recipient of aggressive behavior or, when the preferred partner was nearby, the recipient herself.

The significant conclusion drawn from these observations is that the

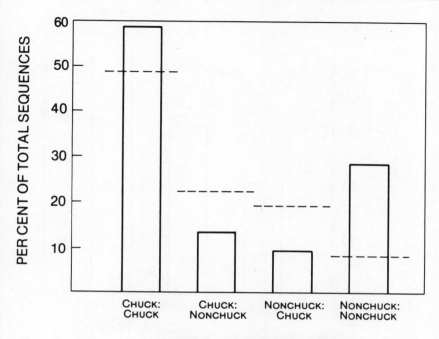

Figure 2.4. Distribution of different types of vocal sequences. Dashed lines indicate expected frequencies.

use of chucks during affiliation (by females) is determined to a great extent by the social relationships of the individuals involved.

Vocal sequences

The distributions of inter-chuck intervals over a sample of several observation days was bimodal, with one peak at short intervals (1 sec or less separation) and another peak at long intervals (10 sec or more separation). In addition, chucks less than 1 sec apart were more likely to represent vocal output by different individuals than chucks at intervals greater than 1 sec. Therefore, we defined chuck sequences as two or more consecutive chucks separated by intervals of 1 sec or less. The same criterion was used to define sequences in which one or both calls were other call types. The percentage of total vocalizations that fit the sequence definition was quite low overall, although 26% of all chucks occurred in sequences. These data show that the vocalization most frequently used in *Saimiri* vocal exchanges is the chuck. Figure 2.4 summarizes the frequency with which other call types were used in vocal sequences, and the

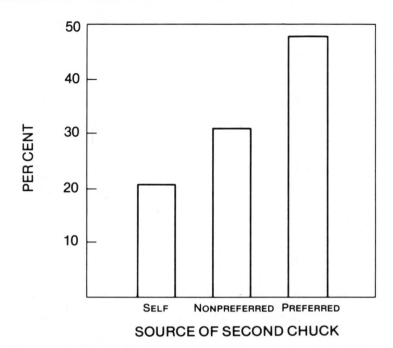

Figure 2.5. Participants in chuck sequences; effects of affiliative relationship on source of second chuck.

differences are significant ($\chi^2(1) = 50.598, p < .001$). Chucks seldom were recorded as components of sequences involving other call types.

To determine whether chuck sequences could be viewed as vocal exchanges, the data were examined for possible dependencies between the origin of the first chuck in a sequence and the origin of the second chuck. Figure 2.5 shows the percentage of all chuck sequences in which the sources were identical (monkey "answers" herself), were nonpreferred partners, or were preferred partners. A χ^2 goodness-of-fit test applied to the data indicated a significant difference in the distribution of chucks under these three conditions ($\chi^2(2) = 8.601, p < .025$). Preferred partners were found to answer each other more frequently than expected, and monkeys answered themselves or nonpreferred partners less frequently than expected.

Vocal sequences involving males ($N = 36$) primarily consisted of the dominant male "answering" himself (61% of all sequences); his second call type was usually different from the first. Of the remaining 14 sequences, 12 consisted of female calls following those of males, 1 of a male call following a female call, and 1 of a male call following that of another male. In contrast to this was the pattern shown by female participants in vocal sequences. Only 32 of 140 cases (23%) consisted of individual females vocalizing immediately after their own calls, whereas in 68% of sequences the vocalizers were different females. We concluded that vocal sequences with female participants may be viewed as vocal exchanges, whereas vocal sequences with male participants may not be interpreted in this manner.

Lack of behavioral response to chucks

Explicit behavioral responses coincident with, or following, chuck sequences were not found. When we reviewed our data on the behavior of both animals immediately following a vocal exchange, we found that the behavior of neither individual was likely to change as a consequence of vocalization. Animals engaged in affiliative behavior during a chuck sequence were about as likely to continue this behavior as they were to change their behavior, and animals whose ongoing behavior was nonaffiliative during a chuck sequence were not more likely to initiate an affiliative interaction than to enter into any other behavior.

SEX DIFFERENCES IN VOCAL BEHAVIOR

The observation that female squirrel monkeys vocalized more frequently and used different vocalizations during affiliation as compared with males may result from several factors. In this connection, we would like to discuss four simple hypotheses relating to male behavior in this species.

1. Males are physically unable to produce chucks. This hypothesis is untenable in view of published work by Jürgens (1979) in which chucks were recorded from males during brain stimulation experiments. In addition, we have observed that males utter a variety of chucklike calls (yaps, keckers), which would seem to require the same phonological capabilities as chucks.
2. Males are generally more silent than females, that is, the rate of vocalizing by males is lower during all behaviors. We have examined rates of vocalizing by males and females in a variety of contexts and have found that male rates are lower during affiliation (males 23/hr vs. females 43/hr), much higher during aggression (males 943/hr vs. females 62/hr), and about the same during mounting (males 117/hr vs. females 86/hr). Thus sex differences in overall rates of vocalizing appear to be context dependent rather than general (see also Gautier and Gautier-Hion, chap. 1, this volume).
3. Males engage in affiliative behavior less frequently than females, resulting in less opportunity for chuck calling. Distributions of time spent by sex in the different types of social behaviors scored in this study are shown in Figure 2.6. Males devoted slightly less time to affiliation than did females, but the difference was not significant. Consequently, the much lower chuck usage by males (no male chucks were recorded in the present study) cannot be explained on the basis of less affiliative behavior.

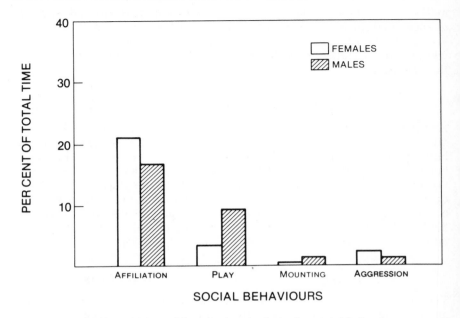

Figure 2.6. Male and female time budgets for social behaviors.

4. Affiliative bonds involving males are weaker and less persistent than those involving females, and chucks are associated with strong, stable bonds. To consider this hypothesis, we must review the published evidence relating to distinct social roles of males and females and their participation in affiliative relationships.

Nearly all authors agree, whether they have studied *Saimiri* in the laboratory (Rosenblum, Levy, and Kaufman 1968; Mason 1971; Coe and Rosenblum 1974; Fairbanks 1974; Strayer, Taylor, and Yanciw 1975; Hopf 1978; Mendoza, Lowe, and Levine 1978; Vaitl 1978), under semi-free ranging (Dumond 1968; Baldwin 1971) or free-ranging conditions (Thorington 1968; Baldwin and Baldwin 1972), that adult female squirrel monkeys are most attractive as a class to all other age and sex classes. Laboratory studies have demonstrated that females prefer one another as affiliative partners and that associations among females may be stable for months or even years and are relatively impermeable to entry by strangers (Fairbanks, 1974). Female subgroups or cliques are characterized by frequent affiliative interactions that include physical contact, resting together, and allomaternal behavior toward the offspring of subgroup members.

Relationships among males living in heterosexual groups are clearly different from those among females. Fairbanks (1974) characterized male relationships by their low rates of social interaction and relative permeability to strangers. Baldwin (1971) and Dumond (1968) proposed that all-male groups functioned primarily to provide travel companions for males (prevented as a group from integrating into the main troop). These all-male groups exhibited both a negligible amount of positive social interaction and large interindividual distances. Our data agree with these findings, as our two males did not display a strong mutual attraction for one another.

Relationships between male and female squirrel monkeys in the laboratory have been described by most of the previously cited authors as relatively weak, temporary, or unrestrictive toward new members. Field work has shown that females tolerated the presence of males nearby primarily during the mating season, when males approached females frequently for sexual, rather than affiliative, interactions (Dumond 1968; Baldwin and Baldwin 1972). In several laboratory studies intersexual preferences have been experimentally assessed. Mason (1971), using a "forced choice" technique, found that female squirrel monkeys preferred to be near a female stranger rather than a familiar male cagemate. After observing a heterosexual group moving freely in a large enclosure, Strayer, Taylor, and Yanciw (1975) concluded that only females with infants completely avoided adult males; several adult females without young did develop affiliative relationships with males. Fairbanks (1974) found that single females introduced into heterosexual groups containing established male and female subgroups were readily accepted by the male subgroup but not by the female subgroup. Finally, Vaitl (1978) found that females easily developed close affiliative relationships with males that were "almost indistinguishable in behavioral content from typical relationships among females," primarily when sufficient female partners were unavailable.

These reports all support the conclusion that males and females may develop cohesive relationships under special circumstances and that the bonds thus established are less stable than those between females. Although no striking differences were observed in either amount or quality of affiliative behavior shown by male–female and female–female partners in the present study, it remains to be seen whether both types of associations are equally stable in the face of "natural" changes in the biosocial environment such as the estrous cycle or the birth of young into the group. The available evidence, in our view, does not justify the rejection of the fourth hypothesis, and we propose that significant sex differences exist in the nature of affiliative bonding.

What might have been the ultimate force behind the evolution of a particularly strong affiliative bond between females? Allomaternal behavior has been well documented in female squirrel monkeys (Long and Cooper 1968; Rosenblum 1968). The ratio of infant birth weight to maternal weight in *Saimiri* is the highest of any primate species in which males do not contribute to parental care (Kleiman 1977). Perhaps the influence of allomaternal support on the successful rearing of offspring provided the stimulus for development of strong bonding among females, thereby ensuring that mothers and "aunts" did not become widely separated. A short-range vocal signal such as the chuck could function both to maintain contact between bonded females and to reiterate the symbol of a particular bond when the individuals were engaged in affiliative behavior. No parallel function for males can be seen.

FUNCTIONAL ROLE OF CHUCK CALLS

We have found no evidence that chucks function to maintain a given minimum spatial distance between individuals as proposed by Winter, Ploog, and Latta (1966). We have concluded from our observations that a major use of this call is to communicate about existing affiliative relationships. The evidence for this conclusion comes from both our behavioral data and our acoustic analysis of chuck structure. The higher rate of chuck calling during affiliative behavior and the significance of the preferred status of a partner, both in rate of calling and in participation in exchanges or dialogues, indicate the association of chucks with affiliative relationships. According to this view, the chuck "answer" may be a substitute for actual physical contact among separated animals, as well as a reinforcement of the bond among interacting individuals. This is clearly different from an undirected contact call that serves to maintain general proximity for all group members (such as those calls commonly uttered during foraging and traveling). The chuck therefore is more specified, and the nature of the response is dependent upon the affiliative relationships between particular individuals.

Statistically reliable individual differences in chuck-call structure as demonstrated in this study raise the possibility that these differences are perceived by the monkeys. It would seem reasonable that the mutual acoustic recognition of partners would facilitate communication between them (particularly in diaglogues). However, only the potential for such recognition was demonstrated here, and the evidence for vocal signature in chuck calls is not yet as strong as that for the isolation peep (Symmes et al. 1979). More data is needed on chuck stability in large groups over extended periods.

In considering the descriptions by others of the class of soft intragroup calls of which the *Saimiri* chuck is a splendid example, we have found in one study an interesting parallel to our findings. Green (1975), working with Japanese macaques in a semi-free ranging habitat, looked closely at vocalization that could not be understood in terms of an association between graded structural variation and changes in internal state. This call, the girney, is an articulated sound, tonal or atonal, with great structural variability. Girneys are relatively soft sounds that occur primarily in contexts of affiliation, although they are common in other contexts as well, as are chucks. Several variants may be given rapidly in sequence, and girneys are the only vocalizations in this species that appear to be used in vocal exchanges. Girneys are often uttered by adult female macaques that are huddling in clusters or are grooming or by females without infants as they approach females with infants.

Chucks and girneys thus present many functional similarities. We have demonstrated, however, that chucks uttered by female squirrel monkeys perform one additional function: communication about the quality of ongoing affiliative relationships. Chuck production may therefore be a form of metacommunication (communication about communication; see Bateson 1955), in which the chuck call "stands for" the level of affiliative relationship between two individuals and in that sense modifies or alters the interpretation of other acts. Chucks are therefore not affiliative behavior; they are communication about affiliative relationships. Altmann (1967) has described three types of metacommunicative messages found in primates: (1) Those that serve to direct messages to particular individuals, (2) those that distinguish between playful and agonistic behavior, and (3) those that indicate the hierarchical status of the communicator. To this list we propose to add a fourth category: Those signals that indicate the quality of social bonds between individuals.

We believe that clues as to the origins and evolution of human language may be uncovered through study of these short-range, socially mediated vocalizations. A likely hypothesis is that such vocalizations have evolved in parallel with affiliative behavior in many primate species. Vocal behavior in nonhuman primates not predictive of motor acts and not indicative of (changing) internal state thus may be general and comprehensible. It is our hope that the emphasis in previous investigations on the defensive, territorial, and agonistic vocal forms may give way to broader analysis of animal sounds and that additional continuities between human and animal language will be found in the study of affiliative behavior and its signals.

REFERENCES

Altmann, S. A. 1967. The structure of primate communication. In *Social communication among primates*, ed. S. A. Altmann, pp. 325–62. Chicago: University of Chicago Press.
Baldwin, J. D. 1971. The social organization of a semifree-ranging troop of squirrel monkeys. *Folia Primatologica* 14:23–50.
Baldwin, J. D., and Baldwin, J. I. 1972. The ecology and behavior of squirrel monkeys (*Saimiri oerstedi*) in a natural forest in western Panama. *Folia Primatologica* 18:161–84.
Bateson, G. 1955. A theory of play and fantasy. *Psychiatric Research Reports A* 2:39–51.
Coe, C. L., and Rosenblum, L. A. 1974. Sexual segregation and its ontogeny in squirrel monkey social structure. *Journal of Human Evolution* 3:1–11.
Dumond, F. V. 1968. The squirrel monkey in a seminatural environment. In *The squirrel monkey*, ed. L. A. Rosenblum and R. W. Cooper, pp. 87–145. New York: Academic Press.
Eisenberg, J. F. 1976. Communication mechanisms and social integration in the black spider monkey, *Ateles fusciceps robustus*, and related species. *Smithsonian Contributions to Zoology* No. 213:1–108.
Fairbanks, L. 1974. An analysis of subgroup structure and process in a captive squirrel monkey (*Saimiri sciureus*) colony. *Folia Primatologica* 21:201–24.
Fossey, D. 1972. Vocalizations of the mountain gorilla (*Gorilla gorilla beringei*). *Animal Behaviour* 20:36–53.

Gautier, J.-P. 1974. Field and laboratory studies of the vocalization of talapoin monkeys (*Miopithecus talapoin*). *Behaviour* 51:209–73.

Green, S. 1975. Variation of vocal pattern with social situation in the Japanese monkey (*Macaca fuscata*): a field study. In *Primate behavior: developments in field and laboratory research*, Vol. 4, ed. L. A. Rosenblum, pp. 1–102. New York: Academic Press.

Green, S., and Marler, P. M. 1979. The analysis of animal communication. In *Handbook of behavioral neurobiology, Vol. 3: Social behavior and communication*, ed. P. Marler and J. G. Vandenbergh, pp. 73–158. New York: Plenum.

Hopf, S. 1978. Huddling subgroups in captive squirrel monkeys and their changes in relation to ontogeny. *Biology of Behaviour* 3:147–62.

Hopf, S., Hartmann-Weisner, E., Kuhlmorgen, B., and Mayer, S. 1974. The behavioral repertoire of the squirrel monkey (*Saimiri sciureus*). *Folia Primatologica* 21:225–49.

Jürgens, U. 1979. Vocalization as an emotional indicator: a neuroethological study in the squirrel monkey. *Behaviour* 69:88–117.

Kleiman, D. G. 1977. Monogamy in mammals. *Quarterly Review of Biology* 52:39–69.

Lillehei, R. A., and Snowdon, C. T. 1978. Individual and situational differences in the vocalizations of young stumptail macaques (*Macaca arctoides*). *Behaviour*, 65:270–81.

Long, J. O., and Cooper, R. W. 1968. Physical growth and dental eruption in captive bred squirrel monkeys, *Saimiri sciureus* (Leticia, Colombia). In *The squirrel monkey*, ed. L. A. Rosenblum and R. W. Cooper, pp. 193–205. New York: Academic Press.

Marler, P. R. 1965. Communication in monkeys and apes. In *Primate behavior: field studies of monkeys and apes*, ed. I. DeVore, pp. 420–38. New York: Holt, Rinehart and Winston.

1976. Social organization, communication and graded signals: the chimpanzee and the gorilla. In *Growing points in ethology*, ed. P. P. G. Bateson and R. A. Hinde, pp. 239–80. Cambridge University Press.

Marler, P. R., and Hobbett, L. 1975. Individuality in a long-range vocalization of wild chimpanzees. *Zeitschrift für Tierpsychologie* 38:97–109.

Mason, W. A. 1971. Field and laboratory studies of social organization in *Saimiri* and *Callicebus*. In *Primate behavior: developments in field and laboratory research*, Vol. 2, ed. L. A. Rosenblum, pp. 107–37. New York: Academic Press.

Mattingly, I. 1972. Speech cues and sign stimuli. *American Scientist*, 60:327–37.

Mendoza, S. P., Lowe, E. L., and Levine, S. 1978. Social organization and social behavior in two subspecies of squirrel monkeys (*Saimiri sciureus*). *Folia Primatologica* 30:126–44.

Moynihan, M. 1976. *The New World primates*. Princeton: Princeton University Press.

Newman, J. D., Lieblich, A., Talmage-Riggs, G., and Symmes, D. 1978. Syllable classification and sequencing in twitter calls of squirrel monkeys (*Saimiri sciureus*). *Zeitschrift für Tierpsychologie* 47:77–88.

Rosenblum, L. A. 1968. Mother–infant relations and early behavioral development in the squirrel monkey. In *The squirrel monkey*, ed. L. A. Rosenblum and R. W. Cooper, pp. 207–34. New York: Academic Press.

Rosenblum, L. A., Levy, E. J., and Kaufman, I. C. 1968. Social behavior of squirrel monkeys and the reaction to strangers. *Animal Behaviour* 16:288–93.

Schott, D. 1975. Quantitative analysis of the vocal repertoire of squirrel monkeys (*Saimiri sciureus*). *Zeitschrift für Tierpsychologie* 38:225–50.

Sebeok, T., ed. 1977. *How animals communicate*. Bloomington: Indiana University Press.

Smith, H. J., Newman, J. D., Hoffman, H. J., and Fetterly, A., In press. Statistical discrimination among vocalizations of individual squirrel monkeys (*Saimiri sciureus*). *Folia Primatologica*.

Strayer, F. F., Bovenkirk, A., and Koopman, R. F. 1975. Social affiliation and dominance in captive squirrel monkeys (*Saimiri sciureus*). *Journal of Comparative and Physiological Psychology* 89:308–18.

Strayer, F. F., Taylor, M., and Yanciw, P. 1975. Group composition effects on social behaviour of captive squirrel monkeys (*Saimiri sciureus*). *Primates* 16:253–60.

Symmes, D., Newman, J. D., Talmage-Riggs, G., and Lieblich, A. K. 1979. Individuality and stability of isolation peeps in squirrel monkeys. *Animal Behaviour* 27:1142–52.

Talmage-Riggs, G., Winter, P., Ploog, D., and Mayer, W. 1972. Effect of deafening on the vocal behavior of the squirrel monkey (*Saimiri sciureus*). *Folia Primatologica* 17:404–20.

Thorington, R. W. 1968. Observations of squirrel monkeys in a Colombian forest. In *The squirrel monkey*, ed. L. A. Rosenblum and R. W. Cooper, pp. 69–85. New York: Academic Press.

van Lawick-Goodall, J. 1968. A preliminary report on expressive movements and communication in the Gombe Stream chimpanzees. In *Primates: studies in adaptation and variability*, ed. P. C. Jay, pp. 313–74. New York: Holt, Rinehart and Winston.

Vaitl, E. 1978. Nature and implications of the complexly organized social system in non-human primates. In *Recent advances in primatology*, ed. D. J. Chivers and J. Herbert, pp. 17–30. New York: Academic Press.

Waser, P. M. 1977. Individual recognition, intragroup cohesion and spacing: evidence from sound playback to forest monkeys. *Behaviour* 60:28–74.

Winter, P. 1968. Social communication in the squirrel monkey. In *The squirrel monkey*, ed. L. A. Rosenblum and R. W. Cooper, pp. 235–53. New York: Academic Press.

Winter, P., Ploog, P., and Latta, J. 1966. Vocal repertoire of the squirrel monkey (*Saimiri sciureus*), its analysis and significance. *Experimental Brain Research* 1:359–84.

3 · A neuroethological approach to the classification of vocalization in the squirrel monkey

UWE JÜRGENS

The squirrel monkey (*Saimiri sciureus*) has an extremely rich vocal repertoire of about a dozen very differently structured call types. However, the call types do not occur as strictly stereotyped forms but manifest themselves as points along continua of vocal variants. Not all variants along these continua occur with the same frequency; some forms clearly occur much more often than others. Nevertheless, there are hundreds of variants forming gradations and intermediates between a few "typical" call types.

In such a situation, to treat each call variant as a distinct call type and look for its specific meaning would require an unrealistic time investment in data collection and statistical analyses. In this chapter I intend to propose a different approach to the problem. The goal of this approach is to group the numerous call variants according to functional *similarity*. More specifically, an attempt is made to devise a precisely defined *system* of classification without knowing the exact meaning of each call, that is, a classificatory scheme in which each call is ordered according to very few overall parameters, comparable to the periodic ordering of the chemical elements. By knowing a few general relationships between function and acoustic structure, together with the relative position of a call within the system, an investigator can extrapolate its meaning, at least roughly, without doing a specific behavioral analysis.

THE CLASSIFICATION SCHEME

Acoustic analysis

The first step in grouping the squirrel monkey's calls according to functional similarity was to classify them according to their frequency–time course by comparing the sonagrams of all calls recorded. The assumption underlying this step is that calls with a similar function have a similar acoustic structure. This assumption seems reasonable but, in practice, raises the problem of how to judge similarity. With such complex phenomena like vocalizations, which are characterized by several parameters, namely, frequency, amplitude, and time, similarity is not a well-defined, specific property but the result of a comparison of complex gestalts. In other words, two calls usually do not differ along only one dimension but along two or three, so that, for instance, there is a problem of how to judge changes of a dimension in one call against changes of a different parameter in a second call. The only way out of this dilemma is to avoid a detailed ordering of the calls and, instead, simply sort the calls into a rough grouping of a few classes according to natural breaks within the vocal gestalt continuum. Five call classes emerge in such a grouping of the squirrel monkey. Within each class, any two calls are related to each other by a continuum of intermediate forms between them. Intermediate forms between classes, in contrast, are sparse.

Transitional frequency-of-occurrence analysis

In those few cases in which a call seems to bridge different classes, the spectrogram alone cannot help to decide into which specific existing class it should be placed or whether it should be treated as a separate class. To deal with this problem, a new criterion was introduced: the transitional frequency of occurrence of the calls. That is, a count was made of how often each call type followed any other in intraindividual call sequences. Such sequences were defined as successions of calls of the same animal (i.e., uninterrupted by a partner) with a time interval between individual calls of less than 10 sec. These intraindividual transitional frequencies were then compared statistically with computed chance frequencies. The working hypothesis underlying this classification criterion assumes that calls with a high probability of following each other intraindividually express more similar motivational states than calls with a low transitional probability. In other words, two calls occurring in the same situational context are probably more closely related functionally than calls occurring in different situations. If this assumption is correct, each call type should have the highest positive intraindividual correlation with itself. This was exactly the case for all call types of the squirrel monkeys we tested.

A concrete example illustrates this second classification step. A typical vocalization of the squirrel monkey is the yapping call. This call is uttered during the mobbing of a potential predator. In its most common form, it consists of a high-pitched component that descends steeply in frequency over several kilohertz and ends in a low-pitched cawinglike component. As Figure 3.1 shows, there is a series of variants in which the cawing component is reduced until, in the extreme, only the high-pitched component is left. Conversely, there are also variants in which the high-pitched component is reduced, so that, at the other extreme of the series, only the cawing component remains. There is thus a continuum of call variants leading from the low-pitched cawing, with a fundamental frequency of a few hundred hertz, to the isolated high-pitched component, the so-called alarm peep, with a fundamental frequency of up to 17 kHz. The question then is whether yapping should be grouped with cawing, with alarm peep, with both, or with neither. In the first description of the squirrel monkey's

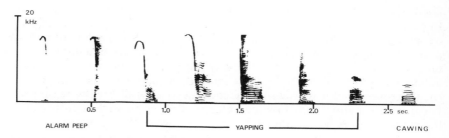

Figure 3.1. Frequency–time diagram of eight representatives of the alarm-peep/yapping/cawing continuum of vocal gestalts.

calls by Winter, Ploog, and Latta (1966), yapping and cawing were placed in the same class; the alarm peep was placed in a different class. The only justification given by the authors for this classification was that yapping was more "similar" in its frequency–time structure to cawing than to any other call. As Figure 3.1 illustrates, this argument is not sound, because it holds only for some yapping variants. Another attempt to classify these calls was published by Schott (1975). He regarded yapping as a separate class. This position is, of course, more defensible. However, it is not the best of all possible classifications, for if the whole vocal repertoire was classified in such a manner, the consequence would be a very large number of isolated call classes, with no indication of the manifold structural and functional interrelationships that in fact exist. In our own classification, we calculated the intraindividual transitional frequencies of yapping, alarm peep, and cawing and found that the alarm peep was the only call of the whole vocal repertoire that showed a statistically significant positive correlation in its occurrence with yapping. Furthermore, different yapping variants were highly correlated with each other. Cawing, in contrast, neither followed nor preceded yapping more often than would be expected by chance. Cawing showed, however, a high positive correlation with the shrieking and groaning calls to which it is closely similar in frequency–time structure. This means that yapping and alarm peep are related to each other both structurally and functionally. On the other hand, cawing shows only a structural, not a functional, relation to yapping but bears a structural, as well as a functional, relation to shrieking and groaning. From these results, it seems appropriate to classify yapping and alarm peep in the same class and cawing in a different class.

Hedonic-quality analysis

Although the two criteria of structural similarity and transitional frequency of occurrence are useful for the formation of call classes and grouping within classes, they do not supply information about the functional significance of the calls. The next step in a classification approach, therefore, is to identify a *functional* indicator. To be suitable as a classification parameter, that is, to allow its application to all levels in the vocal repertoire, such an indicator should be as general as possible. Such a universal indicator, in our opinion, is the hedonic (or aversive) quality of the emotional state accompanying vocalization. In other words, we assume that vocalizations are expressions of specific emotional states and that all emotional states can be characterized (to some degree) by the aversive or pleasurable quality they bear. Therefore, the third step in our classification procedure required the measurement of the aversive or pleasurable emotional concomitants of different call types. Earlier studies had found that electrical stimulation of specific brain structures produced vocalizations (Jürgens, Maurus, Ploog, and Winter 1967; Jürgens and Ploog 1970). Therefore, depending on the brain structure stimulated and the electrical stimulus parameters, a great number of different call types could be obtained. More recent studies (Jürgens 1976) showed that most electrically elicitable vocalizations were not triggered directly by the electrical stimulation but represented secondary reactions to stimulation-induced motivational changes. That is, in most brain structures from which vocalizations can be obtained, the vocalizations must be understood as expressions of stimulation-controlled emotional states. A measurement of the hedonic, or aversive, quality of these emotional states therefore could be attained by giving the animals themselves the opportunity to switch on and off the vocalization-inducing brain stimulation.

Monkeys, implanted with intracerebral electrodes located at points yielding different call types, were put into a self-stimulation cage, which consisted of two compartments. Presence of the animal in one compartment was followed automatically by electrical stimulation of a vocalization-eliciting electrode, whereas presence in the other compartment was not associated with stimulation. As the animals were free to move from one compartment to the other, they could switch the vocalization-eliciting stimulation on and off at will. Every few minutes, the stimulation was switched externally from one compartment to the other to observe whether the animal followed the stimulation, escaped it, or remained passively in one compartment. The stimulation was balanced in such a way that the total amount of exposure was equal in both compartments. Data were evaluated by calculating for each electrode the percentage of the total session time during which the animal received stimulation. Then the mean for all electrodes yielding the same call type was calculated for each call that had been elicited from at least five loci. Five categories were set up: 0–20% – highly aversive; 20–40% – slightly aversive; 40–60% – neutral; 60–80% – slightly rewarding; 80–100% – highly rewarding. Altogether, 251 electrode positions yielding 47 call types were tested.

The results of these self-stimulation tests not only confirmed the classifications made on the basis of the other two criteria but, in addition, allowed a general characterization of the various call types. A concrete example may illustrate this.

Figure 3.2 shows four call variants that, according to their structural similarity and intraindividual transitional frequencies of occurrence, have been grouped into one class. It can be seen that twitter is more similar to

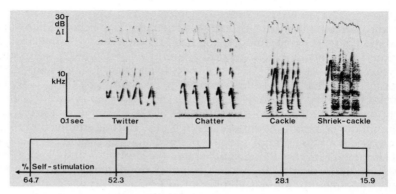

Figure 3.2. Four representatives of the twittering/chattering/cackling class with their self-stimulation scores.

chatter than to cackle and shriek-cackle; chatter is more similar to twitter and cackle than to shriek-cackle; and cackle is most similar to shriek-cackle; so that, according to structural similarity, the order should be: twitter, chatter, cackle, shriek-cackle. If the transitional frequencies of occurrence are calculated, it is found that the only other call showing a significant positive correlation with twitter is chatter; chatter shows its highest positive correlation with cackle, and cackle is correlated most closely with chatter (shriek-cackle did not occur in the sequences evaluated for transitional-frequency calculations). The order according to transitional frequencies of occurrence thus parallels that of structural similarities. The self-stimulation test, finally, yielded the highest self-stimulation score for twitter, a somewhat lower score for chatter, a still lower score for cackle, and the lowest score for shriek-cackle; it thus led to the same ordering as the other two criteria. Moreover, it revealed that twitter signals a pleasurable state of the animal; that chatter lacks a clear pleasurable or aversive component; that cackle expresses a slightly unpleasant emotional state; and that shriek-cackle appears to be uttered only in highly aversive situations, from which the animal tries to escape if at all possible.

APPLICATION OF CLASSIFICATION SCHEME TO REPERTOIRE

A comparison of the self-stimulation scores across the different call classes reveals that highly aversive, slightly aversive, and neutral emotional states can be expressed by call variants of all five classes. This is illustrated by Figure 3.3. Here, for each self-stimulation category, one representative is shown for each class. The innermost circle includes only calls expressing highly aversive states. The next circle includes representatives of the slightly aversive category; the third circle represents neutral

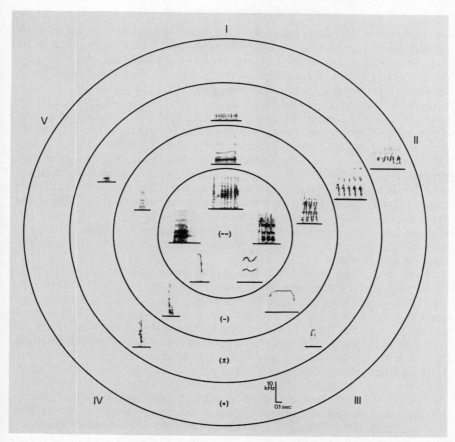

Figure 3.3. Vocalization system of the squirrel monkey. For reasons of simplicity, only one call type per class and self-stimulation category is shown.(−−), highly aversive calls (self-stimulation: <20%); (−), slightly aversive calls (self-stimulation: 20–40%); (±), neutral calls (self-stimulation: 40–60%); (+), pleasure calls (self-stimulation: >60%). The names of the 16 vocalizations shown are, beginning with the innermost circle: I, spitting, growling, purring; II, shriek-cackling, cackling, chattering, twittering; III, squealing, isolation peep, chirping; IV, alarm peep, yapping, clucking; V, shrieking, cawing, groaning.

calls. Thus, for instance, a highly aversive state can be represented by very different call types. This means that the vocal repertoire of the squirrel monkey is multidimensional. Or, more specifically stated, the functional significance of a specific call cannot be characterized exclusively by its self-stimulation scores; in addition the call class to which it belongs must be considered. The following examples may illustrate this in greater detail.

Spitting/growling/purring (class I)

Let us first consider call class I of Figure 3.3. The calls in this class consist of very short (<30 msec) clicklike elements that follow each other in series. One series can have up to 60 elements, which are repeated in a 50 ± 20-Hz rhythm and are always nonharmonic. Figure 3.3 only shows three variants from a continuum. The category highly aversive is represented by spitting (I(− −)). This call typically accompanies dominance gestures like head grasping, genital display, and food stealing and is always carried out by the sender, never by the recipient, of the dominance gesture.[1] A representative of the category slightly aversive is growling (I(−)). Growling occurs when a strange monkey is put into an established group and dominance gestures are exchanged. Occasionally, mothers with infants on their backs use this call against other monkeys that try to touch the infant. It is also uttered by males pursuing females during the reproductive period. Finally, as a representative of the category neutral, purring is shown. Purring is uttered by infants during and immediately before suckling; it is also heard from adults during huddling and between copulatory activities.

If one compares the situations in which these calls occur, it becomes obvious that their common denominator is a state of self-assertiveness, that is, a challenging attitude. This is especially evident for spitting and growling but also holds to a lesser degree for purring. The intensity of challenge, that is, the probability that the animal will start an attack, is reflected directly by the self-stimulation score. Neutral self-stimulation behavior (purring) is associated with a state between contentment and self-confident demanding (infant before and during suckling). Slightly aversive self-stimulation behavior (growling) is typical for low-level threatening (mother keeping other females at a distance from her infant). Highly aversive self-stimulation behavior (spitting) occurs when an animal is aggressively threatening and is accompanied by actual aggressive intervention (head grasping).

These relationships between the functional significance of a call and its place within the hedonic system allow an approximate understanding of the significance of a call by just knowing a very few general characteristics of the class to which a call belongs, together with its self-stimulation score. The latter is not indispensable. As can be seen from Figure 3.3, the stepwise decrease in self-stimulation score from purring to spitting is paralleled by a systematic change in the acoustic structure of the calls. The lower the self-stimulation score, that is, the greater the aversiveness of a call, the greater its frequency range. Whereas purring has a frequency range of about 3 kHz, growling shows appreciable energy up to 7 kHz, and spitting spans more than 16 kHz. Consequently, if the self-stimulation score of a particular call is not known, it can be extrapolated from the scores of other calls of the class by simply considering its acoustic

[1] The situations in which this and the other calls being described occur are taken from Winter et al. (1966), Ploog, Hopf, and Winter (1967), Schott (1975), Marriott and Salzen (1978), and my own observations.

structure and the special kind of relationship between self-stimulation scores and the acoustic structure within that class.

Groaning/cawing/shrieking (class V)

A relationship between self-stimulation score and acoustic structure like that of the purring/growling/spitting class is found in the groaning/cawing/shrieking class. In Figure 3.3, class V again shows one representative for each self-stimulation category. The highly aversive representative, shrieking, is a nonharmonic call with a frequency range of more than 16 kHz. It occurs immediately before and after fights and can be heard from animals closely pursued and attacked by others as well as from higher ranking animals in connection with dominance gestures or during mock attacks against predators. Cawing (Figure 3.3, V (−)), a less aversive call type of this class, shows a more or less harmonic structure with a fundamental frequency of a few hundred (200–700) Hertz. Cawing regularly occurs in a group of animals when a strange monkey is introduced. It also occurs during food stealing, occupation and defense of a favorite resting place, unwanted body contact, before and after fights, and during attempts by a weaning mother to push her resisting baby from her back. The neutral self-stimulation category is represented by groaning, a low, nonharmonic call with a frequency range of about 3 kHz. Groaning is uttered by pregnant females during labor, during unwanted body contact, and at the very beginning or end of agonistic interactions. Groaning, cawing, and shrieking thus have in common the expression of protest; that is, the animals have a somewhat pronounced aggressive tendency but are clearly less self-confident than during the utterance of purring/growling/spitting calls. In other words, groaning, cawing, and shrieking are motivated by a mixture of aggression and flight. In a state of low excitement (neutral self-stimulation behavior), this mixture means slight uneasiness (groaning). In a more aroused state (slightly aversive self-stimulation behavior), cawing is used to signal objections made by the vocalizer to actions of other group members. Shrieking, finally, with its very low self-stimulation score, represents a defensive threat call of high intensity, that is, it has a high probability of being followed by either attack or flight. An examination of the relationships between self-stimulation score and acoustic structure reveals that, in correspondence with the purring/growling/spitting class, the lower the self-stimulation score, the greater the frequency range. A frequency range of about 3 kHz is typical for neutral calls; frequency ranges of up to 10 kHz and more than 16 kHz represent slightly and highly aversive calls, respectively.

Clucking/yapping/alarm peep (class IV)

A class very different in acoustic structure from the two just described is the clucking/yapping/alarm-peep class (Figure 3.3, IV). The calls of this class are characterized by a steep descent of their fundamental frequency over several kilohertz. The maximal fundamental frequency, that is, the point at which the downward sweep starts, is highest for alarm peep (average maximal frequency: 13 kHz) and lowest for clucking (7 kHz), with yapping taking an intermediate position (11 kHz). Alarm peep (Figure 3.3, IV (− −)) is a typical alarm call and is followed immediately by flight. It is elicited by the sudden appearance of a large bird or a snake. Yapping, as has been mentioned, is also used against potential predators. In contrast to the alarm peep, however, it is not followed by flight; in other words, it occurs only when there is a safe distance between monkey and predator. As yapping is always uttered in chorus, it may be called a mobbing call against potential predators. Clucking (Figure 3.3, IV (±)), finally, is uttered by huddling animals who are becoming restless or are trying to change the huddling order. It is also used by mothers calling their infants to mount their backs, and, conversely, by weaning mothers evading infants who are trying to return to their backs. Disputes about food are also accompanied occasionally by clucking (for a more detailed behavioral analysis see Smith, Newman, and Symmes, chap. 2, this volume; their chuck calls include our clucking and chattering calls). Clucking, yapping, and alarm peeps are similar in the sense that they are produced by a vocalizer who is worried about something but at the same time has a very low aggressive tendency. The low aggressive motivation in the case of alarm peep and yapping is due to the obvious superiority of the predator; in the case of clucking, it is due partly to positive social bonds (mother against infant) and partly to a low level of arousal (huddling animals). In anthropomorphic terms, we might say that alarm peep corresponds to a cry of alarm, yapping to a shouting or a scolding, and clucking to a curse.

The highly aversive self-stimulation behavior found for alarm peep corresponds well with its function in natural situations, namely, to indicate a high flight motivation. Yapping, with a clearly lower flight motivation, belongs to the category slightly aversive. Clucking, with the lowest flight motivation, shows neutral self-stimulation behavior. A comparison of the self-stimulation data with the call structures reveals that the greater the aversiveness of a call, the higher its maximum fundamental frequency.

Chirping/peeping/squealing (class III)

All the calls of the chirping/peeping/squealing class (Figure 3.3, III) have a harmonic structure with a fundamental above 2 kHz. In contrast to the clucking/yapping/alarm-peep class, their structure is not dominated by a steep frequency descent. Figure 3.3 again shows one representative for each self-stimulation category. The highly aversive representative of this class, namely, squealing (III (− −)), is characterized by an irregular frequency course with an average fundamental frequency of about 4 kHz and several intense overtones. This call has been observed in a masturbating, mature, low-in-rank male that has been threatened severely immediately prior to the call and in a dominant male spatially separated from, but in acoustic contact with, his group, wherein vigorous sexual intercourse was taking place. A somewhat higher pitched version of this call type is also heard from animals being chased and from an infant hindered from regaining its mother's back.

The slightly aversive self-stimulation category is represented by the isolation peep (isolation peep is only one out of five call types of this class that have been found to be slightly aversive). The isolation peep is a long, drawn-out call with a relatively constant frequency course, except at the beginning and end, and has a fundamental frequency of about 10 kHz. It is emitted as soon as one or several animals lose visual contact with the group. Hearing this sound, members of the group answer antiphonally. The isolation peep thus may be called a long-distance contact call.

Chirping, finally, which belongs to the neutral self-stimulation category, consists of a short ascending and then descending frequency modulation in the region of about 5 to 8 kHz. It is used only if there is visual and acoustic contact between the animals, that is, over small distances. Chirping serves to draw the attention of other group members to the vocalizer in nonagonistic situations, for instance during play and exploration of the environment. It is most often uttered by younger animals.

Chirping, isolation peeps, and squealing signal a desire to enter into social contact. In squealing, this desire is superimposed on, or in conflict with, a simultaneous, strong tendency to escape the momentary situation. Squealing, therefore, may be called a frustration call. Isolation peep, a classic long-distance contact call, functions to maintain the integrity of the social group. Chirping, a short-distance contact call, serves to focus a partner's attention to a specific situation.

A comparison of the acoustic structures and self-stimulation scores within this class reveals again that the total frequency range correlates positively with the degree of aversiveness. More specifically, the more aversive the calls showing a lack of overtones, the higher their average fundamental frequency; calls with high-reaching overtones are more aversive than calls without overtones even if their fundamental frequency is below that of the latter (compare squealing with isolation peep).

Twittering/chattering/cackling (class II)

The calls of the twittering/chattering/cackling class, which were mentioned briefly earlier (figures 3.2 and 3.3, II), are characterized by a periodic repetition of one or two elements in a 14 ± 3-Hz rhythm. These elements are chirpinglike in the twitter call, cluckinglike in the chatter call, chirpinglike and cawinglike or yappinglike in the cackle call, and chirpinglike and shriekinglike in the shriek-cackle. Twittering is uttered during feeding and in expectation of feeding, that is, if food or a person associated with food is detected. It also occurs when the sun breaks through the clouds and the animals gather for sunbathing and during the reunion of group members who have been separated for some time. Thus, although twittering occurs in a variety of situations, it seems to accompany situations that have in common a pleasurable quality. The positive self-stimulation behavior at twitter-eliciting loci corroborates this suggestion. Chattering is heard during feeding after prolonged food deprivation and during the feeding of an animal that is spatially isolated from, but in visual and acoustic contact with, other animals. Cackling and shriek-cackling typically occur before and during agonistic interactions between different groups. All four calls have a high *interindividual* frequency of occurrence, that is, they (and the motivation underlying them) have a contagious character. This high interindividual correlation points to the general function of this class, namely, to confirm social bonds between the vocalizers. Such a reassurance of social integration or "companionship" serves to recruit fellow combatants in the case of cackling and shriek-cackling and to draw the attention of other group members to pleasurable events of common interest in the case of twittering.

The relationships between call structure and self-stimulation score are somewhat more complex than in the other classes. Twittering is the only call type of the whole vocal repertoire showing positive self-stimulation. Therefore we cannot determine which of its acoustic features must be regarded as typical for the vocal expression of pleasure. It is of interest, however, that the single elements of twittering, which are practically identical with chirping, do not have this quality. As for chattering, cackling, and shriek-cackling, there is no distinct increase of the total frequency range with the degree of aversiveness. This results because the clucking and chirping elements, which make up the highest-pitched part of the call, reach about equal heights in all three call types. However, if the frequency range of the cawing and shrieking element is considered, a clear correlation emerges. Chattering, which lacks such a component, belongs to the neutral self-stimulation category, to which its constituent element, clucking, also belongs. Cackling, which is made up of chirpinglike and cawinglike elements, is a member of the same self-stimulation category as cawing. Shriek-cackling, with its combination of chirping and shrieking elements, belongs to the same category as shrieking.

CONCLUSIONS AND SUMMARY

The present study shows that the extensive vocal repertoire of the squirrel monkey can be arranged in such a way that a meaningful system emerges. The criteria applied in the formation of this system are: similarity of call structure, transitional frequency of occurrence, and hedonic quality of the emotional state underlying the calls. The system thus obtained consists of five call classes. Within each class a large number of call variants are found, which form a continuum of vocal gestalts. However, intermediate forms between different classes are sparse. Each class can be characterized by specific acoustic and functional features. Within each class, the aversiveness criterion allows a linear ordering of the diverse calls. The characteristics of the five classes can be summarized as:

1. *Purring/growling/spitting (class I).* The calls of this class consist of click series in a 50 ± 20-Hz rhythm. Functionally, they express a challenging (or self-asserting) attitude. The more aversive the call, the

more intense is its threatening character; that is, there is an increasing probability that it will be accompanied by an aggressive action.

2. *Groaning/cawing/shrieking (class V)*. This class consists of nonrhythmic, noiselike or low-pitched harmonic calls with major acoustic energy below 3 kHz. Functionally, they act as protest calls. With increasing aversiveness, their function changes from an expression of slight uneasiness to one of intensive defensive threat.
3. *Clucking/yapping/alarm peep (class IV)*. The calls of this class are short, loud, and show a steep descent of main energy from higher to lower frequencies. They function as a warning signal. The less aversive versions announce disagreement with a conspecific; the more aversive versions serve as alarm calls against predators.
4. *Chirping/isolation peep/squealing (class III)*. This class contains all high-pitched, harmonic, nonrhythmic calls with the exception of clucking and alarm peep. Its function is to express the desire for social contact. The less aversive versions are used to draw another group member's attention to the vocalizer in nonagonistic situations; the more aversive versions indicate a conflict between contact-seeking and flight motivation, that is, submission or social frustration.
5. *Twittering/chattering/cackling (class II)*. The calls of this class consist of rhythmic repetitions of elements similar to chirping, clucking, cawing, and shrieking in various combinations; the repetition rate is 14 ± 3 Hz. Their function is to confirm social bonds, that is, to create a feeling of companionship. Calls without an aversive quality are used to announce pleasurable events of common interest; calls with an aversive quality announce intraspecific mobbing against one or several outsiders.

In a recent study (Jürgens 1979), an attempt was made to compare the squirrel monkey's calls and their associated situations with those of other primates as given in the literature. This comparison revealed a high degree of conformity between situations in which calls with a comparable acoustic structure were uttered by different species. Consequently, the classification system described here, with its general relationships between acoustic structure and functional significance, is valid not only for the squirrel monkey but, in an adapted form, for a much broader group of primates. Its validity even may encompass nonprimate mammals. As Morton (1977) has shown in a comparative study, a tendency exists for very different mammalian and avian species to use relatively low-frequency, harsh sounds when aggressive and purer tones of higher frequency when frightened, appeasing, or approaching in a friendly manner. Such a trend corresponds well with the findings described in this chapter.

As regards the measurement of the emotional state (aversiveness) underlying a specific call type, it is clear that the method used here is of very limited applicability to other species. As a neuroethological method, it requires a number of neurophysiological and neuroanatomical prerequisites (stereotaxic brain atlas, map of the vocalization-eliciting brain structures, etc.) not available for most other primate species. Nevertheless, we think

that this handicap is less severe than it appears and that the establishment
of a call classification system like that described here for another species
is quite feasible. The present study has shown a general correlation between aversiveness and frequency range. Consequently, even calls whose
absolute aversiveness scores are not known can be ordered according to
relative aversiveness by considering their frequency range. However, as
can be seen from Figure 3.3, such a procedure demands that only calls
from within the same class be compared. Clucking (IV (±)), for instance,
is less aversive than cawing (V (−)), despite its wider frequency range,
because of physiological variables. Low-pitched calls are produced by
different laryngeal muscles than those used to produce high-pitched calls
(Jürgens, Hast, and Pratt 1978). As the aversiveness of a call seems to
depend upon the global muscular effort invested in its production, it is un-
derstandable that calls so different in structure, like high-pitched clucking
and low-pitched cawing, which involve different muscle groups, cannot
be compared.

REFERENCES

Jürgens, U. 1976. Reinforcing concomitants of electrically elicited vocalizations.
    Experimental Brain Research 26:203–14.
    1979. Vocalization as an emotional indicator. A neuroethological study in the
    squirrel monkey. *Behaviour* 69:88–117.
Jürgens, U., and Ploog, D. 1970. Cerebral representation of vocalization in the
    squirrel monkey. *Experimental Brain Research* 10:532–54.
Jürgens, U., Hast, M., and Pratt, R. 1978. Effects of laryngeal nerve transection
    on squirrel monkey calls. *Journal of Comparative Physiology A: Sensory,
    Neural, and Behavioral Physiology* 123:23–9.
Jürgens, U., Maurus, M., Ploog, D., and Winter, P. 1967. Vocalization in the
    squirrel monkey (*Saimiri sciureus*) elicited by brain stimulation. *Experimental
    Brain Research* 4:114–17.
Marriott, B. M., and Salzen, E. A. 1978. Facial expressions in captive squirrel
    monkeys (*Saimiri sciureus*). *Folia Primatologica* 29:1–18.
Morton, E. S. 1977. On the occurrence and significance of motivation–structural
    rules in some bird and mammal sounds. *American Naturalist* 111:855–69.
Ploog, D., Hopf, S., and Winter, P. 1967. Ontogenese des Verhaltens von Toten-
    kopf-Affen (*Saimiri sciureus*). *Psychologische Forschung* 31:1–41.
Schott, D. 1975. Quantitative analysis of the vocal repertoire of squirrel monkeys
    (*Saimiri sciureus*). *Zeitschrift für Tierpsychologie* 38:225–50.
Winter, P., Ploog, D., and Latta, J. 1966. Vocal repertoire of the squirrel monkey,
    its analysis and significance. *Experimental Brain Research* 1:359–84.

PART II

Social and environmental determinants of primate vocalizations

The structural complexity of primate vocal systems is likely to parallel interspecific differences in the size, organization, and distribution of the social unit. For example, species with small social units, consisting of three to six individuals, as is characteristic of de Brazza's monkey (*Cercopithecus neglectus*), may require a less sophisticated vocal system for the mediation of social exchanges than species with large social units of a hundred or more individuals, as exhibited by talapoin monkeys (*Miopithecus talapoin*). Furthermore, the precise acoustic structure, or the design features, of many vocalizations may have evolved to favor accurate communication in acoustically nonoptimal environments. The perspective advanced by Part II assumes that primate vocalizations are not arbitrary or accidental in their acoustic structure; rather, it is likely that they have been selected because they satisfy the communicative demands of the social unit with respect to ecological forces in the habitat. Thus both ecological and social factors are pertinent to the structure and organization of vocal displays.

Many primate vocalizations are short, telegraphic, pulsed signals. However, investigators are beginning to appreciate and describe remarkably elaborate vocal sequences. Deputte describes and analyzes one of the most elaborate primate vocalizations discovered to date. Song bouts of the white-cheeked gibbon are sung by mated pairs as a duet. These duets, which may last 20 or more minutes, combine various components, including intense vocalizations, brachiation, hugging, and so forth. These remarkable signals have been implicated in the preservation of monogamy in the mated pair by repelling potential intruders from the social unit. Furthermore, these signals may communicate family size, their relative location, the size of their territory, and their socionomic composition. These displays may be critical in strengthening the pair bond, intergroup spacing, and ultimately the solicitation of potential mates for adolescent offspring. The sequencing and timing of the various components of the duet implies that the duet is governed by a syntactic structure in which the female may direct the timing, pace, and intensity of the male song. The question of syntax in primate vocalizations is gaining increased attention (see Smith, Newman, and Symmes, chap. 2, this volume; Snowdon, chap. 9, this volume), and Deputte's description of the gibbon song indicates that this remarkable display is syntactically governed.

The distribution and spacing of individuals is as important within the social unit as it is between groups. Social groups presumably form and persist because they provide a better biological environment for members, who live longer and reproduce in greater numbers than their solitary counterparts. Social behaviors and communication systems have likely evolved to orchestrate the behavior of individuals to heighten the benefits of group living and to attenuate the costs of high-density living. Accordingly, communication systems may be designed to monitor and to modify the distribution and density of individuals within the group. Robinson describes a cohesion-spacing vocal system in a New World monkey, the wedge-capped capuchin. Robinson's field observations reveal that some calls act to reduce interindividual distances, whereas other calls act to increase interindividual spacing. Vocally mediated changes in group density may be important for optimizing foraging strategies and reducing individual or group vulnerability to predation. These observations provide firm evidence that the maintenance and adjustment of interindividual space is of prime significance to primates within groups. The calls described by Robinson indicate that a relatively complex system is based on quite subtle vocal differences, a finding that underscores other investigative work showing that communicatively significant distinctions may be based on acoustically minor vocal features (Green 1975; Brown, Beecher, Moody, and Stebbins 1979). In this respect primate vocal communication may resemble speech perception in that the perceptual distance between various sounds may not be paralleled by their acoustic difference. One classical problem in the analysis and description of primate vocalizations has been the identification and categorization of the communicatively salient attributes of the vocal repertoire. This problem is certain to receive intensive research in the future, and Robinson's chapter illustrates one approach that investigators may employ in the field.

What factors govern the structure of primate vocalizations? Most primatologists would argue that the vocal apparatus of any species is a product of natural selection and consequently vocal production and acoustic displays should exhibit characteristics that reveal the course of natural selection. Indeed, vocal behavior may be taxonomically diagnostic (Struhsaker 1970). Yet the extent to which vocal elements may have been shaped in response to habitat-specific ecological factors is unclear. For example, the reverberation time of one habitat may hinder the perception of rapid acoustic sequences, and the background noises in a second habitat may mask the perception of sounds in a certain frequency band. In this context Waser surveys the loud calls of a number of mangabeys and baboons. Primate species belonging to the tribe Papionini are distributed over a wide range of African habitats from rain forests to arid scrub, and their respective social organizations are thought to vary in response to ecological factors. Waser, however, details how interspecific variations in the use of loud calls appears to reflect differences in social structure, not differences in the environment. His analysis, suggesting that the acoustic structure of loud calls has been governed primarily by phylogenetic and social rather than ecological factors, is sure to evoke additional research into this question.

The regulation and maintenance of spatial relationships both within and between groups may be mediated by special classes of vocalizations. Type 1 loud calls have been implicated in regulating spacing between groups (Gautier and Gautier 1977). Within the social unit a variety of position marking, cohesion, dispersal, and contact vocalizations may regulate interindividual distances. In either case, the mediation of spatial relationships is dependent upon the production of signals that effectively communicate the position of the vocalizer. Brown describes a number of primate vocalizations that vary in their suitability for sound localization and hence in their suitability for communicating location. Remarkably subtle adjustments of acoustical elements are important for sound localization, and the design features of vocalizations that are significant for sound localization also may be important semantically (Green 1975). Brown's chapter illustrates the fruitful interplay between field and laboratory methodologies in the study of primate communication. Until recently, the flow of hypotheses has been from the field to the laboratory. Brown's investigation has generated hypotheses that have influenced field (see Snowdon, chap. 9, this volume) as well as laboratory research.

The chapters in Part II illustrate a variety of methodological approaches in the study of communication. Robinson and Waser demonstrate the importance of field research conducted in the species' natural habitat. Waser's chapter documents the utility of the comparative approach, through which he examines the vocal behavior of several closely related species. In contrast, Robinson's chapter is focused on a single species. However, Robinson's focus is not simply descriptive; he formulates a family of competing hypotheses regarding the function of several capuchin (*Cebus nigrivittus*) vocalizations and structures his field observations so that he may test and discriminate between these hypotheses. In the same tradition as Robinson and Waser, Deputte employed ethological techniques, but his study was conducted on a captive colony. A number of forest primates, like the white-cheeked gibbon (*Hylobates concolor*), are difficult to study in their natural habitat. However, important observations may be conducted on a captive colony as testified by Deputte's research. Brown's research adapts techniques from the human psychoacoustics laboratory for studying auditory perception in primates. The approach exemplified by the primate psychoacoustics laboratory is far more manipulative and considerably less natural than that utilized by ethologists, yet the animal's perceptual capacities and the organization of its vocal system becomes much more amenable to intensive and precise analysis than might be possible in the field.

The methodologies adopted by these four investigators illustrate that there is no single preferred approach to the study of primate communication, for each approach has inherent advantages and liabilities. Furthermore, the interplay between field-centered and laboratory-centered approaches provides the most compelling perspective for analyzing communication systems. Collectively, these four chapters document several social and environmental factors pertinent to the organization and evolution of communication systems. The syntactic and semantic sophistication of the natural languages of nonhuman primates is an unresolved question. Yet the communicative sophistication of all primate species may be a consequence of the historical exchange between the social and environmental factors introduced in this section.

REFERENCES

Brown, C. H., Beecher, M. D., Moody, D. B., and Stebbins, W. C. 1979. Locatability of vocal signals in Old World monkeys: design features for the communication of position. *Journal of Comparative and Physiological Psychology* 93:806–19.

Gautier, J.-P., and Gautier, A. 1977. Communication in Old World monkeys. In *How animals communicate,* ed. T. A. Sebeok, pp. 890–964. Bloomington: Indiana University Press.

Green, S. 1975. Variations of vocal pattern with social situation in the Japanese monkey (*Macaca fuscata*). In *Primate behavior: Developments in field and laboratory research,* Vol. 4, ed. L. A. Rosenblum pp. 1–102. New York: Academic Press.

Struhsaker, T. T. 1970. Phylogenetic implications of some vocalizations of *Cercopithecus* monkeys. In *Old World monkeys: evolution, systematics and behavior,* ed. J. R. Napier and P. H. Napier, pp. 365–403. London: Academic Press.

4 · Duetting in male and female songs of the white-cheeked gibbon (*Hylobates concolor leucogenys*)

BERTRAND L. DEPUTTE

The emergence and development of sociability has been accompanied by the development of sophisticated communicative processes. These are especially well developed in primates, whose social life displays a great variety of patterns. In all cases, the integrity of the group must be maintained against disturbances coming from either the group itself or external sources. These two conditions present different constraints on communication. At short interindividual distances, several different communicative channels can be used. In intergroup exchanges, communication is generally over large distances and essentially relies on the acoustic channel.

Forest-dwelling primates in particular rely heavily on acoustic signals. Within the group, vocalizations carry short distances. They are often given in conjunction with visual signals and may assume different functions, the essential one being to maintain bonds among all individuals in the group but also between some preferred partners (Gautier and Gautier-Hion, chap. 1, this volume; Smith, Newman, and Symmes, chap. 2, this volume). On the other hand, intergroup communication vocalizations must be loud at the source and of precise structural characteristics to carry long distances despite the high attenuation rates in the forest environment. Such vocalizations have reached a high degree of specialization through selective pressure to keep their high information-transfer abilities (Waser, chap. 6, this volume; Brown, chap. 7, this volume). The principal characteristics of these loud calls are the stereotypy of both the physical form and the delivery patterns. They are widespread in arboreal primates, including New World monkeys, Old World monkeys, and apes, and are assumed to play a major role in intergroup spacing (howlers, Altmann 1959 and Southwick 1962; lemurs, Petter 1962, Jolly 1966, and Klopfer 1977; guenons and mangabeys, Gautier 1969, Gautier and Gautier 1977; Waser 1977b; and Quris 1980; langurs, Jay 1965; colobus, Marler 1972; apes, van Lawick-Goodall 1968 and MacKinnon 1974).

This study was completed under the direction of Dr. M. Goustard. I thank him for his collaboration on portions of this research, for helpful discussions on drafts of this manuscript, and for introducing me to gibbons. I also thank A. Gautier-Hion and J.-P. Gautier for their help in making corrections and for other suggestions.

In intragroup vocal communication, individuals differentially use specific parts of the vocal repertoire, depending upon their age, sex, and status (Gautier and Gautier-Hion, chap. 1, and Smith, Newman, and Symmes, chap. 2, this volume). In the intergroup communication of most primate species the loud calls are a male attribute. They appear in the male at sexual and/or social maturity either as new acquisitions to the repertoire or as a differentiation of an already existing vocal type (Gautier 1975; Gautier and Gautier 1977). In gibbons these loud calls are generally referred to as songs (Carpenter 1940; Chivers 1972; Ellefson 1974; Tenaza 1976). In contrast to most monkey species, both male and female adult gibbons utter such songs, but there are clear differences in the songs of each sex. Except in Kloss's gibbon, the songs are given in duets by mated pairs.

Vocal duetting is a well-known and well-documented phenomenon, especially in tropical birds where males and females establish long-lasting bonds, live in forest habitats, and show territorial behavior (Thorpe 1963; Grimes 1966; Diamond and Terborgh 1968; Hooker and Hooker 1969; Watson 1969; Payne 1971; Thorpe 1972; Vencl and Soucek 1976; Harcus 1977; Wiley and Wiley 1977; Todt and Fiebelkorn 1980; Wickler and Seibt 1980). Duetting and antiphonal calling has also been studied in the horseshoe bat, particularly in the case of mother–infant dialogue when the mother returns to its young (Matsumura 1979). Alarm duetting has been described in a desert antelope *Oreotragus oreotragus* (Tilson 1977), where mated pairs duet in response to potential predators. Among primates this duetting phenomenon has been mentioned and studied only in South American Cebidae (*Callicebus moloch*, Robinson 1979; *Callicebus torquatus*, W. Kinzey quoted by Robinson 1979), tree shrews (Sorensen and Thompson 1969), and Asian gibbons, where it is assumed to be "unique among vocalizations of higher primates" (Marler and Tenaza 1977).

The structure of the duets in gibbons depends upon the structure of both male and female songs. In gibbon species where sexual differences in song patterns are slight, male and female contributions to duets tend to overlap (e.g., the siamang, *Symphalangus syndactylus,* Lamprecht 1970; the capped gibbon, *Hylobates lar pileatus,* and the hoolock gibbon, *H. hoolock,* Marshall, Ross, and Chantharojvong 1972) and are chorused in unison. In contrast, when songs show strong sexual differences in structure and organization, they tend to be uttered antiphonally (e.g., the white-handed gibbon, *H. lar,* Marshall et al. 1972; and Marler and Tenaza 1977; the red-cheeked gibbon, *H. concolor gabriellae,* Goustard, unpublished). In the white-cheeked gibbon (*H. c. leucogenys*), male and female songs are highly dimorphic, especially in their delivery pattern. They are also antiphonal.

The captive population of 12 white-cheeked gibbons at the Laboratory of Etho-Primatology in Saint Cheron (France) provided us with the material to answer some questions about the organization of duets and their inferred functions. For example, what are the mutual influences of male and female songs on the structure of the song, its temporal pattern, and its

Table 4.1. *Temporal characteristics of song and song bout in female white-cheeked gibbon*

	Sample size	Mean	Coefficient of variation	Range
Song: phase I (whôô)				
Number of units	74	5.6	110.4	1–17
Duration (sec)	73	51.7	90.8	2–208
Rhythm (unit/sec)	72	.14	62.0	
Song: phase II (whuit)				
Number of units	73	25.2	11.2	18–32
Duration (sec)	73	14.6	19.6	6.4–28
Rhythm (unit/sec)	72	1.8	22.2	
Total duration (sec)	158	77.3	74.5	11–309
Song bout				
Number of songs	213	2.67	101.6	1–15
Total duration (min)	39	9.04	36.07	4–14
Rhythm (song/sec)	39	.012	18.6	.008–.016

associated behaviors? Is the information transferred by a duet the mere sum of information of male and female songs? What may be the selective pressures that have favored male and female singing in duets?

The results are based on the study of one captive group composed of three adults (two females and one male) and four subadults (three males and one female). There is only one mated pair in that group (Deputte and Goustard 1978).

VOCAL AND TEMPORAL PATTERNS OF FEMALE AND MALE SONGS

Female song

Three phases may be distinguished in the female song (Figure 4.1). The first consists of long "whôô" calls with moderately linear increases in frequency over time. This call belongs to the type 2 class of the common repertoire (Deputte and Goustard 1978). The emission rate gradually increases from the beginning to the end of this first introductory phase, and sound frequencies display a simultaneous gradual increase. The second phase, a "buildup" or "paroxymal" stage, is composed of 18 to 32 calls with a strong rising inflection. This element belongs to the type 4 class of the common repertoire. These "whuit" calls are uttered at a fast, steady rate (Table 4.1). The second phase ends with a mixed-type unit, which indicates the climax of the song (Figure 4.1). The song itself ends with a sequence of "thiô" calls, which belong to the type 3 class of the common repertoire (Deputte and Goustard 1978), uttered at a decreasing rate.

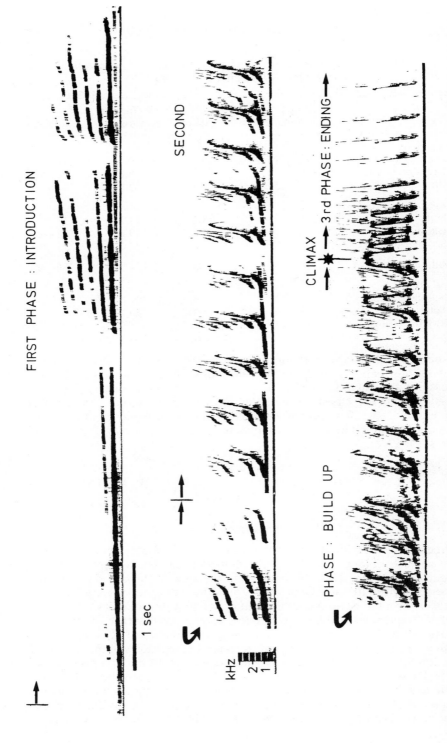

Figure 4.1. Frequency–time display of female song.

Table 4.2. *Influence of the temporal structure of the beginning of the female song bout on the occurrence of the male song*

	♂ does not ever sing[a]	♂ sings after 2nd song[a]
Delay (in sec) between 1st and 2nd female songs, $\bar{x}$	166.6	108.9
cv	60.1	42.4
	$t(30) = 2.43, p < .05$	

[a] $N = 16$.

The whole song lasts anywhere from 11 to more than 300 sec (Table 4.1). This variability in duration depends mainly on the variability of the duration of the first phase (duration: song vs. first phase, $r (70) = +.930$, $p < .001$; song vs. second phase, $r (70) = -.173, p > .10$). One or several songs (up to 15), constituting a song bout, may be uttered by the female at a mean rate of 1 song/100 sec (Table 4.2). In a sample of 197 songs and song bouts, 46% were uttered "spontaneously," and 37% followed different calls (especially types 2B, 4, and 4B, Deputte and Goustard 1978) uttered by group members either in harassment situations or agonistic episodes. Of the female songs, 6% were given in such aggressive interactions where the female herself gave aggressive calls (type 1B).

During the paroxymal phase, the female becomes progressively more agitated, jumping, shaking, and brachiating quickly in many directions. At the climax of her song, the female comes to her partner or another group member and embraces him. At the onset of the paroxymal phase, juveniles or subadults of both sexes may utter long whôô calls. These calls are at first given in an asynchronous pattern with the female calls, but as the phase progresses, the juvenile whôô calls evolve into whuit calls, which then become synchronized with the female sounds. This constitutes a chorus, which climaxes with the individuals jumping around simultaneously and embracing each other.

Male song

The male's song progresses through a succession of calls of increasing complexity. When the song has reached maximum complexity it is said to be fully developed. The typical series is then repeated up to 40 times. At the very beginning of the male song, a long unitary whôô call is uttered (Figure 4.2A). It is rapidly followed by a pair of long whôô calls (Figure 4.2B). Then two other sounds appear: (a) a low-pitched sound associated with the filling of the laryngeal sac (this sound is a new acquisition in the mature, adult male vocal repertoire) (Figure 4.2C and Da) and (b) short, high-pitched "ha" sounds (Figure 4.2C and Db) ("eek," Marshall and Marshall 1976; "*sons brefs*," Demars and Goustard 1978) belonging to the type 2 class of the common repertoire. These two sounds preceding the

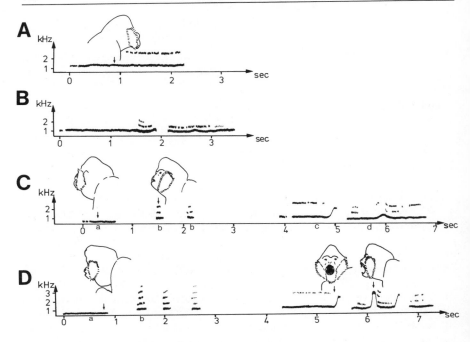

Figure 4.2. Frequency–time displays of different calls in the male song: (A) monotonous unitary long "whôô" and (B) binary long "whôô." Both calls are uttered mainly at the beginning of the song. (C) Typical series: (a) air-filling sac sound, (b) short "ha" sounds. The next long calls present (i) on the first unit (c), a type 4 "whuit" ending and on the second unit (d), the earliest form of the type 4B subunit. (D) Typical series of greater complexity than C, with (a) and (b) as in C. The series of long calls present three units with a fully developed type 4B subunit plus a type 4 ending on the second unit. The arrows indicate the place of the illustrations in the call. Illustrations are taken from photographs.

long whôô calls constitute a sequence that I call a typical series (Figure 4.2C and D). As the male song progresses there is a parallel elaboration of the sequences of short ha sounds and the series of long calls (see figure on p. 7 in Demars and Goustard 1978 for the whole representation of a song). Frequency modulations gradually appear, first a type 4 whuit ending on the first whôô unit and then a modulated subunit (type 4B) on the second whôô unit. In the course of the song, this latter modulation is never uttered before the former. When the type 4B subunit has reached a maximum frequency of about 4 kHz, the second unit ends, also with a type 4 modulation, and then a third monotonous whôô unit appears (Figure 4.2C and D). At the same time there is an increase in the number of short ha sounds in the sequence preceding the long calls as they increase in complexity (Figure 4.3A). The proportion of long calls uttered in a typical series decreases as the complexity of modulation increases (Figure 4.3B).

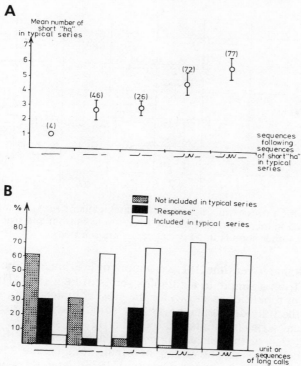

Figure 4.3. (A) Relation between structural complexity of series of long calls (horizontal axis) and the number of short ''ha'' sounds in the sequence that precedes a series of long calls in a typical series. (B) Proportions of different structural forms of the male song, uttered in or out of typical series or as ''responses'' to female song. The diagrams on the horizontal axis represent prototypical sonagrams, arranged in order of increasing complexity from left to right.

In the male song, the transition from early, simple monotonous forms to more complex ones is graded. If we consider the matrix of the first-order transitional frequencies (see Figure 4.4) between successive long calls (the other types of sound in a typical series are not taken into account) arrayed in order of increasing complexity, we observe that the maximum transition frequencies appear on the descending diagonal of the matrix. To test the degree of heterogeneity of the matrix, the G statistic (Sokal and Rohlf 1969) is used. This log-likelihood ratio test reveals whether the observed heterogeneity of transitional frequencies between the different calls is likely to have arisen by chance from a homogeneous distribution of transition. A high value of G means a strong association (or lack of independence) between some variables of the matrix. Thus each form of long call is most frequently followed by a form like itself ($G(16) = 423.4$, $p < .001$; Figure 4.4). That is, each type is repeated a certain number of times before the transition to the next more complex type takes place.

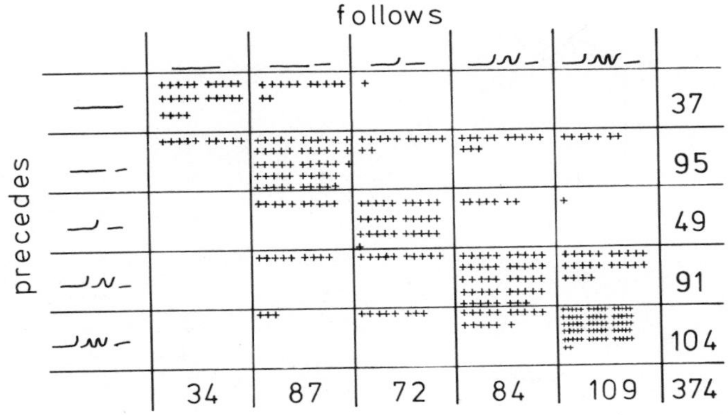

Figure 4.4. First-order transition matrix of long calls or series of modulated long calls arranged in increasing complexity from top to bottom and from left to right, respectively (other sounds in a typical series are omitted).

This means that, over time, there is a more or less linear elaboration of the male song from a simple to a complex series of long calls. The maximum frequency of repetition is shown for the most complex form, which corresponds to the fully developed part of the song. The mean succession rate of the typical series is 6 units/100 sec ($N = 12$, $cv = 18$), that is, six times faster than that of the female songs. The delay between the successive series is about 16 sec. The ending of a male song is characterized by increasing delays (Demars and Goustard 1978), and its whole duration varies from 350 to 860 sec ($N = 34$, $\bar{x} = 618$ sec; $cv = 22.4$). Together the song bout of a mated pair may last longer than 1 hr.

INFLUENCE OF THE FEMALE SONG ON THE MALE SONG IN A DUET

During daytime, once the female of a mated pair has sung, the male may begin to sing. The two then sing in duet. The female song is intercalated between the male typical series (or calls). The song of the female elicits the male song 90% of the time at sunrise (Goustard 1979) but only 30% of the time during the rest of the day. Nevertheless, one may assume that the female song has a strong eliciting effect on the male song. The male song begins in the interval between the first two songs of a female bout ($\chi^2(1) = 17.4$, $p < .001$). The shorter the delay between the first two female songs, the more likely the utterance of the male song will occur after the second female song (Table 4.2). Once his song has started, the male utters a long whôô call or a series of modulated calls after 98% of the female songs in a female song bout. This temporal association suggests that the male call is a *true response* to the female song. When the male sings alone in the interval between the female songs, the complex, modulated calls are always uttered in a typical series (Figure 4.3B). The responses

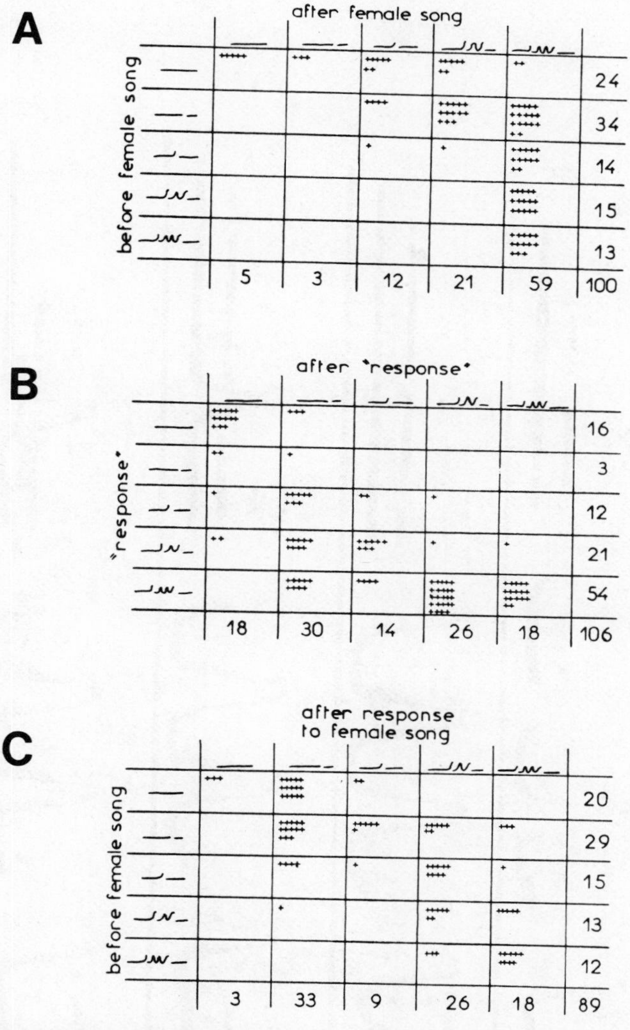

A — after female song (columns) / before female song (rows)

before \ after	—	— —	⌐ —	⌐N—	⌐w—	total
—	+++++	+++	+++++ ++	+++++ ++	++	24
— —			++++	+++++ +++++ +++	+++++ +++++ ++	34
⌐ —			+	+	+++++ +++++ ++	14
⌐N—					+++++ +++++ +++++	15
⌐w—					+++++ +++++ +++	13
total	5	3	12	21	59	100

B — after "response" (columns) / "response" (rows)

response \ after	—	— —	⌐ —	⌐N—	⌐w—	total
—	+++++ +++++ +++	+++				16
— —	++	+				3
⌐ —	+	+++++ +++	++	+		12
⌐N—	++	+++++ +++	+++++ +++	+	+	21
⌐w—		+++++ +++	++++	+++++ +++++ +++	+++++ ++	54
total	18	30	14	26	18	106

C — after response to female song (columns) / before female song (rows)

before \ after	—	— —	⌐ —	⌐N—	⌐w—	total
—	+++	+++++ +++++ +++++	++			20
— —		+++++ +++++ +++	+++++	+++++	++++	29
⌐ —		++++	+	+++++ ++++	+	15
⌐N—		+		+++++ ++	+++++	13
⌐w—				+++	+++++ +++	12
total	3	33	9	26	18	89

Figure 4.5. First-order transition matrix of male song calls. In each matrix the calls are arranged as in Figure 4.4. (A) Series of long calls (or a long call) preceding the female song versus male "response." (B) "Response" versus calls or series of long calls just following the response (if they are present, the other sounds in the typical series are omitted). (C) Male calls just preceding the "buildup" phase of the female song versus male calls just following the male response.

are the only complex calls that are *never uttered* in these typical series (Figure 4.3B). Examination of the first-order transitional frequencies between the male long calls preceding the female song and the male "response" reveals that the male responds to the female with his most com-

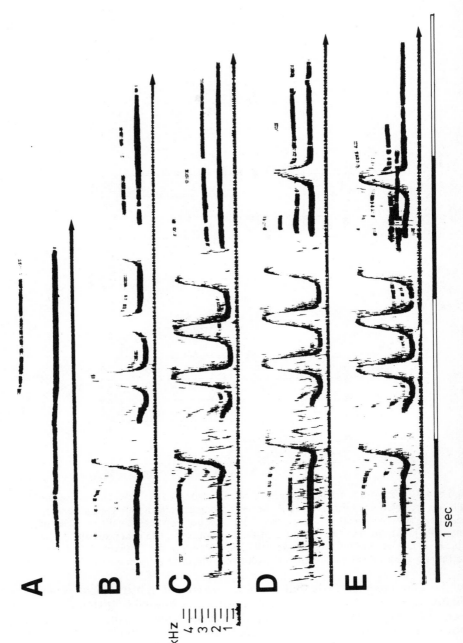

Figure 4.6. Frequency–time displays of successive male responses to successive female songs during a duet. From top to

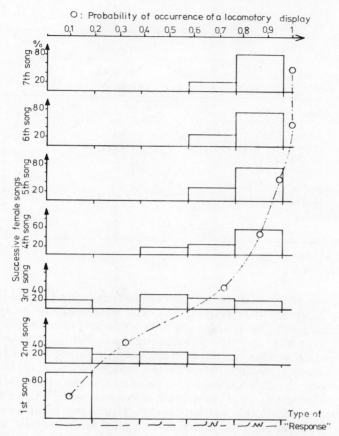

Figure 4.7. Development of the structural forms of response during the female song bout. On the horizontal axis, the "responses" are arranged in increasing order of complexity from left to right. The interrupted-line/open-circle curve shows the development of the male locomotory display, associated with his response, during the female song bout.

plex calls ($G(16) = 77.15, p < .001$; Figure 4.5A). In addition, the "after response" calls are less complex than the response itself ($G(16) = 106.7$, $p < .001$; the "response" calls are more complex than the "after response": $\chi^2(1) = 53.4$, $p < .001$; Figure 4.5B). Hence, the female song causes a modification in the development of male calls consisting of a transitory increase in complexity. This contrasts with the repetition pattern leading to a progressive increase in complexity, as mentioned earlier, for the normal course of the male song, that is, when the female does not sing. Furthermore, the complex call that follows the response is significantly more complex than the one that precedes the female song ($G(16) = 78.6$, $p < .001$; Figure 4.5C). This suggests that the increased complexity induced by the female song lasts even after the response itself.

Table 4.3. *Influence of female song on temporal structure of male song*[a]

| | Male songs | | | |
Delay of male calls (sec)	Number of songs	Number of delays	Mean	t'_s
In absence of female song				
D_1	25	85	16.3	
D_2	25	409	16.9	
D_3	25	400	4.7	1.53*
				4.55**
When separated by female song in duet				35.2**
D'_1 or D'_2	26	71	14.8	
D'_3	25	79	14.9	

* ns, $p > .05$; ** $p < .001$
[a] See Figure 4.8.

The response of the male to the successive female songs becomes progressively more complex (Figure 4.6). The percentage of complex responses increases as the female song bout progresses (Figure 4.7). Simple responses even after the third female song mean that the male may start his song as late as after the third female song. In the course of the male song, the female "buildup" phase of the song may occur: (a) after long unitary whôôs; (b) after a sequence of short ha sounds; or (c) after a typical series (Figure 4.8). At the beginning of the male song, when unitary or binary whôôs are uttered, the temporal pattern is not influenced by the utterance of the female song (Table 4.3; Figure 4.8). In contrast, once the typical series has begun, the female song causes a significant modification of the temporal pattern of the male song (Table 4.3; Figure 4.8). The duration of the pause in the male song, wherever it occurs, depends only on the duration of the buildup phase of the female song (male D'_1 vs. D'_3: $t(148) = .102$, $p > .90$; male D'_1 and D'_3 vs. duration of female buildup phase: $t(142) = .55$, $p > .90$ and $t(142) = .655$, respectively; Figure 4.8). It is conceivable that the introductory whôô phase of the female song has an "alerting" and synchronizing effect on the occurrence of the male pause and hence on the adjustment of the utterance of the response.

INFLUENCE OF THE FEMALE SONG ON THE LOCOMOTORY BEHAVIOR OF THE SINGING MALE

During his song, the male does not move while uttering his calls. But when he utters a response to the female song, he executes locomotory displays, violent brachiation, shaking, or prancing ($\chi^2(1) = 929.5$, $p < .001$; Table 4.4). These displays are associated with the utterance of the whuit

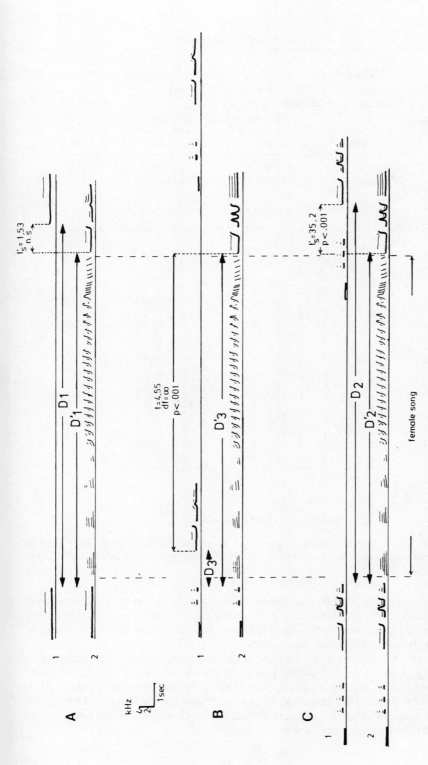

Figure 4.8. Influence of the female song on the temporal structure of the male song; (A) at the beginning of the male song, when typical series are not yet uttered; (B) after a sequence of short "ha" sounds; (C) after a typical series. Male song alone is on line 1 of each row. Male plus female song is on line 2 of each row. Female song in light outline occurs between male call series in thick outline.

Table 4.4. *Association between male response and
male locomotory displays*

| | Male locomotory behavior | | |
Male song	Display	No display	Total
Response	169 (28.5)	66 (206.5)	235
Other calls	10 (150.5)	1,229 (1088.5)	1,239
Total	179	1,295	1,474

Note: $\chi^2(1) = 929.5$; $p < .001$. Figures in parentheses are expected values.

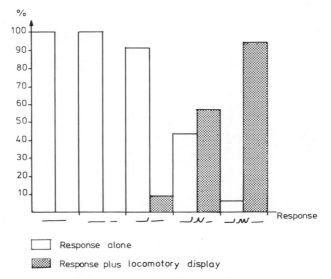

Figure 4.9. Relation between locomotory displays and structural complexity of the response in the male (increasing structural complexity from left to right on the horizontal axis).

modulation (Figure 4.9). Actual locomotory behavior does not occur until the modulation appears. The probability of occurrence of a locomotory display increases with the complexity of the response, and both increase as the female song bout progresses (Figure 4.7).

To summarize, in a mated pair, when the female sings, the male starts to sing. During each buildup phase of the female song, the male stops singing. After the climax of the female song, he resumes his song by uttering a response. This response is characterized by its structural complexity, loudness, and associated locomotory displays. Therefore, in the white-cheeked gibbons, duets are antiphonal.

Table 4.5. *Influence of the male song on the temporal structure of the female song bout*

Delay between songs in female song bout (sec)	Female alone	Duet
Between 1st and 2nd songs, $\bar{x}$	137.6	83.3
N	32	27
cv	59.7	32.3
$t(57) = 3.9, p < .05$		
Between 2nd and 3rd songs, $\bar{x}$	105.6	80.9
N	5	13
cv	18.9	33.8
$t(16) = 2.25, p < .05$		

INFLUENCE OF THE MALE SONG ON THE FEMALE SONG IN THE DUET

The male song does not modify either the acoustic or temporal structure of the female song. Nevertheless, when the male fails to sing, the female utters significantly fewer songs than when he sings (number of songs in a song bout: female alone, $N = 120$, $\bar{x} = 1.2$, $cv = 54.4$; female plus male, $N = 44$, $\bar{x} = 5.9$, $cv = 29.6$; $t = 17.5$, $p < .001$). When the male sings, the female may utter up to nine songs. Hence the male song influences the duration of the female song bout. The delays between the first, second, and third female songs are significantly shorter when the male sings (Table 4.5). In addition, the later the first typical series with modulated long calls appears in the male song, the longer the female song bout lasts ($r = .558$, $N = 13$, $p < .05$). This suggests that the song of the male stimulates the song of the female and prolongs her song bout. After the third female song, the successive delays between songs increase as the song bout progresses, regardless of the length of the song bout (i.e., number of successive songs). Examination of the different calls uttered by the male between successive female songs reveals that the frequencies of the more complex calls and, correlatively, the number of short ha sounds increase as the delay between the female songs increases (Figure 4.10). Hence the female song bout comes to its end progressively and inexorably when the male song has reached its full development. The female sings most often after the monotonous binary long calls of the male song but not after the complex calls ($\chi^2(1) = 9.8$, $p < .001$; Figure 4.5). Thus it seems that it is an additive effect that slows up the self-ending process of the female song bout – the discrepancy between the temporal development pattern of the songs of the two partners prolongs the duet.

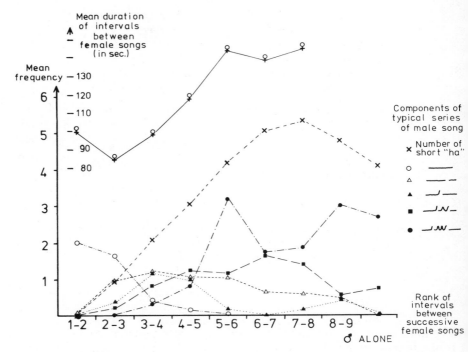

Figure 4.10. Development of the mean frequency of the different sounds or calls uttered by the male between the successive female songs during a duet. The male responses are not taken into account. The upper curve (♀) shows the parallel development of the mean delay between the female songs and so on; "♂ alone" corresponds to the part of the male song uttered after the end of the female song bout. This figure is based on 15 male songs.

INFLUENCES OF MALE AND FEMALE SONG ON THE AROUSAL LEVEL OF THE VOCALIZERS

In the common vocal repertoire of the crested gibbon, a graded structural transition exists between the monotonous type 2 calls and the highly modulated type 4 calls. There is a one-way transition from the type 2 call to the type 4 call. The utterance of type 4 (or 4B) calls stops with the cessation of the releasing context. An individual uttering the type 4 whuit calls is generally agitated, often jerking its hands and feet (Deputte and Goustard 1978). The sounds uttered in male and female songs also belong to the type 2 and the type 4 classes of the common repertoire. The difference is that only when uttered as part of a song, the units are louder, are more structurally stereotyped, are emitted at more stable intensity levels, and are part of prescribed sequences or patterns of sounds. The transitions between types 2 and 4 in both male and female songs closely parallel the transition from the type 2 to the type 4 seen in the common repertoire. This elaboration in songs has three characteristics: a gradual increase in

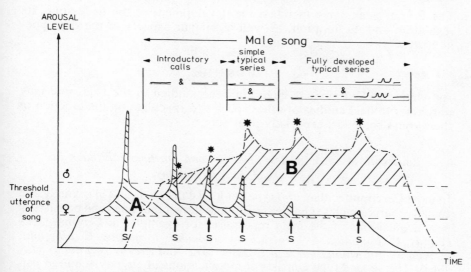

Figure 4.11. Speculative and schematic representation of the evolution of the female and male arousal level during a duet: (A) female song bout, (B) male song, (s) female song, (*) male "response" to female song.

frequency modulation, an increased rate of utterance, and an increased locomotor activity. These characteristics suggest a parallel increase in the arousal level of the vocalizers. Indeed, Jürgens (1979 and chap. 3, this volume) points out that a high arousal level is reflected in the tension of the sound-producing muscles, leading to an increase in the amount of frequency modulation and in loudness. In addition Green (1975) assumes that shifts in physiological states underlying the Japanese macaque's social behavior are also reflected in sound morphology. He cites, for example, the transition from "deep-late high" and "smooth-late high" coos to "*oui*" and "rise" calls. This is very similar to the transition from whôôs to whuit calls in the white-cheeked gibbon. Green (1975) also points out that the more agitated state linked with utterance of rise calls is associated with an increase in the rate and intensity of locomotor activity. In their songs, male and female gibbons reached high arousal levels with different latencies. The female has a greater readiness to sing. She may utter as many as 10 songs, or song bouts, in a day. The male produces only one-third as many and, then, generally after the female has sung. In the female the peaks of arousal level are more transient and phasic than in the male, who may sustain a high arousal level for a longer period of time. The mere occurrence of the male song and its gradual development slows the decrease in the female's arousal level. The female song, on the other hand, seems to increase the male arousal level until it reaches the threshold of song utterance. Then, successive female songs create high and short-lasting peaks in male arousal level, which are reflected in the structural characteristics of the response and its associated behaviors (Figure 4.11).

These brief peaks may facilitate the development of the male song, its regulation and its duration, as well as similar aspects of the duet phenomenon.

DISCUSSION

As white-cheeked gibbons have not been studied in the field, one may only refer to field studies of other species to speculate on the function of their vocal behavior, especially their songs.

Loud calls in other species: Structure and function, utilization by age–sex classes, locatability

In most Cercopithecidae, two types of loud calls are found in sexually and socially mature males. Gautier (1975) has shown that one type consists of new acquisitions of the male repertoire, for example, in *Cercopithecus cephus, C. pogonias, C. neglectus, Cercocebus torquatus, C. albigena*, as well as in *Colobus guereza* (Marler 1972) and *Presbytis antellus* (Jay 1965). The second type consists of modifications of already acquired patterns seen in the alarm and aggressive calls of *Cercopithecus nictitans, C. cephus* (Gautier 1975), and *Papio* (Waser, chap. 6, this volume). In the agile mangabey (*Cercocebus galeritus agilis*), the male loud call represents a modified "summary" of the whole species-specific vocal repertoire (Quris 1980 and personal communication). It seems that the two categories previously mentioned in Cercopithecidae are also present in at least one new World monkey, *Alouatta palliata* (Altmann 1959; Baldwin and Baldwin 1976). The roars or howls may correspond to the type 1 loud calls and the "whoofs" (or barks) to the type 2. However, there is no evidence that howls are new acquisitions of the male repertoire. In apes, the loud calls of all species contain components ordinarily given in aggressive contexts (chimpanzee, van Lawick-Goodall 1968, Marler and Tenaza 1977; gorilla, Fossey 1972; orangutan, MacKinnon 1974). In Cercopithecidae, it appears that the loud calls with aggressive-context components play a role in the active spacing of the groups and antipredator defense, whereas loud calls without those components only maintain intergroup distance (Gautier 1969; Marler 1972; Waser 1975). In howlers, the loud calls that do not contain aggressive-context components serve as a "proclamation of an occupied area" (Altmann 1959) and as cues for intergroup avoidance (Southwick 1962; Baldwin and Baldwin 1976). This is also the function of the loud "wails" of the lemur *Indri indri* (Petter 1962) and of the howls of *Lemur catta* (Jolly 1966). Furthermore, the nonaggressive loud calls also serve to rally group members (Gautier 1969; Struhsaker 1970; Baldwin and Baldwin 1976). This function of reinforcing cohesion in the vocalizer's group has also been demonstrated in mangabeys and baboons (*Cercocebus albigena*, Waser 1977b; *Papio*, Byrne 1980; Waser, chap. 6, this volume). An attracting effect of this type of loud call is assumed for *C. g. agilis* (Quris 1980). In contrast, the loud calls that contain aggressive components, such as the "pant-hooting" of the chimpanzee and the "hoot" series of the gorilla, serve in territorial defense (Marler and Tenaza 1977). This is also the function of male orangutan long calls, "each male vocally defending a mobile territory" (MacKinnon 1974).

In most Cercopithecinae and Colobinae, some lemurs, and orangutans, the loud calls are uttered only by males. On the other hand, in howlers, chimpanzees, and to a lesser extent the gorilla, the loud calls are uttered by all age–sex classes. However, clear, distinctive structural differences are apparent in the male and female roars of howlers (Baldwin and Baldwin 1976). More subtle differences exist in male and female chimpanzees (pant-hooting, Marler and Hobbett 1975). Within a species, loud calls possess individually distinctive features (Marler and Hobbett 1975; Tenaza 1976; Waser 1977b; Quris 1980). This is generally true except for the black-and-white colobus, where the range of intraindividual variability is as great as the range of interindividual variability (Marler 1972). The age–sex class specificity of loud calls, or lack thereof, allows the transmission of information about specific elements of a group's social structure (e.g., uni-male vs. multi-male) and the socionomic composition of the calling group. In addition, information about the individual identity of group members or group leaders is broadcast. This led Marler and Tenaza (1977) to assume that in chimpanzees, where there are many different combinations of sub-groupings, "sexual differences in pant-hooting are presumably as much an issue of communication within groups as between them."

All loud calls are highly locatable vocalizations. Brown (chap. 7, this volume) shows that wide bandwidth calls are best suited for vertical and horizontal localization, but that narrow bandwidth calls are better localized in noisy environments. The low-pitched, frequency-modulated pure sounds delivered in a sequential pattern show the best locatability in forest environments (Waser 1977a; Brown, Beecher, Moody, and Stebbins 1979; Quris 1980). Thus it appears that most sounds produced by the white-cheeked gibbons would be highly locatable.

Function of songs in gibbons in relation to their physical structure

In gibbons, loud calls are given by both the male and female of a mated pair and even by members of all other age–sex classes. Ellefson (1968) asserts that the loud calls of howlers and gibbons (white-handed gibbons) serve opposite functions. Thus, howler roars elicit avoidance, whereas conflict whôôs of adult male white-handed gibbons function as active spacing mechanisms, which elicit confrontation rather than avoidance. Opposite in function to the male conflict whôô is the female gibbon's "great" call, which functions in both within- and between-group social communication. It is a location signal meaning "here we are" (Ellefson 1968). Tenaza (1976) suggests the same location-signaling function for the male Kloss's gibbon song. In this species the more agitated behavior of the singing (or countersinging) female suggests a more active territorial defense. As mentioned previously, the white-cheeked gibbons' songs contain calls derived from the type 2 and type 4 calls of the common vocal repertoire (except for one male sound). These types do not have any ag-

Figure 4.12. Singing white-cheeked gibbons: Subadult female (left) is uttering a whuit call, juvenile male (right) a whôô call.

gressive connotations. However, the female song contains a sequence of type 3 calls, which are usually uttered in aggressive social situations (Deputte and Goustard 1978). Thus, it is possible that the function of the male song is not exactly the same as that of the female. The absence of aggressive components and the general calm state of the emitter could mean that the male song is merely a locale marker. This contrasts with the female song, whose aggressive components may have a repulsive effect on neighboring groups and especially on other nonmated females. This suggests that the white-cheeked gibbons' songs are functionally analogous to songs of Kloss's gibbons but functionally different from those of the white-handed gibbon (Ellefson 1968; Tenaza 1976).

Ontogeny of male song and ontogeny of pair

During ontogeny, juvenile males chorus with juvenile females by using a femalelike song that resembles the terminal portions of the typical adult female song. Hence, as a result of uttering the femalelike song, they participate in the rather active intergroup spacing activities. When they reach subadulthood, their femalelike songs are gradually replaced by malelike songs, but they pass through a stage in which they use both malelike and femalelike utterances (Figure 4.12). This stage very likely precedes the peripheralization of the males. Then, when given alone, the male song may signal the presence of a nonmated male and could have an attracting effect on the surrounding nonmated females of the population, leading to the creation of a new gibbon group (Aldrich-Blake and Chivers 1973; Tenaza 1976; MacKinnon and MacKinnon 1977).

Hence, the white-cheeked gibbon duet is more than a mere female plus male song. It appears to signal the existence and location of a still-established mating pair. The duet, in gibbons, seems to act to protect the monogamous structure. Nonmated females surrounding a duetting pair are informed of the presence of a mated female and are likely to be maintained at a distance by the aggressive components of the female song. They are also informed that the male in the vicinity is already mated, because of the precise, temporally adjusted response to *his* female partner's song. In addition, the chorus of the juveniles may transmit a message about the size of the family group and perhaps even its socionomic composition.

Apart from this intergroup communicative function, it is likely that the duet has an intragroup function. Within the group, cohesion may be reinforced by the interindividual embraces that occur repeatedly during the duets. Furthermore, the individuality of the songs could maintain kinship ties in a gibbon population (Marler and Tenaza 1977) and also serve as a vocal mechanism for incest avoidance.

Structure of male and female song versus structure of duets: Taxonomic implications

The structure of the male and female song seems to influence the organization of the duets. In the siamang there is only a small sexual difference in songs and both songs consist of long sequences of calls (Lamprecht 1970). The duet is precisely organized in spite of a large amount of overlapping. In the hoolock gibbon, male and female songs are only slightly different, and they too are composed of long sequences of calls. Their joint utterance produces a unison duet (Marler and Tenaza 1977). In other species, male and female songs are much different in organization and the duets are generally antiphonal (Marshall et al. 1972; Marler and Tenaza 1977). At the end of the female song, the males utter a "coda" analogous to the "response" described in the white-cheeked gibbon. According to Marler and Tenaza (1977), the differences in the organization of the duets are linked to male and female song structure, but there is no apparent correlation between sexual dimorphism in color and sexual dimorphism in song structure. These authors also point out that there is more similarity among female songs than among male songs in the gibbon species. The male and female songs in white-cheeked gibbons fit well with this general pattern. Female song shows some pattern analogy with the patterns of the Bornean and pileated gibbons (see Figure 23 in Marler and Tenaza 1977) and *H. c. gabriellae* (Goustard, unpublished). Male song in the white-cheeked gibbon is somewhat similar in organization to that of the red-cheeked male song (Demars and Goustard 1978) and, to a lesser extent, the Kloss's gibbon song (Tenaza 1976). The analogies with *concolor* gibbons may be related more to past contact between populations than to present contact because, currently, white-cheeked gibbons are located primarily in Laos and northern Vietnam to the west of the Black River, whereas the red-cheeked gibbon is located in southern Vietnam and southernmost Laos (Groves 1972).

In spite of specific differences in male and even female songs, one may note the close analogy of duetting organization in *H. c. leucogenys* and *H. lar lar*. The latter's song is assumed to be similar to those of *H. l. agilis* (Marler and Tenaza 1977). The antiphonal character of the duet, by preventing interference with calling, permits the broadcasting of a minimally ambiguous message.

Duetting in another primate

Duetting occurs in at least one other primate species, the *Callicebus moloch,* a South American species (Robinson 1979). This species utters long sequences of calls, which Moynihan (1966) has termed "songs." Male and female songs are derived from the common vocal repertoire but are organized differently from gibbon song (Robinson 1979). The duet reflects these differences. Duetting in titis stimulates the approach of neighboring groups. But the function of male and female songs uttered in the duet shows some close analogies with those inferred for white-cheeked gibbons. In titis, Robinson (1981) assumes that the male's contribution to the duet functions to reinforce the pair bond and to maintain conventional boundaries. Such a close analogy in the organization and function of gibbon and titi calling raises the question of whether other behavioral analogies might exist. "Vocal duetting is a complex phenomenon which ideally should be viewed in relation to the total behavior and ecology of these species which duet" (Harcus 1977).

Duet versus male-only loud call: Monogamy versus polygamy? Socioecological and phylogenetic trends in age–sex class participation in loud calling

The common features of duetting titis and gibbons are monogamy and territoriality. These two characteristics are also shared by numerous species of Lemuridae and Cebidae (P-type, Itani 1977), but the titis and gibbons are much more vocal. Territorial behavior is generally associated with monogamy and lack of sexual dimorphism. The absence of sexual dimorphism is often accompanied by a lack of differentiation in social roles. Even if behavior at the boundaries differs in male and female titis, as in gibbons, the two partners nevertheless participate in the regulation of spacing. One monogamous species, *Cercopithecus neglectus,* does not fit with these assumptions because it shows a prominent sexual dimorphism, no territoriality, and the loud calls are male-specific (Gautier-Hion and Gautier 1978).

In polygamous species (T-type, Itani 1977) there is a clear-cut differentiation of the social roles. In such groups, either uni- or multimale, the spacing role is adopted by the males and they alone emit type 1 loud calls. In chimpanzees, the complex and fluctuating group organization perhaps helps "to explain the lack of any vocalizations that are unique to adult males" (Marler and Tenaza 1977).

The "chorus," a vocal phenomenon associated with group cohesion, is characteristic of mangabeys (Deputte 1973; Waser 1977b) and baboons (DeVore and Hall 1965). In mangabeys, the chorus often precedes (perhaps elicits?) the male whoopgobble (Waser 1977b; personal observation). This vocal phenomenon is seen also in *Lemur catta* (Jolly 1966), howlers (Baldwin and Baldwin 1976), and chimpanzees. Talapoin monkeys lack the male loud calls, and choruses are the sole group vocal manifestation (Gautier 1974). Perhaps the chorus is a vocal adaptative feature linked to multimale group organization. The chorus is also present in white-cheeked gibbons and occurs in association with the duet.

Thus two trends are evident: (1) The development of vocal participation and diminution of age–sex specificity in space-related calling is associated with aggregation tendencies (T-type), and (2) the development of participation by all group members in vocal proclamation is associated with phylogenetic development, exemplified in T-type by the chimpanzee and in P-type by the gibbon. The countersigning and the all-male and all-female choruses (not in a group but in a population) may be another adaptative feature in Kloss's gibbon (Tenaza 1976).

Duetting in birds: Evolutionary convergence with primates?

Duetting also occurs in many tropical bird species. Todt and Fiebelkorn (1980) have reviewed the social functions of antiphonal duets in birds. Many of them are shared by the gibbons and the titis; for example, the development and maintenance of pair bond (see, e.g., Thorpe 1963, 1972; Watson 1969; Wickler and Seibt 1980) and territorial display (see, e.g., Grimes 1966; Diamond and Terborgh 1968; Seibt and Wickler 1977; Wiley and Wiley 1977). The duetting birds are supposed to be present primarily in tropical habitats with dense vegetation and, hence, lack visual contact. However, there are exceptions. The duetting of nocturnal owls inhabiting temperate zones confirms the correlation between this specialized vocal display and the absence of visual contact (Diamond and Terborgh 1968), but only three clear examples of duetting bird species (plus four occasional and one rarely) may be found among the 217 species living in New Guinea. Moreover, New Guinea duetters may be found in every possible habitat. The only common features possessed by the duetting bird species are their sociability and their maintenance of pair bonds for long periods of time, even beyond their breeding season (Thorpe and North 1966; Diamond and Terborgh 1968; Payne 1971). This may reflect the influence of the tropical habitat more than the lack of visual contact. Vencl and Soucek (1976) outline a correlation between sexual monomorphism and duetting, but it does not seem to fit the duetters in New Guinea (Diamond and Terborgh 1968). Moreover, Payne (1971) reports that more than half the monomorphic African birds are not duetters. Thus, duetting may be a case of evolutionary convergence evolved in monogamous, territorial species as one of the influences (direct or indirect) of the structure of the habitat.

Birds and gibbons: Neural control differences?

Duetting implies a precise timing between the songs of the two partners. Thorpe (1963) argues that in birds the delay between the two calls in a duet represents the auditory reaction time, but according to Payne (1971), the timing in duets "appears to be more complex than a simple reflex." In white-cheeked gibbons the complex antiphonal organization suggests that in addition to expressing emotion, the call may be produced under the volitional control of the male. But, as Aitken and Wilson (1979) point out, this control does not necessarily imply that the monkeys (nor even the gibbons) "also possess a dual system of neural control as found in humans."

In this discussion I have tried to find some general trends in species that sing in duets despite great variability in social organization and vocal behavior. I am aware that many exceptions *must* exist. Such discrepancies simply express the variety of potential adaptations available to different species as "communication is both an instrument for organizing societies and a mirror of social organization" (Green and Marler 1979).

REFERENCES

Aitken, P. G., and Wilson, W. A., Jr. 1979. Discriminative vocal conditioning in rhesus monkeys: evidence for volitional control? *Brain and Language* 8:227–40.
Aldrich-Blake, F. P. G., and Chivers, D. J. 1973. On the genesis of a group of siamang. *American Journal of Physical Anthropology* 38:631–6.
Altmann, S. A. 1959. Field observations on a howling monkey society. *Journal of Mammalogy* 40:317–30.
Baldwin, J. D., and Baldwin, J. I. 1976. Vocalizations of howler monkeys (*Alouatta palliata*) in southwestern Panama. *Folia Primatologica* 26:81–108.
Brown, C. H., Beecher, M. D., Moody, D. B., and Stebbins, W. C. 1979. Locatability of vocal signals in Old World monkeys: design features for the communication of position. *Journal of Comparative and Physiological Psychology* 93:806–19.
Byrne, R. W. 1980. Uses of long-range barks during ranging by Guinea baboons (*Papio papio*) in Senegal. *Antropologia Contemporanea* 3:176.
Carpenter, C. R. 1940. A field study in Siam of the behavior and social relations of the gibbon (*Hylobates lar*). *Comparative Psychological Monographs* 16(5):1–212.
Chivers, D. J. 1972. The siamang and the gibbon in the Malay Peninsula. In *Gibbon and siamang,* Vol. 1, ed. D. M. Rumbaugh, pp. 103–35. Basel: Karger.

Demars, C., and Goustard, M. 1978. Le "grand chant" d'*Hylobates concolor leucogenys*. Comparaison avec les émissions sonores homologues d'*H. concolor gabriellae* et d'*H. klossii* (Iles Mentawei, ouest Sumatra). *Behaviour* 65:1–26.

Deputte, B. L. 1973. Etude d'un type de comportement vocal chez un groupe captif de mangabeys (*Cercocebus albigena* Gray). Méthode télémétrique d'enregistrement individuel des vocalisations. *D.E.A. Biologie Animale: Eco-Ethologie*, Université de Rennes I.

Deputte, B. L., and Goustard, M. 1978. Etude du répertoire vocal du gibbon à favoris blancs (*Hylobates concolor leucogenys*). Analyse structurale des vocalisations. *Zeitschrift für Tierpsychologie* 48:225–50.

DeVore, I., and Hall, K. R. L. 1965. Baboon ecology. In *Primate behavior: Field studies of monkeys and apes*, ed. I. DeVore, pp. 20–110. New York: Holt, Rinehart and Winston.

Diamond, J. M., and Terborgh, J. W. 1968. Dual singing by New Guinea birds. *Auk* 85:62–82.

Ellefson, J. O. 1968. Territorial behavior in the common white-handed gibbon *Hylobates lar* Linm. In *Primates: studies in adaptation and variability*, ed. P. C. Jay, pp. 180–99. New York: Holt, Rinehart and Winston.

1974. A natural history of white-handed gibbons in the Malayan peninsula. In *Gibbon and siamang*, Vol. 3, ed. D. M. Rumbaugh, pp. 1–136. Basel: Karger.

Fossey, D. 1972. Vocalization of the mountain gorilla (*Gorilla gorilla beringei*). *Animal Behaviour* 20:36–53.

Gautier, J.-P. 1969. Emissions sonores d'espacement et de ralliement par deux cercopitheques arboricoles. *Biologia Gabonica* 5:117–45.

1974. Field and laboratory studies of the vocalizations of talapoin monkeys (*Miopithecus talapoin*). *Behaviour* 51:209–73.

1975. Etude comparée des systèmes d'intercommunication sonore chez quelques cercopithécinés forestiers africains. Mise en évidence de corrélations phylogénétiques et socio-écologiques. *Thesis*, Université de Rennes I.

Gautier, J.-P., and Gautier, A. 1977. Communication in Old World monkeys. In *How animals communicate*, ed. T. A. Sebeok, pp. 890–963. Bloomington: Indiana University Press.

Gautier-Hion, A., and Gautier, J.-P. 1978. Le singe de Brazza: une stratégie originale. *Zeitschrift für Tierpsychologie* 46:84–104.

Goustard, M. 1979. Les interactions acoustiques au cours des émissions sonores de gibbons (*Hylobates concolor leucogenys*). *Journal de psychologie normale et pathologique* 77:133–56.

Green, S. 1975. Variation of vocal pattern with social situation in the Japanese monkey (*Macaca fuscata*): a field study. In *Primate behavior*, Vol. 4: *Developments in field and laboratory research*, ed. L. A. Rosenblum, pp. 1–102. New York: Academic Press.

Green, S., and Marler, P. 1979. The analysis of animal communication. In *Handbook of behavioral neurobiology*, Vol. 3: *Social behavior and communication*, ed. P. Marler and J. G. Vandenbergh, pp. 73–158. New York: Plenum Press.

Grimes, L. G. 1966. Antiphonal singing and call notes of *Laniarius barbarus barbarus*. *Ibis* 108:122–6.

Groves, C. P. 1972. Systematics and phylogeny of gibbons. In *Gibbon and siamang*, Vol. 1, ed. D. M. Rumbaugh, pp. 1–89. Basel: Karger.

Harcus, J. L. 1977. The functions of vocal duetting in some African birds. *Zeitschrift für Tierpsychologie* 43:23–45.

Hooker, T., and Hooker, B. I. 1969. Duetting. In *Bird vocalizations*, ed. R. A. Hinde, pp. 185–205. Cambridge University Press.

Itani, J. 1977. Evolution of primate social structure. *Journal of Human Evolution* 6:235–43.

Jay, P. C. 1965. The common langur of north India. In *Primate behavior: field studies in monkeys and apes,* ed. I. DeVore, pp. 197–249. New York: Holt, Rinehart and Winston.

Jolly, A. 1966. *Lemur behavior.* Chicago: University of Chicago Press.

Jürgens, U. 1979. Vocalization as an emotional indicator: a neuroethological study in the squirrel monkey. *Behaviour* 69:88–117.

Klopfer, P. H. 1977. Communication in prosimians. In *How animals communicate,* ed. T. A. Sebeok, pp. 841–50. Bloomington: Indiana University Press.

Lamprecht, J. 1970. Duettgesang beim Siamang, *Symphalangus syndactylus* (Hominoidea, Hylobatinae). *Zeitschrift für Tierpsychologie* 27:186–204.

MacKinnon, J. 1974. The behaviour and ecology of wild orang-utan (*Pongo pygmaeus*). *Animal Behaviour* 22:3–74.

MacKinnon, J., and MacKinnon, K. 1977. The formation of a new gibbon group. *Primates* 18:701–8.

Marler, P. 1972. Vocalizations of East African monkeys. II. Black and white colobus. *Behaviour* 42:175–97.

Marler, P., and Hobbett, L. 1975. Individuality in a long-range vocalization of wild chimpanzees. *Zeitschrift für Tierpsychologie* 38:97–109.

Marler, P., and Tenaza, R. 1977. Signaling behavior of apes with special references to vocalization. In *How animals communicate,* ed. T. A. Sebeok, pp. 965–1033. Bloomington: Indiana University Press.

Marshall, J. T., and Marshall, F. R. 1976. Gibbons and their territorial songs. *Science* 193:235–7.

Marshall, J. T., Jr., Ross, B. A., and Chantharojvong, S. 1972. The species of gibbons in Thailand. *Journal of Mammalogy* 53:479–86.

Matsumura, S. 1979. Pattern of mother–infant vocal communication in a horseshoe bat. *Abstracts of XVIth International Ethological Conference,* Vancouver.

Moynihan, M. 1966. Communication in the titi monkey *Callicebus. Journal of Zoology (London)* 150:77–127.

Payne, R. B. 1971. Duetting and chorus singing in African birds. *Ostrich, Supplement* 9:125–46.

Petter, J. J. 1962. Ecological and behavioral studies of Madagascar lemurs in the field. *Annals of the New York Academy of Sciences* 102:267–81.

Quris, R. 1980. Emission vocale de forte intensité chez *Cercocebus galeritus agilis:* structures caractéristiques spécifiques et individuelles. Modes d'émission. *Mammalia* 44:35–50.

Robinson, J. G. 1979. An analysis of the organization of vocal communication in the titi monkey *Callicebus moloch. Zeitschrift für Tierpsychologie* 49:381–405.

1981. Vocal regulation of inter- and intragroup spacing during boundary encounters in the titi monkey *Callicebus moloch. Primates* 22:161–72.

Seibt, V., and Wickler, W. 1977. Duettieren als Revier-Anzeige bei Vögeln. *Zeitschrift für Tierpsychologie* 43:180–7.

Sokal, R. R., and Rohlf, F. J. 1969. *Biometry: the principles and practice of statistics in biological research.* San Francisco: Freeman.

Sorensen, M. W., and Thompson, P. 1969. Antiphonal calling of the tree shrew *Tupaia palawanensis. Folia Primatologica* 11:200–5.

Southwick, C. 1962. Patterns of intergroup social behavior in Primates with special references to rhesus and howling monkeys. *Annals of the New York Academy of Sciences* 102:436–54.

Struhsaker, T. T. 1970. Phylogenetic implications of some vocalizations of *Cerco-pithecus* monkeys. In *Old World monkeys: evolution, systematics and behavior*, ed. J. R. Napier and P. H. Napier, pp. 365–403. London: Academic Press.

Tenaza, R. R. 1976. Songs, choruses and countersinging of Kloss' gibbons (*Hylobates klossii*) in Siberut Island, Indonesia. *Zeitschrift für Tierpsychologie* 40:37–52.

Thorpe, W. H. 1963. Antiphonal singing in birds as evidence for avian auditory reaction time. *Nature* 197:774–6.

1972. Duetting and antiphonal song in birds: its extent and significance. *Behaviour, Supplement* 18:1–197.

Thorpe, W. H., and North, M. E. W. 1966. Vocal imitation in the tropical bou-bou shrike *Laniarius aethiopicus major* as a means of establishing and maintaining social bonds. *Ibis* 108:432–5.

Tilson, R. L. 1977. Duetting in Namib desert klipspringers. *South African Journal of Science* 73:314–15.

Todt, D., and Fiebelkorn, A. 1980. Display timing and function of wing movements accompanying antiphonal duets of *Cichladusa guttata*. *Behaviour* 72:82–106.

van Lawick-Goodall, J. 1968. The behaviour of free-living chimpanzees in the Gombe Stream Reserve. *Animal Behaviour Monographs* 1(3)·161–311.

Vencl, F., and Soucek, B. 1976. Structure and control of duet singing in the white-crested laughing-thrush (*Garrulax leucolophus*). *Behaviour* 57:206–26.

Waser, P. M. 1975. Experimental playbacks show vocal mediation of intergroup avoidance in a forest monkey. *Nature* 255:56–8.

1977a. Sound localization by monkeys: a field experiment. *Behavioral Ecology and Sociobiology* 2:427–31.

1977b. Individual recognition, intragroup cohesion and intergroup spacing: evidence from sound playback to forest monkeys. *Behaviour* 60:28–74.

Watson, M. 1969. Significance of antiphonal song in the eastern whipbird *Psophodes olivaceus*. *Behaviour* 35:157–78.

Wickler, W., and Seibt, U. 1980. Vocal duetting and the pair bond. II. Unison duetting in the African forest weaver *Symplectes bicolor*. *Zeitschrift für Tierpsychologie* 52:217–26.

Wiley, R. H., and Wiley, M. S. 1977. Recognition of neighbor duets by stripe-backed wrens *Campylorhynchus nuchalis*. *Behaviour* 62:10–34.

5 · Vocal systems regulating within-group spacing

JOHN G. ROBINSON

This chapter concerns vocalizations that control spacing within primate groups. The spatial consequences of different calls have been described in a number of species. But to understand the vocal determinants of the characteristic spatial configurations of primate groups, the analysis must consider not only how the initial spacing of animals affects the probability of use of a call and the spatial responses to it but also how it affects the probabilities of use of, and spatial responses to, a group of interrelated calls. In other words, one must ask how calls, as a system, regulate spacing.

A vocal system is a set of vocalizations that are related to one another. In the classification of a vocal repertoire, calls are usually assigned to the same vocal type on the basis of structural similarity. Vocal types are then grouped using some *relationship* among calls, such as similarities or differences in the physical structure, use, or effect of the calls on conspecifics. Relationships have been posited on the basis of similarity of physical structure (Rowell and Hinde 1962; Winter, Ploog, and Latta 1966), transition frequencies between calls in the same animal (Green 1975; Jürgens 1979 and chap. 3, this volume) and in different animals (Maurus, Pruscha, Weisner, and Geissler 1979), and equivalent places in vocal sequences (Newman, Lieblich, Talmage-Riggs, and Symmes 1978; Maurus et al. 1979; Robinson 1979). Vocal types have been grouped using correlation between the vocalizations and behaviors (Smith 1977) and inferred functions (Collias 1960).

However, the relationship that an investigator chooses to use in structuring a vocal repertoire is not trivial. If vocalizations that are linked have similar causes or effects, then the method of structuring the vocal repertoire defines how the vocal repertoire is organized and specifies which variable underlies the vocal system. By grouping calls in different ways, different investigators have argued that the morphology of calls varies

Field work was supported by a Smithsonian Institution grant to J. F. Eisenberg and NIMH Research Grant MH 28840 to J. F. Eisenberg and R. Rudran. T. Blohm generously provided accommodation, facilities, and the opportunity to work at Hato Masaguaral. D. Moskovits and C. Snowdon constructively criticized earlier drafts of the manuscript. My wife, Linda Cox, was an irreplaceable help in the field and an unflagging judge of the result.

with the animal's motivation (Rowell and Hinde 1962; Jürgens, chap. 3, this volume), internal state (Green 1975), the characteristics of the signaler and its tendencies (Smith 1977), and the functional contexts in which the vocalizations are used (Collias 1960).

A second step frequently is to specify the *rules* that relate variation in vocal morphology to variation in one or more of these postulated independent variables. For example, vocalizations might follow the rule that noisiness indicates high arousal (Green 1975), that harsh, low-frequency sounds indicate hostility, and that high-frequency, pure-toned sounds indicate friendliness (Morton 1977; Jürgens, chap. 3, this volume) and that pure-toned calls allow vocalizers to avoid detection by predators (Marler 1955). These rules are quite distinct from syntactic rules (see Marler 1977; Robinson 1979; Snowdon, chap. 9, this volume).

In this chapter I discuss the vocal systems that operate to regulate the spacing within social groups. These systems are not defined by similarity in the calls' physical structure nor by similarity in use or response to the calls but, assuming that vocalizations reflect the patterns of social interactions, by the axis along which the vocal types vary continuously in physical structure, use, and the response of conspecifics.

MOTIVATIONAL, FUNCTIONAL, OR OPERATIONAL VOCAL SYSTEMS?

Different determinants of vocal output and thus different classifications of vocal repertoires have been suggested. Can one predict vocal output from the supposed motivational state and vice versa? From the observation that there are structural similarities and intergradations among vocalizations in *Macaca mulatta,* Rowell and Hinde (1962) suggested that this relatedness among vocalizations reflects the underlying motivational substrate. If vocalizations are external manifestations of motivational conditions, then similar vocalizations provide evidence of a similarity of motivation.[1] Green (1975) reported concordance between the use of vocalizations and the animal's demeanor in *Macaca fuscata.*

This concept has been extended in two directions. First, reflecting the continuing discussion over the relative influence of specific and nonspecific factors in motivation, Eisenberg (1974) distinguished calls reflecting a specific mood from those reflecting a general arousal. The primacy of specific mood (e.g., Moody and Menzel 1976) or arousal (Kiley 1972; Baldwin and Baldwin 1976) apparently varies among species. Second, Green (1975) pointed out that in addition to demeanor and arousal, extrinsic factors such as social spacing and orientation contribute to internal state. Wiley (1976) felt that in common grackles a change in spatial relationships between animals was more important than a change in locomo-

[1] This has been questioned by Fentress (1976) who points out that (1) discontinuities in overt expression do not imply discontinuities in cause, (2) different causal factors may "blend" to produce apparently continuous expressions or (3) may "interleaf" to produce sequences for which it is difficult to separate the components.

tor tendencies in eliciting certain vocalizations. The existence of both extrinsic and nonspecific factors in internal state means that the vocal system cannot be strictly mapped onto intrinsic motivational continua (e.g., aggression, fear, sex, etc.), although the match still might be sufficiently precise to be useful.

There is another consideration. Motivational systems perhaps should be defined in part in terms of their interactions with other systems (Fentress 1976), with boundaries between systems not clearly demarcated. If this is the case, clearly defined motivational clusters and rigid motivational axes may be illusory, and correlation between vocal output and inferred internal state might be quite poor. This is not to say that vocalizations are not expressions of internal state, only that internal state is a composite of a number of factors, both external and internal, whose boundaries and relative influences are difficult to specify.

Do social or functional circumstances predict vocal usage? Certain calls do occur in certain contexts (Fossey 1972; Gautier 1974, 1978; Green 1975). When vocalizations occur in a narrowly defined set of circumstances, the responses of other animals will be less variable and a restricted function can be inferred (Eisenberg 1974, 1976). Calls given only on the approach of a predator may elicit quite specific avoidance responses (Seyfarth and Cheney, chap. 10, this volume), for example. However, many calls are not contextually restricted (Kiley 1972) and consequently elicit quite varied responses (Smith 1977, Waser, chap. 6, this volume). This does not invalidate functional analysis, but it does make it difficult. Green (1975), for instance, concluded that in *M. fuscata* the correlation between vocal types and functionally defined contexts was poorer than between vocal types and contexts defined by the animal's demeanor, arousal, social spacing, and orientation.

Nevertheless, the effects of calls are often of interest. In an attempt to determine the consequences of vocalizations in West African cercopithecines, Gautier (1974, 1978; Gautier and Gautier 1977) has described vocal systems that regulate social interactions within and between groups. If, during social interactions, particular sets of circumstances frequently lead into, or are followed by, other sets of circumstances, transitions between vocalizations associated with those contexts will be frequent and perhaps intergrade. The relationships between vocalizations define vocal systems that operate to regulate the species' social interactions. Gautier has suggested two such systems: the aggression–flight system and the cohesion system. Vocalizations are not grouped by similarity of motivational state nor by function but to map the patterns of social interaction. In this chapter I follow this approach and discuss the vocal systems regulating spacing within primate groups.

VOCAL SYSTEMS OF SPACING

Calls that are uttered when group members are separated but within auditory and often visual communication of one another frequently grade into calls given when animals are isolated from others. Noting this relation in the vocalizations of the West African guenons, Gautier (1967, 1974, 1978; Gautier and Gautier 1977) proposed the existence of a vocal "cohesion" system, with vocal morphology varying with increasing spatial separation and isolation of the calling animal from other group members. The system has the consequence of maintaining a characteristic spacing. A similar gradation between "contact" calls and the characteristic "lost" calls of Neotropical primates has been noted by Oppenheimer (1977). In the marmoset *Callithrix jacchus* the contact "phee" becomes longer, louder, and more frequent when animals are losing contact (Epple 1968). The gradation between "contact" and "isolation" calls is apparently not so complete in the tamarins, although the short contact "pe" is structurally similar to the longer isolation "whee" in *Leontopithecus rosalia*, and in *Saguinus oedipus*, calls resembling the short contact "te" are delivered at a rapid rate when animals are losing contact with one another (Epple 1968). *Saimiri* isolation peeps are structurally similar to, but discrete from, contact peeps (Winter, Ploog, and Latta 1966). Transitions between these calls within individual sequences are significantly greater than expected (Jürgens 1979). Finally, in *Ateles*, "slow whinnies," which are contact calls frequently given in feeding contexts, grade into the long calls that maintain contact between subgroups (Eisenberg 1976).

For the remainder of this chapter I will discuss the vocal system of spacing in the wedge-capped capuchin. For this species I posit the existence of a "cohesion-spacing" system, with calls spacing out other group members and other calls reducing interindividual separations.

THE COHESION–SPACING SYSTEM OF *CEBUS NIGRIVITTATUS*

From June 1977 through July 1979 I studied a population of wedge-capped capuchins, *Cebus nigrivittatus*, inhabiting gallery forest in the savanna *llano intermedio* of Venezuela. The study site was a cattle ranch, Fundo Pecuario Masaguaral, about 45 km south of the town of Calabozo. Vocalizations were recorded using a Uher 4000-L tape recorder and a Sennheiser MKH 816T directional microphone. Recordings were opportunistic, and the additional information I noted included the identity and behavior of the signaler, the identity and distance to the nearest neighbor, and other animals' vocal and behavioral responses.

Three calls that have consequences for the spacing of animals are structurally related (see later discussion). The "arrawh" (the following terminology after Robinson 1979) is uttered in phrases that normally comprise 2 to 5 calls (see Oppenheimer and Oppenheimer 1973) that have a single, low-frequency, generally tonal, dominant band, possibly with some overtones. The duration of each call ($\bar{x} = 0.75$ sec, $cv = .303$, $n = 52$) generally increases through the phrase, as does the intercall interval ($\bar{x} = 0.28$ sec, $cv = .272$, $n = 40$), which is measured from the end of one call to the beginning of the next. There are two common variants, each occurring in a different set of circumstances. Loud arrawhs, with some atonality (Figure 5.1H), are given by animals separated from other group

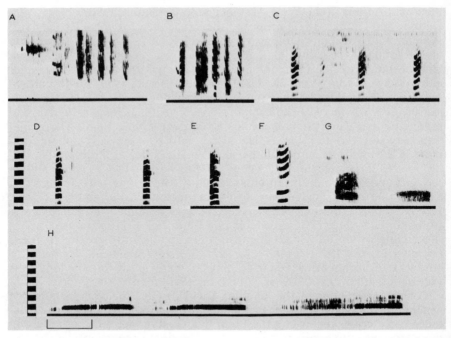

Figure 5.1. Spectrograms of *Cebus nigrivittatus* vocalizations analyzed with a broadband filter (effective resolution, 300 Hz): (A) hehs from very excited animal; (B) hehs, less excited; (C) huhs delivered in a rapid cadence; (D–F) huhs; (G) calls intermediate between huhs and arrawhs; (H) loud arrawh phrase. Frequency scale indicates kilohertz; time scale is 0.5 sec.

members. The response of the latter if within earshot (~300 m) is to increase their "huh" delivery rate and sometimes to reply with arrawhs. The isolated animals can then approach the body of the group. Quiet arrawhs, which are pure toned and much softer (range ~20 m), are given by animals within the group who are dropping behind during group progression or who are moving in a direction different from other group members.

The huh call (Figure 5.1C–F) of capuchins is a short ($\bar{x} = 0.10$ sec, $cv = .31, n = 30$), largely tonal call. In some examples atonality occurs at the end of the call. Frequency modulation is slight. Typically, the frequency of the bands increases gradually through the course of the call, often ending with a slight drop in pitch. The plaintive quality of the call is responsible for one of the common English names of *C. nigrivittatus:* the weeper capuchin. This is the most commonly occurring call in the repertoire. It is highly repetitious and apparently has a "tonic" effect (Schleidt 1973) on other group members, for it rarely elicits an obvious response. Oppenheimer (1977) terms it a contact call.

The "heh" call is a noisy, atonal call of 3 to 7 syllables (Figure 5.1A and B), although at lower excitation levels (Figure 5.1B) some tonal struc-

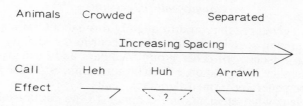

Figure 5.2. Operation of *C. nigrivittatus* cohesion-spacing vocal system showing the association between vocalization and characteristic spacing and the effect of the vocalization.

ture is discernible. The syllables are rapidly pulsed with a mean intersyllable interval (measured from the beginning of one syllable to the beginning of the next) of 0.21 sec ($cv = .43$, $N = 78$). The calls are a component of the threat display of capuchins, and the threatened animal usually moves away rapidly.

The three calls are related by both their structure and their proximity in sequences. Hehs and huhs intergrade, and intermediates are not uncommon. Figure 5.1C illustrates four short huh calls delivered in a cadence similar to, but not as rapid as, hehs. Intergradations between huhs and arrawhs are less common, but Figure 5.1G illustrates one example. The first call of the pair is huhlike but with fewer bands and a longer duration. The second, which follows at an interval similar to the temporal patterning of arrawhs, is a short single arrawh call. Moving along the axis from arrawhs through huhs to hehs, the changes in physical structure are consistent: The perceived pitch increases, the frequency span widens, and the call becomes more atonal. Heh calls are composed of a number of pulsed syllables. Monosyllabic huh calls are given singly but are frequently repeated. Arrawhs are composed of a number of arrawh calls.

Relatedness among vocal classes is evident from the transitions from one class to another during vocal sequences. I constructed a matrix of pairwise transitions for all the vocalizations in the repertoire from a random selection of vocal sequences. The number of transitions between arrawhs and huhs was significantly higher than expected from a random model, as was the number of transitions between huhs and hehs.

The calls differ in physical form and in the spacing associated with their delivery, characteristics that vary predictably (see Brown, chap. 7, this volume). Hehs occur when the animals are crowded, such as when they are feeding in fruit trees or when the distance between two individuals diminishes, for whatever reason, during invertebrate foraging. Arrawhs are uttered by an animal that is isolated or separated from other group members. The spatial response of neighbors tends to bring animals into a more characteristic spacing (Figure 5.2). Huhs are given at intermediate separations and in all circumstances, but the frequency of their use is variable, a characteristic that can be examined to determine the call's spatial consequences.

Contact calls in general

Calls that are given regulary and frequently throughout daily activity are widespread among social primates. They occur in the vicinity, and often in the visual presence, of other group members and in many contexts. They seem to elicit little response from group members. These low-amplitude calls are generally termed "contact" calls.

It is not clear what elements are common to the circumstances associated with these calls, nor whether there are any common responses to them. Gautier and Gautier (1977) felt that the frequency of calling in forest monkeys increased with the "potential risk of losing contact with congeners." The calls often increase in frequency during contexts that are potentially disruptive. The frequency with which they are delivered increases at the start of group progressions, at times when the habitat or light intensity makes contact more difficult, and after disturbances. The Gautiers concluded that an important consequence of calling and counter-calling was to permit "reciprocal individual localization" and to allow "the general cohesion of the social unit." Oppenheimer (1977) concluded from his survey of Neotropic primates that such calling "probably functions to maintain contact" among group members. This implies that animals give, and respond to, these calls so as to identify their positions and thus allow each animal to position itself relative to other group members.

Another possibility is that animals call to increase distance between neighbors when group density is high. In addition, calling could deflect the approach of other animals. Gautier and Gautier noted that animals frequently call when coming into the presence of a conspecific animal.

Frequency of calling also increases when animals enter fruiting trees (Gautier and Gautier 1977). *Cebuella* give contact calls as they approach major sap trees (C. T. Snowdon, personal communication). Oppenheimer (1977) noted that the huh call of *Cebus capucinus* usually occurred "at fruit trees where several individuals feed at once." There was also an association between frequent calling and approach by other group members, suggesting that calls might "draw nearby troop members to the food source." Frequent calls announce the presence and location of a fruit tree, and this allows others to exploit the food source. Izawa (1979) suggested further that when *Cebus apella* enter a fruit tree "one or a few capuchins that had been at the head of group movement would give a soft clear vocal signal to let the following members of their group know about the presence of food."

The consequences of these calls include maintaining group cohesion, possibly increasing distance between nearest neighbors, and attracting other group members to fruit trees. The consequences of a behavior cannot always be equated with its function, for this would imply that all effects are adaptive. Some effects necessarily will be neutral or disadvantageous, on average, to the signaler.

Predictions

Under what circumstances is the use of huh calls advantageous, on average, to the signaler? The four hypotheses proposed below are based on the suggested advantages, and each predicts the circumstances in which huhs should be uttered. Vocalizations can have three spatial consequences: Animals can approach, maintain their distance, or move away. The hypotheses are grouped by the spatial effects that frequent calling might have on other group members.

I. *Calling decreases spacing.* (a) Calls bring in other group members to food sources. This hypothesis makes the following predictions: Animals should call more frequently when foraging than when moving or resting and more frequently when foraging successfully than unsuccessfully. Neighbors should respond to frequent calling by approaching.

Approach by neighbors will decrease the separation between individuals, and proximity in capuchins increases competition for food items (Robinson, 1981). Calling thus potentially lowers the amount of food available to the caller. Such de facto sharing of food items is only likely to occur when the cost to the caller is minimal or when other group members are rather closely related (Hamilton 1964). The cost decreases with the abundance of food items at the source. So animals should call more frequently when feeding on clumped, abundant resources such as fruit trees than when foraging on discrete, discontinuous food items such as invertebrate prey.

As most, if not all, males emigrate from their natal troop (J. G. Robinson, unpublished data) but few females move, females are, on average, more closely related to other group members than males. If calls summon others, then females in the adult, subadult, and juvenile age classes should call more at food sources than the equivalent male classes, who regularly move between groups. Females might also call more often when the nearest animal is one of their own offspring.

(b) Calls counteract a scattering of animals during the course of daily activity. An animal should call more when it is likely to lose contact with other members or when it is separated from them. Reciprocal countercalling and appropriate spatial responses should bring animals back together. Moving, especially when the group is traveling rapidly, and foraging are activities that should elicit calling. Females should call more than males, especially in the presence of kin. But one would not expect variation in the calling rate with success in foraging, food item, and so on.

II. *Calling increases spacing.* (c) Calls space out group members or deter the approach of neighbors. The call rate should be high when the subject is in proximity to neighbors or when other group members approach or are likely to approach. Like the preceding hypothesis, this predicts no dramatic shifts in spatial pattern, but neighboring animals should tend to move away from a frequently calling animal.

Circumstances in which neighbors approach or are likely to approach include those in which the caller is foraging, especially if foraging successfully. If the approach of neighbors can be deterred, potential competition for food sources or preferred positions could be avoided. As processing time for prey is rapid when foraging on invertebrates, an animal is unlikely to be approached and displaced by its neighbor. Thus, call rate

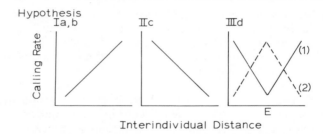

Figure 5.3. Relations between calling rate and interindividual distance, as predicted by the three hypotheses. For hypothesis III there are two possibilities, and E marks the "expected" position.

under these circumstances should be lower than when an animal is feeding in fruit trees, where a competitor is more likely to approach. If calls space out animals, and animals call more when foraging, subjects foraging next to kin should call less and thus allow those individuals preferential access to resources. Animals should call more frequently when they are moving, a circumstance in which animals are likely to bunch together, and the rate should increase with speed of group movement. Finally, in contrast to the previous hypotheses, this predicts that animals will call more frequently when in proximity to neighbors (Figure 5.3).

III. *Calling maintains spacing.* (d) Calls maintain interindividual distance. This distance may vary with the activity of the animals, environmental conditions, and so on. Calls increase spacing when the density of animals is too high and decrease it when animals are too scattered. Response to the call must vary with the distance between animals. The system could operate with either call rate increasing or decreasing when spacing departs from some "expected" distance (Figure 5.3). If animals call and countercall, neighbors can adjust their distances from one another accordingly.

At separations below the expected distance, animals should call more in all circumstances proposed in hypothesis (c). At separations above this distance, animals should call in all circumstances proposed in (b).

Table 5.1 summarizes the predictions from three of the four hypotheses.

THE STUDY

The frequency of huh calls in different contexts was sampled intermittently from March 1978 through July 1979. Both the onset and termination of a sample were determined by the sampled animal, for I ended the sample when the subject switched activities or as soon as the spatial configuration of neighboring animals changed significantly (see later discussion). No activity was included in the sample unless its total duration exceeded 15 sec. This restriction allowed a fairly accurate estimate of calling fre-

Table 5.1. *Summary of calling rate and response*
predictions from three hypotheses

	Hypotheses				
	I(a)	I(b)	II(c)	III(d)[a]	Results
Contexts					
Foraging vs. resting	+	+	+		+
Foraging vs. moving	+	0	0		0
Moving vs. resting	0	+	+		+
With increasing speed of group movement	0	+	+		+
Foraging successfully vs. unsuccessfully	+	0	+		+
Foraging on fruit vs. invertebrates	+	0	+		+
Females vs. males	+	0	0		+
Close kin vs. others	+	0	−		0
Nearest neighbor dominant vs. subordinate	0	0	+		+
With increased interindividual separation	+	+	−		−
Responses	Approach	Approach	Move away	Maintain distance	

Note: A plus indicates that the hypothesis predicts a higher calling rate in the first context relative to the second. A zero indicates that the hypothesis either makes no prediction or does not distinguish a priori in which context the rate should be higher. A minus indicates that the hypothesis predicts a lower rate of calling in the first context.

[a] See text.

quency. In addition to the number of times the animal called and the total duration of the sample, I noted the time, the identity of the individual, the distance to, and identity of, the nearest neighbor, and the subject's activity. If the subject was foraging, each sample recorded the food items taken, if any.

A change in spatial configuration was recorded as significant if the nearest neighbor moved at least 1 m directly toward or away from the subject when initially 5 m or less from the subject or at least 2 m when initially more than 5 m away. Such moves were scored as spatial responses and terminated the sample. A total of 415 samples were taken.

For each sample I also estimated the overall speed of group movement: stationary; stirring, group displacement less than 75 m/hr; drifting, between 75 and 150 m/hr; moving, between 150 and 225 m/hr; and racing, more than 225 m/hr.

For each sample I calculated the rate of calling (huhs/10 sec) by dividing the number of huhs recorded by the duration in seconds of the sample and (for clearer presentation) multiplying by 10. In the following discus-

Table 5.2. *Rate of calling during three activities*

Activity	Rate of calling		
	Mean	*SD*	*N*
Foraging	.56	.81	316
Resting	.22	.50	39
Moving	.74	.84	58

Note: Differences among distributions were tested with the nonparametric median test, which examines whether a pair of groups differs in central tendencies. Foraging vs. resting: $\chi^2(1) = 8.77, p < .01$; resting vs. moving: $\chi^2(1) = 12.11, p < .001$; foraging vs. moving: $\chi^2(1) = 2.47, p > .05$.

sion I present mean and standard deviations of calling rates in various contexts. As there is no reason to suppose that the distribution of call rates approximates a normal distribution, I tested differences with the nonparametric median test (Siegel 1956).

Association of calling frequency with context

Variation in calling frequency with activity. Table 5.2 presents the mean calling rate and the standard deviations of the distributions in the three contexts that were sampled: foraging, moving, and resting. Animals called significantly more frequently when they were moving than when resting and more frequently when foraging than when resting, but differences in median values between moving and foraging were not significant. The high calling rate when moving was not predicted by hypothesis (a).

The other hypotheses also predict that huh-call frequency will increase with the overall speed of group movement. Although calling rate does not correlate significantly with group speed ($r_s = .70, p > .05$), Figure 5.4 nevertheless indicates a trend: Animals tended to call more when the group was moving rapidly.

Possibly this trend is not a direct consequence of group speed but the result of the proportions of different activities in the five group speed categories. When the group was stationary, animals were more likely to be resting or feeding in large fruit trees. During slow group movements (stiring and drifting), animals tended to forage for invertebrates or feed in small fruit trees. During rapid group progressions (moving and racing), most animals were moving. To test this possibility I examined the subsample in which all animals were moving. The same trend is evident if the single stationary sample is ignored (Table 5.3).

Variation in calling frequency with success in foraging. Successful foraging was scored when the subject animal ingested at least one food item during

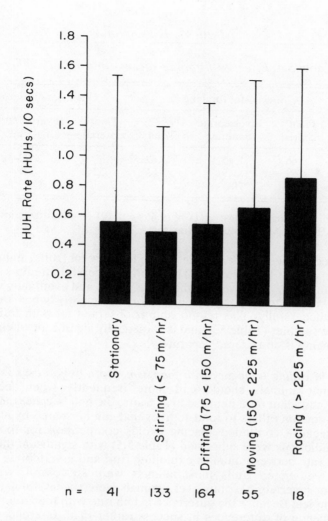

Figure 5.4. Correlation between call rate and speed of group movement. Top of darkened bar represents mean, and one standard deviation is enclosed by the brackets. *n* is the number of samples.

Table 5.3. *Rate of calling at different group speeds*

Group speed	All samples			Samples in which subject was moving		
	Mean	*SD*	*N*	Mean	*SD*	*N*
Stationary	.56	.98	41	1.20	—	1
Stirring	.49	.71	133	.58	.76	13
Drifting	.55	.81	164	.67	1.13	20
Moving	.66	.86	55	.76	.64	12
Racing	.87	.73	18	.87	.62	13

Table 5.4. *Effect of foraging success on rate of calling*

Rate of calling	Successful foraging			Unsuccessful foraging		
	Plant material	Invertebrate material	Total	Plant material	Invertebrate material	Total
Mean	.66	.53	.62	.37	.23	.23
SD	.81	.94	.85	.52	.48	.47
N	163	70	233	2	45	47

Note: Total foraging samples: $\chi^2(1) = 10.65$, $p < .01$, median test; invertebrate foraging: $\chi^2(1) = 4.65$, $p < .05$, median test.

each 15-sec segment of the sample. When foraging for fruits, animals were successful almost by definition. The two unsuccessful attempts involved caruto fruit, *Genipa americana*, that were unripe and eventually were discarded. Animals called significantly more frequently when they were foraging successfully. This relationship is observed for both fruit and invertebrate samples (Table 5.4) and is statistically significant when invertebrate foraging is considered separately.

Variation in calling frequency with food item. Both hypotheses (a) and (c) predict that animals should call more frequently when feeding on clumped, abundant food items such as fruits, the first because callers supposedly summon others to food sources that can be shared by all and the second because continued spacing avoids competition for food items. These predictions were supported (Table 5.5) with significant differences in calling rate between animals exploiting fruit and invertebrate sources.

When feeding on fruits most animals were successful; when they foraged for invertebrates, however, animals were successful in only 60% of the samples. Because the difference in huh rate with food item might be a consequence of differences in success rather than differences in food item, I compared calling rate in animals feeding successfully on fruits with the rate of those feeding successfully on invertebrates. Mean calling rate was still higher when feeding on fruits (Table 5.4), and differences were statistically significant ($\chi^2 = 6.11$, $p < .05$, median test).

Variation in calling frequency with the sex of the vocalizer. Figure 5.5 illustrates the differences in call rate among age–sex classes, as defined by Robinson (1981) for *C. nigrivittatus*. I have excluded the two infant classes as my sample size for these animals was too small. Hypothesis (a) predicts that females of all ages should call more frequently than males of comparable age. Adult females called significantly more frequently than adult males ($\chi^2 = 20.68$, $p < .001$, median test), and subadult females called significantly more frequently than subadult males ($\chi^2 = 11.65$, $p < .001$). These results support hypothesis (a). But among juveniles, the median rate was not significantly different, being slightly higher for males.

Table 5.5. *Effect of food item on calling rate*

Rate of calling	Food item	
	Fruit	Invertebrates
Mean	.65	.43
SD	.81	.79
N	166	122

Note: $\chi^2(1) = 9.72$, $p < .01$, median test.

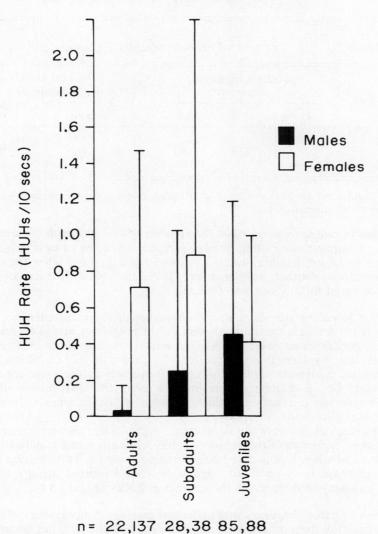

n = 22,137 28,38 85,88

Figure 5.5. Differences in the rate of calling of the sexes, divided into age–sex classes.

Table 5.6. *Effect of kin presence on rate of calling*

Subject female	Rate of calling: nearest neighbor daughter			Rate of calling: nearest neighbor nonkin		
	Mean	SD	N	Mean	SD	N
HI	.61	.82	15	.66	.85	33
BE	1.11	.75	9	.91	.83	19
BU	.37	.28	4	.32	.43	15

Table 5.7. *Effect of rank of nearest neighbor on rate of calling*

Rate of calling	Nearest neighbor dominant		Nearest neighbor subordinate	
	Foraging	Total	Foraging	Total
Mean	.79	.75	.53	.51
SD	.96	.90	.76	.79
N	98	132	128	168

Note: Total samples: $\chi^2(1) = 8.71, p < .01$, median test; foraging samples: $\chi^2(1) = 5.01, p < .05$, median test.

Variation in calling frequency with the presence of kin. For each of three females I compared the calling rates of samples when her nearest neighbor was her youngest daughter under 2 years of age with those when the closest animal was another group member (Table 5.6). No differences of rate were detected for any of these females.

Variation in calling frequency with the presence of dominant animals. Hypothesis (c) predicts that animals will call if neighbors are likely to approach. As dominant neighbors are more likely than subordinates to approach and displace the subject from a food site (J. G. Robinson, unpublished), subjects should call more frequently in the presence of a dominant. Group members were ranked using the distribution of all dyadic agonistic interactions, which included all displacements, threats, and chases (Robinson, in press).

Mean rates of calling were significantly higher when the nearest neighbor ranked higher than the subject (Table 5.7) both when I included all samples and when I considered only foraging samples. This relation held at a significant level ($p < .01$, sign test) when I plotted mean rates in every category of distance to the nearest neighbor (Figure 5.6).

Variation in calling frequency with individual spacing. If increasing isolation and separation from other group members elicits huh calls, the mean call rate should *increase* with distance to the nearest neighbor (see Figure

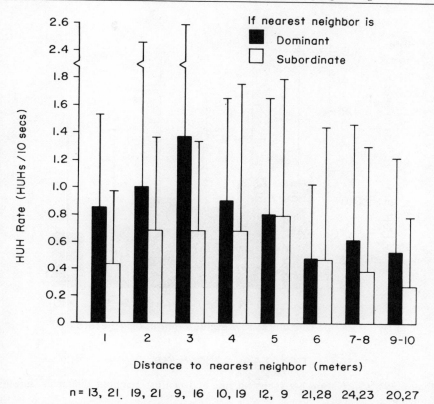

Figure 5.6. Effect of the presence of a dominant neighbor on call rate at all nearest-neighbor distances.

5.3). Hypothesis (c) predicts that mean call rate will *decrease* with increasing distance from the nearest neighbor. Figure 5.7 plots the mean and standard deviation of call rate with distance to the nearest neighbor ($r_s = .89, p < .05$). I restricted the sample to foraging to control for variability in calling frequency with activity. Animals uttered huh calls more frequently when they were close to neighbors, and the rate decreased with greater interindividual distance.

Conclusions. This examination suggests that animals called more frequently when in close proximity to other group members. They called when approach by a neighbor was likely, especially when such approach would increase competition for resources: when the subject was foraging, especially if foraging successfully, or on fruits, and when the nearest neighbor was a dominant animal. Animals called frequently when the group was moving, especially if it was moving rapidly.

Whereas some of the predictions from hypotheses (a) and (b) were sup-

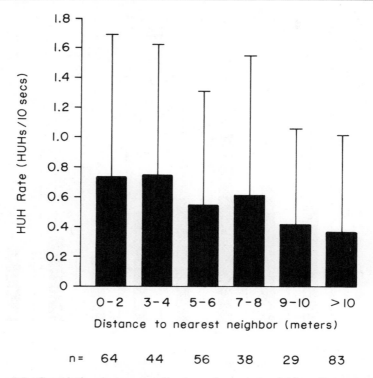

Figure 5.7. Correlation between call rate and nearest-neighbor distances during foraging.

ported, many were not. Not predicted was the increase in calling rate when animals were close together. The results support hypothesis (c). This spacing hypothesis failed to predict that females would call more frequently than males. But males and females are often subject to dissimilar selective forces: Costs and benefits of their actions are potentially different. This might be true for calling. Frequent calling, for instance, might differentially increase male susceptibility to predators, for this class forages frequently on the ground (Robinson 1981).

Responses of other group members to huh calls

What are the spatial responses of neighboring animals to huhs? I divided the foraging samples into two categories: "frequent" calling samples in which the subject called more frequently than the mean rate and "infrequent" samples in which the call rate was less than, or equal to, the mean. Table 5.8 summarizes how the spatial response of neighbors varied with the frequency of calling when the subject was foraging. The majority of animals in both frequent and infrequent groups did not respond spatially

Table 5.8. *Spatial response of neighbors to frequent and infrequent calling of foraging animal*

Call rate of foraging animal	Spatial response of neighbors		
	Approach	No move	Move away
Frequent	37(39%)	55(57%)	4(4%)
Infrequent	35(18%)	151(77%)	11(6%)
Total	72	206	15

Note: $\chi^2(2) = 15.03$, $p < .001$.

Table 5.9. *Relation between spatial response of nearest neighbor and rate of calling*

Response of nearest neighbor	Rate of calling		
	Mean	SD	N
Approach	.92	1.02	86
No change	.45	.71	272
Move away	.73	1.03	19

Note: Approach vs. no change: $\chi^2(1) = 15.22$, $p < .01$, median test; approach vs. move away: $\chi^2(1) = 1.65$, $p > .05$, median test: no move vs. move away: $\chi^2(1) = 0.04$, $p > .05$, median test.

during the sample period, as predicted by the "tonic" hypotheses (b) and (c). But when the subject was calling frequently, a significantly higher proportion of neighbors approached as predicted by hypotheses (a) and (b).

If this is a real response, one would expect that animals approached by their neighbors would show significantly higher call rates than animals whose neighbors maintained the same distance or moved away. Comparisons of mean call rates based on the spatial response of nearest neighbors conform to these predictions (Table 5.9). Animals who were approached called at significantly higher rates than animals whose nearest neighbors did not move, but differences in rate between individuals whose neighbors approached or moved away and between no-move and move-away neighbors were not significant.

These results suggest that although the most frequent spatial response is inertial, neighbors are more likely to approach frequent callers. This contradicts the conclusions derived from the contextual associations of calling.

Are neighbors approaching because the subject is calling or in spite of the calling? Neighbors could be approaching the food source, not the call-

Table 5.10. *Spatial response of neighbors to frequent and
infrequent calling of moving animal*

Call rate of moving animal	Spatial response of neighbors		
	Approach	No move	Move away
Frequent	5(24%)	14(67%)	2(9%)
Infrequent	3(12%)	23(88%)	0(0%)
Total	8	37	2

Note: $\chi^2(2) = 4.20$, $p > .05$.

ing animal, either cueing in directly to the source or to the "unintentional" advertisement from animals already feeding.

This possibility can be tested in two ways. First, do neighbors still approach frequently calling animals when resources are not involved? Table 5.10 presents the spatial response to frequent versus infrequent calling when the subjects were moving. Neighbors did not approach frequent callers significantly more. Nor was the calling rate significantly higher for those subjects who were approached than for those whose neighbors did not move ($\chi^2 (2) = 1.21, p > .05$, median test).

Second, I examined whether the rank of the nearest neighbor affected the probabilities of approach. If huhs are meant to deter approach, one would expect that lower-ranking animals would approach less readily. I compared the frequencies of approach to the calling animal by the dominant adult male, BM, and the dominant adult female, HI, who ranked first and second in the group, respectively, with that by all other group members. The most aggressive animals, BM and HI, were involved in more than 36% of all dyadic aggressive interactions in the group and usually displaced subordinates from food resources or preferred positions. They approached frequently calling animals (Table 5.11) significantly more than did other group members. But when the subject animal was calling infrequently, BM and HI were no more likely to approach than others.

This suggests that the tendency of animals to approach frequently calling animals does not imply a cause–effect relationship. Animals appear to approach *in spite of* high rates of calling during foraging, for (1) they did not approach frequent callers during moving and (2) high-ranking animals approached more than low-ranking animals. This conclusion supports the observation that huh calls have the effect of increasing spacing between animals or maintaining a minimal spacing within the group.

CONCLUSIONS

The cohesion–spacing vocal system of *C. nigrivittatus* operates to maintain the characteristic spacing. The huh, a highly repetitive call, appears to elicit little response from neighbors. Animals deliver these calls when

Table 5.11. *Comparison of spatial response of high-ranking BM and HI and other group members to frequent and infrequent calling of another animal*

Spatial response	Frequent calling			Infrequent calling		
	BM and HI	Other group members	Total	BM and HI	Other group members	Total
Approach	10(71%)	27(33%)	37	10(25%)	25(16%)	35
No move	3(21%)	52(63%)	55	28(70%)	123(78%)	151
Move away	1(7%)	3(4%)	4	2(5%)	9(6%)	11

Note: Frequent calling: $\chi^2(2) = 8.63$, $p < .05$; infrequent calling: $\chi^2(2) = 1.80$, $p < .05$.

in auditory and often visual contact with neighbors. The circumstances of their use indicate that huh calls discourage neighbors from approaching and encourage them to move away. The continual production of these calls counteracts a tendency of animals to bunch together, especially around resources. Arrawhs and hehs are given following more extreme spatial perturbations. Arrawhs bring animals together and hehs space them out.

The possibility cannot be excluded that, instead of only tending to space out group members, huhs attract others when animals are widely spaced and as a result maintain spacing within certain limits or at some expected distance, as predicted by hypothesis (d). Some observations are suggestive. In response to arrawh calls from animals separated from the group, other animals sometimes arrawh, in turn, but more often increase their frequency of huh calls. Once a separated animal has reestablished vocal contact with the group, it usually moves toward the group giving huh calls. In addition, when a group separates into subgroups moving in different directions, animals in each subgroup increase their rate of huh calls. Eventually the subgroups coalesce. Thus, whereas huhs space out members within the group context, they might elicit approach from animals outside the immediate subgroup.

Depending on the characteristics of the food resource or the vulnerability of the species to predators, animals tend to aggregate or disperse. Foraging techniques and social organization affect these tendencies. Thus, to regulate the characteristic spatial configuration, the use of calls and the responses to them might differ within and between species. Waser (chap. 6, this volume) makes a similar argument for the different spatial responses to male loud calls in different species. Contact calls might continually promote spacing *or* aggregation.

A final word on the usefulness of examining vocalizations as operational systems seems appropriate. A call like the huh, which occurs in many contexts and seems to elicit no obvious response, is difficult to examine motivationally or functionally. However, once the call is defined as an element in a spacing system, one can pose testable hypotheses relative to its role in spacing.

REFERENCES

Baldwin, J. D., and Baldwin, J. I. 1976. Vocalizations of howler monkeys (*Alouatta palliata*) in southwestern Panama. *Folia Primatologica* 26:81–108.
Collias, N. E. 1960. An ethological and functional classification of animal communication. In *Animal sounds and communication,* ed. W. E. Lanyon and W. N. Tavolga, pp. 368–91. Washington, D.C.: American Institute of Biological Science.
Eisenberg, J. F. 1974. The functional and motivational basis of hystricomorph vocalizations. *Symposium of the Zoological Society of London* 34:211–47.

1976. Communication mechanisms and social integration in the black spider monkey, *Ateles fusciceps robustus*, and related species. *Smithsonian Contributions to Zoology* No. 213:1–108.

Epple, G. 1968. Comparative studies on vocalizations in marmoset monkeys (Hapalidae). *Folia Primatologica* 8:1–40.

Fentress, J. C. 1976. Dynamic boundaries of patterned behaviour: interaction and self-organization. In *Growing points in ethology,* ed. P. P. G. Bateson and R. A. Hinde, pp. 135–69. Cambridge University Press.

Fossey, D. 1972. Vocalizations of the mountain gorilla (*Gorilla gorilla beringei*). *Animal Behaviour* 20:36–53.

Gautier, J.-P. 1967. Emissions sonores liées a la cohésion du groupe et aux manifestations d'alarme dans les bandes de talapoin (*Miopithecus talapoin*). *Biologia Gabonica* 3:17–30.

1974. Field and laboratory studies of the vocalizations of talapoin monkeys (*Miopithecus talapoin*). *Behaviour* 51:209–73.

1978. Repertoire sonore de *Cercopithecus cephus. Zeitschrift für Tierpsychologie* 46:113–69.

Gautier, J.-P., and Gautier, A. 1977. Communication in Old World monkeys. In *How animals communicate,* ed. T. A. Sebeok, pp. 890–964. Bloomington: Indiana University Press.

Green, S. 1975. Variation of vocal pattern with social situation in the Japanese monkey (*Macaca fuscata*): a field study. In *Primate behavior: developments in field and laboratory research.* Vol. 4, ed. L. A. Rosenblum, pp. 1–102. New York: Academic Press.

Hamilton, W. D. 1964. The genetical evolution of social behaviour, I, II. *Journal of Theoretical Biology* 7:1–52.

Izawa, K. 1979. Foods and feeding behavior of wild black-capped capuchin (*Cebus apella*). *Primates* 20:57–76.

Jürgens, U. 1979. Vocalization as an emotional indicator: a neuroethological study in the squirrel monkey. *Behaviour* 69:88–117.

Kiley, M. 1972. The vocalizations of ungulates, their causation and function. *Zeitschrift für Tierpsychologie* 31:171–222.

Marler, P. 1955. The characteristics of certain animal calls. *Nature* 176:6–7.

1977. The structure of animal communication sounds. In *Recognition of complex acoustic signals: report of the Dahlem Workshop,* ed. T. H. Bullock, pp. 17–36. Berlin: Abakon.

Maurus, M., Pruscha, H., Weisner, E., and Geissler, B. 1979. Categorization of behavioral repertoire with respect to communicative meaning of social signals. *Zeitschrift für Tierpsychologie* 51:48–57.

Moody, M. I., and Menzel, E. W., Jr. 1976. Vocalizations and their behavioral contexts in the tamarin *Saguinus fuscicollis. Folia Primatologica* 25:73–94.

Morton, E. S. 1977. On the occurrence and significance of motivation–structural rules in some bird and mammal sounds. *American Naturalist* 111:855–69.

Newman, J. D., Lieblich, A. K., Talmage-Riggs, G., and Symmes, D. 1978. Syllable classification and sequencing in twitter calls of squirrel monkeys (*Saimiri sciureus*). *Zeitschrift für Tierpsychologie* 47:77–88.

Oppenheimer, J. R. 1977. Communication in New World monkeys. In *How animals communicate,* ed. T. A. Sebeok, pp. 851–89. Bloomington: Indiana University Press.

Oppenheimer, J. R., and Oppenheimer, E. C. 1973. Preliminary observations of *Cebus nigrivittatus* (Primates: Cebidae) on the Venezuelan llanos. *Folia Primatologica* 19:409–36.

Robinson, J. G. 1979. An analysis of the organization of vocal communication in the titi monkey, *Callicebus moloch. Zeitschrift für Tierpsychologie* 49:381–405.

———— 1981. Spatial structure in foraging groups of wedge-capped capuchin monkeys *Cebus nigrivittatus. Animal Behaviour* 29:1036–56.

Rowell, T. E., and Hinde, R. A. 1962. Vocal communication by the rhesus monkey (*Macaca mulatta*). *Proceedings of the Zoological Society of London* 138:279–94.

Schleidt, W. M. 1973. Tonic communication: continual effects of discrete signs in animal communication systems. *Journal of Theoretical Biology* 42:359–86.

Siegel, S. 1956. *Nonparametric statistics for the behavioral sciences.* New York: McGraw-Hill.

Smith, W. J. 1977. *The behavior of communicating.* Cambridge, Mass.: Harvard University Press.

Wiley, R. H. 1976. Communication and spatial relationships in a colony of common grackles. *Animal Behaviour* 24:570–84.

Winter, P., Ploog, D., and Latta, J. 1966. Vocal repertoire of the squirrel monkey (*Saimiri sciureus*), its analysis and significance. *Experimental Brain Research* 1:359–84.

6 · The evolution of male loud calls among mangabeys and baboons

PETER M. WASER

How labile, evolutionarily, are the components of the communicative process? We view the abilities of nonhuman primates as interesting when we can demonstrate specialization of signals for the transmission of particular classes of information (Brown, chap. 7, and Seyfarth and Cheney, chap. 10, this volume) or the detection and processing of information by specialized perceptual mechanisms (Petersen, chap. 8, and Snowdon, chap. 9, this volume). This chapter assumes that our ability to discover such specializations in primates, and our ability to reconstruct the selective processes leading to more complex systems of communication, will be aided by examination of the ways in which the behaviors of sender and recipient of homologous signals vary (or fail to vary) across species.

Kummer (1970:30), discussing the value of primate behavior patterns to taxonomists, concluded that "action patterns of intraspecific communication . . . provide the most reliable taxonomic characters. In adaptation to their communicative function, they generally vary little in form." Many primatologists today would hesitate to argue, as Kummer did, that ecological factors exert little influence on signal form (see, e.g., Michelson 1978; Wiley and Richards 1978), yet few would question the accuracy of his generalization. The Gautiers' recent review amply catalogs the conservatism of basic signal forms across taxonomic groups (Gautier and Gautier 1977). Does this mean that signal form is, for some reason, relatively resistant to selection? Kummer concluded that, in contrast to their form, the rates of action patterns (including social signals) are highly labile, both within and between species. Long *sequences* of behavior patterns, including responses to the relatively invariant signals, are also labile in form. Does this indicate that these components of the communicative process are subject to stronger, or more divergent, selective pressures?

I thank D. McKey, K. Homewood, T. T. Struhsaker, and M. S. Waser for assistance in the field; S. Gartlan, A. Horn, R. Quris, and T. Struhsaker for use of recordings; N. Sallea and E. Rodewald for assistance in data analysis; and R. W. Byrne, R. S. Harding, G. Hausfater, K. Homewood, J. F. Oates, C. Packer, R. Quris, J. Sabater-Pi, R. Seyfarth, M. S. Waser, and R. H. Wiley for comments on the manuscript and/or access to unpublished data. The preparation of this paper was supported in part by NSF Grant SER-77-6731.

Traditional views, arguing to various degrees that selection in communication systems maximizes the efficiency of information transfer, have recently been challenged (e.g., Dawkins and Krebs 1978). I will argue that the data reviewed here are consistent with the position taken by Wiley and Richards (forthcoming) that selection generally increases the efficiency with which the recipient extracts information from the signal. Further, the data support the following interpretations:

First, the value to a signaler of broadcasting information to recipients, and thus the degree to which selection favors specialized "information-transfer" abilities, depend on the social system. Information on the signaler's future behavior may be available in a variety of motor patterns; who the recipients are, and their evolutionary interests relative to the signaler's, should determine whether these motor patterns become specialized for the transmission of that information. The acoustic environment may indeed place constraints on signal form; for instance, where it is advantageous to a male to identify himself, the habitat influences how that information is best encoded. However, the major effects of habitat differences on signaling may be secondary, that is, through their effects on social structure.

Second, the use to which a recipient can put the information contained in a signal varies strongly with the sex and social status of the recipient and with the nature of the social system. As a consequence, selection should favor specialized perceptual abilities tuned to those particular signal characteristics from which the recipient can potentially extract information advantageous to himself. Also as a consequence, responses should be more variable across populations and across age–sex classes than the signals themselves.

THE ANIMALS

Cercocebus and *Papio* represent a cercopithecine radiation into a variety of African habitats, from rain forest to arid scrub; it can be assumed, therefore, that selective factors influencing acoustic communication vary within this group. Data are available for the following populations (Figure 6.1):

Cercocebus albigena. An arboreal, rain-forest species, the gray-cheeked mangabey has been the subject of extensive field studies in eastern Uganda, 1964–5 (Chalmers 1968); Río Muni, 1967–8 (Cashner 1972); and the Kibale forest, western Uganda, 1971–7 (Waser 1975, 1976, 1977a, b; Wallis 1979). My work on *C. albigena johnstoni* in western Uganda has concentrated on the function of male loud calls and includes experimental playbacks of recorded vocalizations. I have also recorded male loud calls of *C. a. albigena* in the Edea forest reserve, southwest Cameroon. T. Struhsaker (personal communication) has recorded *C. a. zenkeri* in the Dja reserve in southeast Cameroon. Sonograms from other *C. albigena* populations have been published by Chalmers (1968; *C. a. johnstoni*) and Gautier and Gautier (1977; *C. a. albigena*).

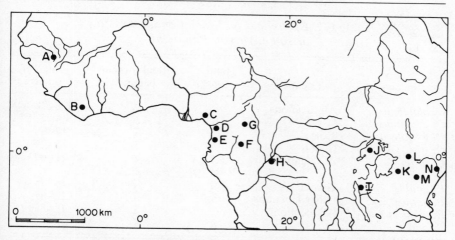

Figure 6.1. Locations of study sites mentioned in text: (A) Niokolo-Koba, Senegal; (B) Troya, Ivory Coast; (C) Korup, Cameroon; (D) Edea, Cameroon; (E) Río Muni; (F) Makokou, Gabon; (G) Dja, Cameroon; (H) Lac Tumba, Zaire; (I) Gombe, Tanzania; (J) Kibale, Uganda; (K) Serengeti, Tanzania; (L) Gilgil, Kenya; (M) Amboseli, Kenya; (N) Tana River, Kenya.

Cercocebus aterrimus. Studied during 1973–4 near Lac Tumba in central Zaire (A. Horn, personal communication), the black mangabey appears ecologically and behaviorally similar to *C. albigena*. Both Horn and S. Gartlan (personal communication) have recorded vocalizations from this population, although no sonograms or descriptions have yet been published.

Cercocebus galeritus. The agile or crested mangabey has a disjunct distribution, occurring in gallery or flooded forest in west Africa as well as in relict populations in east Africa. Recordings are available from *C. galeritus agilis* near Makokou in northeast Gabon, where Quris (1973, 1975, 1976, 1980) worked in 1972–3 and 1976, and from *C. g. galeritus* in the Tana River forests of Kenya, where I recorded groups studied by Homewood (1976, 1978) between 1973 and 1978. Homewood and I have also investigated responses to recorded male loud calls in the Tana mangabeys (Waser and Homewood 1979). Other than the sonograms and descriptions published by Quris (1973, 1980), I am aware of no data in the literature on the structure of these calls in *C. galeritus*.

Cercocebus torquatus. Like *C. galeritus,* this mangabey is primarily a terrestrial species, often found in seasonally flooded forests. No long-term field studies of this species are yet available, but I have recorded *C. torquatus torquatus* from the Korup reserve, west Cameroon, and Struhsaker (personal communication) has recorded this subspecies near the Edea forest reserve, Cameroon. Cashner (1972) has described the calls of this same subspecies in Río Muni.

Cercocebus atys. Struhsaker (personal communication) recorded vocalizations of these mangabeys near Troya, Ivory Coast. I am aware of no

Table 6.1. *Ecological and social characteristics
of some African Papionini*

Species	Location[a]	Habitat type[b]	Modal foraging height[c]	Range of group sizes[d]
C. albigena	All locations	R	A	6–28
C. aterrimus		R	A	14–22
C. g. galeritus	Tana	G	T	17–36
C. g. agilis	Gabon	R/F	T	8–25
C. torquatus	All locations	R/F	T	11–40+
C. atys		R	T	?
P. anubis	Gombe	M	T	26–75
P. anubis	Other locations	S	T	12–87
P. cynocephalus	All locations	S	T	12–85
P. papio		S	T	?

[a] Location is specified when more than one ecologically different population is mentioned in the text.
[b] (R) rain forest (moist evergreen forest); (G) gallery (riverine) forest; (R/F) moist evergreen forest, often flooded at least seasonally; (M) mixed deciduous and evergreen forests, woodland; (S) savanna and woodland.
[c] (A) arboreal; (T) terrestrial.
[d] Range of group sizes as reported in all references cited.

published descriptions or sonograms for any vocalizations of *C. atys*.

Papio. Baboons inhabit a wide range of habitats, including gallery forest (Rowell 1966) and, in western Uganda, moist evergreen forest (Struhsaker 1975), in addition to more characteristic savanna and woodland areas. This variation in habitat provides the raw material for a study of intraspecific variation in communication, but, despite the large number of field studies of baboons, little has been published on vocalizations since Hall and DeVore's early (1965) catalog of *Papio anubis* signals. I have recorded *P. cynocephalus* in the Tana River district and Amboseli National Park, Kenya. Descriptions of male vocalizations are also available for *P. anubis* in the forests fringing Lake Tanganyika in the Gombe National Park, Tanzania (Ransom 1971; C. Packer, personal communication), and *P. papio* in the Niokolo-Koba National Park, Senegal, (R. W. Byrne, personal communication).

Taxonomically, all these species are closely allied in the tribe Papionini (Jolly 1967). Recent immunological and anatomical studies suggest that *C. albigena* and *C. aterrimus* have diverged more recently from *Papio* than have *C. galeritus* or *C. torquatus* (Barnicot and Hewitt-Emmett 1972; Cronin and Sarich 1976; Groves 1978). Nevertheless, all share many social and ecological similarities, including, as discussed here, the presence of homologous male vocalizations. An overview of some ecological and social characteristics of these species is given in Table 6.1.

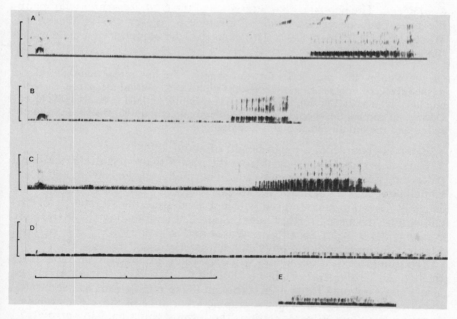

Figure 6.2. *C. albigena* whoopgobbles, Kay 6061B Sonagraph at double speed, wide band. Scale divisions are at 1000 Hz (vertical) and 1 sec (horizontal). (A) *C. albigena johnstoni*, Kibale forest, Uganda, male hk. (B) *C. a. johnstoni*, Kibale, male rf. (C) *C. a. johnstoni*, Kibale, male wr: note aberrant spectral character (A–C illustrate the extremes of call structure in a local area). (D) *C. a. albigena*, Edea Reserve, Cameroon (whoop is very brief); (E) gobble only of *C. a. zenkeri*, Dja Reserve, Cameroon. Lack of harmonic structure in D and E indicates attenuation of higher frequencies in calls recorded at a distance.

THE VOCALIZATIONS

The "whoopgobble"

Within the repertoires of the African Papionini, certain calls are given predominantly or solely by adult males. The most striking of these fall into the category termed "type 1 loud calls": calls notable for "their intensity . . . for their structural originality . . . and for their slight variability in structure, which is stereotyped and discrete" (Gautier and Gautier 1977:895). the *C. albigena* whoopgobble, the first to be described (Chalmers 1968) can serve as the paradigm (Figure 6.2): A low-pitched, relatively tonal unit (the whoop), is followed by a delay of several seconds and terminated by a repetitive series of broader-band pulses (the gobble). Following Quris (1973), I will refer to the tonal unit as part A of the call, the repetitive pulses as B1, and the barks that follow in some species as B2.

C. albigena. The acoustic characteristics of the *C. a. johnstoni* whoopgobble have been described in detail elsewhere (Waser 1977a; Waser and Waser 1977). Chalmers (1968:276) describes the behavior of a male giving the vocalization:

On the whoop, the body jerks forward slightly, the lips remain closed and are pushed forward in a pout, and the cheeks billow out. During the pause . . . the monkey sits tensed, the shoulders shake very slightly and the eyes are half closed. During the gobble, the shoulders shake violently up and down, the head is slightly raised and the mouth remains almost closed.

The call is given from a mean height of about 20 m.

The call is radically different from the rest of the repertoire in temporal patterning and audible distance (which normally reaches 600 m and occasionally exceeds 1 km inside the forest). Figures 6.2A and B illustrate the range of variation encountered within a local area (the stereotypy of this call will be discussed further later). Source sound level is also extremely constant (74.9 ± 1.1 db at 5 m, Waser and Waser 1977). Table 6.2 and Figure 6.3 summarize the physical characteristics of the call.

Recordings of *C. a. johnstoni* whoopgobbles are indistinguishable from those of *C. a. albigena* (Figure 6.2D) and *C. a. zenkeri* (Figure 6.2E; see also Figure 6.3 and Table 6.2), although these populations are separated by more than 2,000 km. The geographic invariance of these calls raises a puzzling question: What constrains the form of this signal so narrowly? I have encountered only one whoopgobble differing substantially from the norm: an extremely noisy, rattling call, ordinary in temporal pattern but lacking the normal resonances, recorded from a male *C. a. johnstoni* in the Kibale forest, Uganda (Figure 6.2C). This call was about 6 db softer than "normal" whoopgobbles and was audible for less than 100 m. Nevertheless, this male was the most frequent vocalizer in his group, responded most intensely to whoopgobble playbacks, and was the most frequent recipient of female grooming – all traits of males who mate most frequently. If females do not discriminate against males with aberrant calls, are there environmental constraints on signal structure that are identical across the species' geographical range?

C. aterrimus. As with many other aspects of its behavior (A. Horn, personal communication), the *C. aterrimus* whoopgobble is essentially identical to that of *C. albigena* (Figures 6.2, 6.3, 6.4A, Table 6.2).

C. galeritus. The loud call of *C. g. galeritus* males consists of "a preliminary 'orientation whoop' [part A] . . . and after a pause a crescendo of whoops of mounting pitch and loudness [B1], followed by a descending cascade of gobbles, cackles, and chuckles [B2]" (Homewood 1976:389). Superficially rather different from the *C. albigena* whoopgobble (Figure 6.4B and C), on closer examination it shares not only the same general temporal pattern but also a homologous "gobble" structure. Each unit in the "crescendo of whoops" has two elements: a softer, tonal element, presumably inspiratory, followed by a louder, high-pitched, broader-band

Table 6.2. *Species and subspecies differences in call parameters presented as mean ± standard error, sample size in parentheses. Second line presents coefficient of variation for all available populations*

	A: length (sec)	A–B1: interval (sec)	B1: length (sec)	B1: number of units	A: frequency range (Hz)	B1: frequency range (Hz)
C. a. johnstoni	0.2 ± 0.01 (40)	5.2 ± 0.25 (38)	2.1 ± 0.08 (53)	14.2 ± 3.24 (73)	40–560 (60)	80–480 (52)
	0.35	0.30	0.28	0.23		
C. a. albigena	0.1 (1)	4.5 (1)	3.1 ± 0.06 (3)	19.7 ± 0.33 (3)	40–440 (1)	40–540 (1)
			0.04	0.03		
C. a. zenkeri	—	—	1.7 (1)	9.0 (1)	—	40–500 (1)
C. aterrimus	0.1 ± 0.01 (6)	4.0 ± 0.09 (6)	3.0 ± 0.02 (7)	12.7 ± 0.28 (7)		
	0.16	0.06	0.02	0.06		
C. g. galeritus	0.2 ± 0.01 (12)	3.8 ± 1.34 (14)	2.2 ± 0.09 (34)	5.8 ± 0.21 (34)	150–440 (12)	100–450 (26)
	0.20	0.35	0.23	0.23		
C. g. agilis[a]	0.2	6.4	1.9	4	150–435	95–410
P. cynocephalus	0.7	2.7	2.0	3.0	50–500 (1)	50–500 (1)

[a] *C. g. agilis* measures, from Quris (1980), are means only.
Note: The dash indicates not measurable from available recordings.

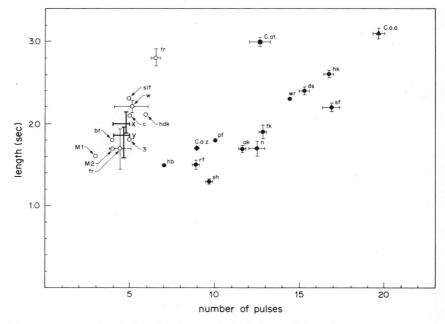

Figure 6.3. Temporal characteristics of B1 (the gobble) in *C. galeritus, C. albigena,* and *C. aterrimus* individuals. Means and standard errors for all individuals for which more than two recordings are available; means only otherwise. Dots: *C. albigena johnstoni,* Kibale forest, Uganda, eleven males in six groups. Triangle: one male *C. a. albigena,* Edea, Cameroon. Diamond: one male *C. a. zenkeri,* Dja, Cameroon. Square: one male *C. aterrimus,* Lac Tumba, Zaire. Open circles: *C. galeritus galeritus,* Tana River, Kenya. Hatched lines: range of measurements for two *C. g. agilis* in Gabon (means not given) from Quris (1980).

expiratory element. According to Chalmers (1968), the units in the *C. albigena* gobble are also double, consisting of both softer, lower-pitched inspiratory and louder, broader-band expiratory pulses. *C. albigena* differs in that the repetition rate of these double-element units is about three times as rapid as in *C. galeritus* (Figure 6.3). In addition, *C. galeritus* appends the "descending cascade of cackles," absent in *C. albigena,* to the rest of the call. The cackles, which are quite variable in form and number, are similar to the "staccato-bark" alarm calls used by both mangabeys (discussed later).

The *C. g. galeritus* loud call is audible up to 800 m in gallery forest (Homewood 1976); its SPL at 5 m from the source is estimated at 77 ± 1.1 db ($N = 4$). Though this species is primarily terrestrial, the call is normally given well above the ground.

The loud call of *C. g. agilis* males in Gabon, West Africa, differs only in minor details from that of *C. g. galeritus* (Table 6.2, Figures 6.3, 6.4D), despite the long separation of these two subspecies – and despite the

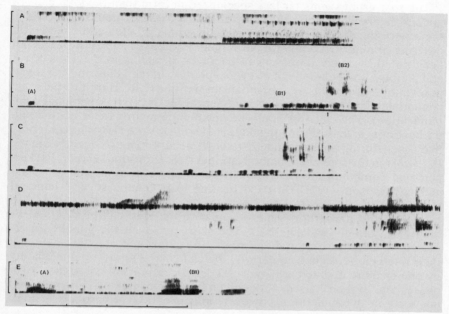

Figure 6.4. Type 1 loud calls of other *Cercocebus* species. Scale and sonagraph settings as in Figure 6.2. (A) *C. aterrimus*, Lac Tumba, Zaire. Note similarity to *C. albigena* calls in Figure 6.2. Arrows at 1,500 and 2,000 Hz indicate insect noise. (B) *C. galeritus galeritus*, Tana River, Kenya, male. Note whoop (A), pause, accelerating repeated double units (B1), and, at arrow, onset of broad-band karau series (B2). Harmonic structure in B1 is not visible because of attenuation. (C) *C. g. galeritus*, Tana River, male. (D) *C. g. agilis*, Gabon; insect noise at 2,700 Hz. (E) *Papio cynocephalus*, Tana River, Kenya. Note very prolonged hum (A) followed by pause and three equally prolonged B1 units.

presence of *C. albigena* in sympatry with *C. g. agilis* but not *C. g. galeritus* (Quris 1973, 1980).

Papio. Baboon males are not generally regarded as having a type 1 loud call in the sense described by Gautier and Gautier (1977). However, in at least some populations they have a vocalization homologous to the whoopgobble (Figure 6.4E). Described as a ''grating roar'' by Hall and DeVore (1965) and as a ''hum-roargrunt'' by Ransom (1971), this call has the same whoop pause repeated-unit structure as its homologs. Ransom (1971:307) describes it as ''a hum (mouth closed) [Part A] followed by a series of . . . inhaled grunts which become deeper in pitch and spaced farther apart [B1]; loud, very low pitch . . . hum may be omitted or inaudible . . . male hunches over during hum, slowly straightens up with jerks of head and shoulders during grunts.'' As with *C. galeritus*, males usually give the calls from an elevated location.

The roar-grunt seems far less specialized in form than the loud calls of male mangabeys. It does not carry noticeably farther than other calls in the repertoire. It resembles other baboon grunts in pitch and bandwidth, differing primarily in temporal patterning. Perhaps, as a result, it has not been noted by many observers of baboons. It occurs in the Tana River and Amboseli populations of *P. cynocephalus,* in the Gilgil and Serengeti populations of *P. anubis* (personal observations; R. Harding, personal communication), and in *P. papio* in Senegal (R. W. Byrne, personal communication). The call appears to be uncommon among these open-country baboons, at least during the day. Hall and DeVore (1965) heard it only five times during their extended study. It occurs more frequently among *P. anubis* in the forested Gombe National Park (Ransom 1971; C. Packer, personal communication). Although quantitative data on call rates are lacking, Ransom (1971) reported hearing it at Gombe several hundred times.

In form the roar-grunt is more similar to the loud call of *C. galeritus* than of *C. albigena* (Table 6.2). The short pulses of the gobble are replaced by long, noisy, "grating" roars; the number of roars is small, and their rate is low. As in *C. galeritus,* the call can be followed immediately by one or more double barks (part B2), apparently identical to the "two-phase" or "wahoo" alarm barks given in other contexts. This trait does not occur in all populations, however, and appears to be particularly common in the Gombe *P. anubis.*

Other species. It is not clear whether *C. torquatus* or *C. atys* have type 1 loud calls homologous with the whoopgobble. A captive *C. torquatus* male in the Gautiers' Paimpont colony frequently gives a loud call similar to that of *C. galeritus* (R. Quris, personal communication). Cashner (1972:204) reports hearing a call "similar to the whoop part of the whoopgobble of *C. albigena*" but not the gobble. These whoops carried long distances. Sometimes, whoops were given "in a series . . . sometimes very fast and descending in tone," usually in conjunction with a two-syllable bark. Although it is conceivable that these descriptions could refer to something resembling a *C. galeritus* loud call, other observers report no such vocalizations from *C. torquatus* in Cameroon (T. Struhsaker, personal communication; J. S. Gartlan, personal communication; personal observations) or Río Muni (J. Sabater-Pi, personal communication), or from *C. atys* in Sierra Leone (J. F. Oates, personal communication). According to Cashner, the loudest vocalization of *C. torquatus* is a two-syllable bark, given by both sexes but distinctly deeper when given by adult males. This is the vocalization most commonly heard by field observers, especially in the early morning and often without any clear eliciting stimuli (Cashner 1972; T. Struhsaker, personal communication; personal observation).

The loud call of a captive *C. atys* male at the National Primate Center, Tigoni, Kenya, is much softer than, but virtually identical in temporal structure to, that of *C. albigena;* this type 1 call is followed immediately by a much louder (>10 db) series of two-syllable barks (P. Waser and C.

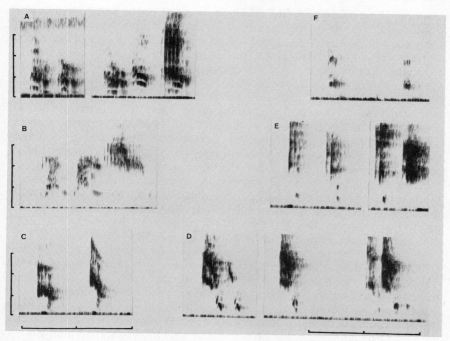

Figure 6.5. Type 2 loud calls. Scale and sonagraph settings as in Figure 6.2. (A) *C. albigena johnstoni*, Kibale, two different sequences. The two staccato barks in the first sequence each consist of a short series of rapidly repeated pulses followed by a much lower-pitched, slightly longer unit. The second call in the second series lacks this lower-pitched unit; the third, higher pitched call is a female staccato bark. (B) *C. aterrimus*, Lac Tumba, two male staccato barks followed by a female's. (C) *C. galeritus galeritus*, Tana River, two male karaus. The initial set of repeated staccato pulses are more clearly differentiated from the low-pitched unit in this species (compare part B2 of Figure 6.4). (D) *C. torquatus*, Korup, Cameroon, two sequences of the male ya-kung. (E) *C. atys*, Troya, Ivory Coast, two male sequences. (F) *Papio cynocephalus*, Tana River, Kenya, male wahoo bark.

Brown, unpublished data). The paucity of fieldwork on *C. atys*, coupled with the low amplitude of the type 1 call, may have prevented its description in the literature. Alternately, it is conceivable that this captive animal includes in his call acoustic elements not given in the wild. In any case, the juxtaposition of the type 1 call and a two-syllable, or "staccato," bark is a familiar combination within the Papionini; what is unusual is the elaboration of the type 2 call.

The staccato bark

In addition to variants of the whoopgobble, the male repertoires of all mangabey species and baboons contain calls resembling the *C. albigena* staccato bark (Figure 6.5). In general, these calls fit the description of the

Gautiers' "type 2 loud calls": they "can occur in the same sequence as [loud calls] of type 1; they are of comparable intensity but they do not have such original structure and are most often related to the aggressive alarm vocalizations of the given species' repertoire" (Gautier and Gautier 1977:900). For example, the *C. albigena* staccato bark "consists of about six pulses of sound of falling pitch . . . in quick succession. Each pulse lasts between 0.04 and 0.09 sec. Several [staccato barks] may be given consecutively, a bout lasting anything up to 5 sec" (Chalmers 1968:271). Staccato barks are given in a wide variety of potentially alarming situations: on appearance of an observer, in response to raptors or other species' alarm calls, or during major intragroup disturbances. Those of adult males are particularly common in the latter context and are distinguished by their overall low pitch and, frequently, by emphasis of the last, lowest-pitched pulse (Figure 6.5A). The form of staccato barks used by the male grades into that used by other individuals, and males rarely append this call to the whoopgobble. The form of the call is virtually identical in all *C. albigena* subspecies and *C. aterrimus* (Figure 6.5B).

In *Papio*, the type 2 loud call is the wahoo or two-phase bark (Figure 6.5F). Given almost exclusively by adult males, it is clearly similar to the "shrill bark" alarm vocalization given by all age–sex classes but appears to be more stereotyped in form, consisting always of two units. Although females and subadults occasionally give the wahoo in full two-phase form (G. Hausfater, personal communication; R. W. Byrne, personal communication), the adult male wahoo is considerably lower-pitched than similar calls given by other age–sex classes and is probably louder; according to Hall and DeVore (1965), it is audible for a half mile or more. As noted earlier, in some populations it frequently follows the type 1 loud call.

In *C. galeritus*, the "karau" call, a rapid, descending sequence of 2–5 staccato pulses, is so characteristic as to serve as the basis for many onomatopoeic vernacular names for this species (Quris 1973; Homewood 1976). Although given by all age–sex classes in situations of alarm, adult males emit a variant. As in *C. albigena*, adult male calls are usually lower-pitched and emphasize the final, lowest-pitched syllable, but they grade into the form used by other individuals. Unlike *C. albigena*, males regularly append these calls to type 1 loud calls. There is no apparent difference in the form of this call in *C. g. galeritus* and *C. g. agilis* (compare Figure 6.5C with Figure 14 in Quris 1973).

In *C. torquatus*, the homologous call (Figure 6.5D) is a two-syllable bark, which has been transliterated "ka-fung" or "ya-kung" (Cashner 1972; J. Sabater-Pi, personal communication). *C. atys* has an almost identical call (Figure 6.5E). As has been noted earlier, the male variant of this call is loud and deep in pitch and may be the only loud call normally given.

Call context

The lack of readily recognizable stimuli for reliably eliciting type 1 loud calls has virtually become one of their diagnostic characters (Gautier and

Gautier 1977). Nevertheless, the lists of conditions under which the whoopgobble and its homologs occur are strikingly similar:

Following a group disturbance. Chalmers (1968:227) first noted this most frequent predecessor of the whoopgobble. "The call would be given . . . a half to one minute after a chase or a fight had finished, either by the chaser or chased, or both . . . [or] after a dog or bird of prey had been in the area for some minutes." Chalmers emphasized that the whoopgobble is neither an immediate nor an invariant response to these conditions. In western Uganda, these two situations characterized 22% of 270 bouts.

In *C. g. galeritus,* the loud call also occurs frequently following group disturbances (personal observation). In *C. g. agilis,* the call "semble suivre une perturbation, mais quelquefois avec une latence telle que la relation n'apparait pas clairement" ("seems to follow a disturbance, but occasionally with a delay such that the relationship seems unclear") (Quris 1973:260).

The *Papio* roar-grunt, as well, is most frequently heard during or following outbursts of alarm or other vocalizations. These may occur when the group is approached by a potential predator; nocturnal outbursts of roar-grunts accompanied by sounds of leopards, wild dogs, or lions are not uncommon among Tana River *P. cynocephalus,* Serengeti *P. anubis,* and Senegalese *P. papio* (personal observations; R. W. Byrne, personal communication). Alternately, roar-grunts may follow agonistic intragroup disturbances. Ransom (1971) lists the primary context of this call as "following aggressive episodes." C. Packer (personal communication) identifies aggressive encounters between males within a group and with single outside males as contexts of the roar-grunt in Gombe *P. anubis.* Similar contexts precede roar-grunts in Amboseli *P. cynocephalus* (G. Hausfater, personal communication).

Packer also reports roar-grunts during intergroup encounters, particularly following chases of females by males (see later discussion). A correlation between rate of calling and intergroup proximity, also reported for *C. galeritus* (Homewood 1976), is particularly obvious in *C. albigena* (Waser 1977a).

Following other vocalizations. In *C. albigena,* 35% of all bouts occurred during or immediately after a characteristic chorus of grunting by other individuals in the group. Soft grunts of *C. albigena* and *Papio* are extremely similar, and a similar chorus sometimes gives rise to the baboon roar-grunt. According to Hall and DeVore (1965:91), chorused soft grunts in baboons occur "when the group [is] congregated near or in sleeping trees [and] when the animals are becoming calm after being frightened."

In both *Cercocebus* species and *Papio,* loud calls tend to be clustered in time. Whoopgobble playbacks elevate the rate of *C. albigena* whoopgobble emission for up to 30 min (Waser 1977a). Loud calls also evoke answers in kind in *C. galeritus,* particularly in *C. g. galeritus.* In some high-density Tana River populations, calls are almost always chorused in the early morning; calls from six to eight sources may be heard within 5 min. Quris (1980) reports the same phenomenon in *C. g. agilis* populations when several males are within earshot of each other. Serengeti and Tana River baboon populations routinely chorus roar-grunts in late evening and at 0400–0500 (personal observation).

After losing contact with conspecifics. C. Packer (personal communication) suggests that the roar-grunt is given most often by males temporarily out of sight of their group. More frequent roar-grunts at Gombe could be interpreted to reflect increased visual isolation of baboons in forested habitats. Solitary or peripheral males give loud calls in both *C. albigena* and *C. galeritus* (personal observations; R. Quris, personal communication), but it is not clear that they call more frequently than males in groups. Packer (personal communication) also suggests that male baboons may give the call when isolated temporarily from consort females, a possibility that has not been investigated in any mangabey species.

Spontaneously. Chalmers (1968) noted that whoopgobbles could be given by males "when sitting quietly," that is, when the observer could detect *no* change in the environment immediately preceding the call. This remains the most common "context" of the *C. albigena* whoopgobble (Cashner 1972; Waser 1977a) as well as of *C. galeritus* homologs (R. Quris, personal communication; personal observation). The tendency for *Papio* roar-grunts to be given "spontaneously" apparently varies among populations; in the forested Gombe habitat, nothing obvious preceded many calls (Ransom 1971; C. Packer, personal communication), whereas among open-country Amboseli baboons, most calls followed male–male agonistic behavior (G. Hausfater, personal communication).

VOCALIZER

In the field, vocalizers have always been identified as adult or "old subadult" males. In captive *C. albigena* and *C. galeritus,* the call appears abruptly, in its final form, in 7-year-old males (Gautier and Gautier 1977). The rate at which males in a group give calls can vary, however, and in *C. albigena,* at least, the rate correlates with other behavioral attributes. Those males who call most frequently are also those termed "rapid approach" (RA) males, because of their response to recordings or naturally occurring whoopgobbles from other groups in their vicinity (Waser 1976, 1977a, b). One male in each group responds to such calls with immediate attention, moving directly to the vicinity of the call source, emitting answering calls of his own, and, often, engaging in a prolonged stay away from his own group but near the original caller. In western Ugandan groups, the RA male was also the male most likely to "win" male–male aggressive interactions, the male who performed the majority of observed copulations, and the male whose own whoopgobble tended to be approached by the other members of his group (see also later discussion).

Following aggressive encounters, the winning male was four times more likely than the loser to give the call.

In *C. g. galeritus,* dominant males also call most frequently (Homewood 1976; personal observation). Hausfater (personal communication) reports that among Amboseli baboons the call following intermale aggression is generally given by the winner; consequently, nearly all calls are given by high-ranking animals. No data are available for other populations.

In *C. albigena* and *C. galeritus* the call can be given by males who are separated by hundreds of meters from the nearest group; these are known to include males who are temporarily absent from a group they otherwise associate with and, probably, males in the course of intergroup transfer.

RESPONSE CHARACTERISTICS

Responses by dominant males

The whoopgobble and its homologs elicit widely different responses in different species. Among Gombe baboons, the most clear-cut male response to roar-grunts occurs when they signal the presence of another group. "Males [hearing calls from outside their own group] usually chased females of their own troop away from the other troop" (C. Packer, personal communication). Though systematic data are lacking, males have not been reported to respond in this way to roar-grunts given by other males within their own group nor has this response been noted during intergroup interactions in other baboon populations. Much more often, males hearing the roar-grunt show little or no visible reaction. Responses resembling the rapid approach of mangabey males apparently do not occur.

Males, particularly high-ranking males, may also respond with an answering roar-grunt to roar-grunts outside the group or to those emitted by less dominant males within the group. Particularly at night, the distribution of roar-grunts is clumped in time, but whether this results from males' responses to common stimuli or to each other's calls is unclear.

In contrast, the rapid-approach responses of certain mangabey males are (as mentioned earlier) so salient and reliable as to serve as an assay of the males' status in the group. Rapid approaches occurred in response to 24 out of 30 recorded whoopgobbles played back to western Ugandan *C. albigena,* and males sometimes covered enormous distances – up to 700 m – during their approaches. Even over such distances, a male's ability to localize the call was extremely precise (Waser 1977b). RA males also responded in this way to calls of males from neighboring groups when these approached within a few hundred meters. RA males rarely approached calls of males in their own group (observed once in four playbacks and never to naturally occurring calls) but did orient toward them and often gave answering calls.

In *C. g. galeritus,* the dominant male in each of two intensively studied groups showed a similar, though somewhat attenuated, response to loud-call playbacks (Waser and Homewood 1979). Approach occurred in 9 of the 15 playback tests during which the dominant male was visible, and immediate attention and answering calls were noted more frequently (all tests used calls recorded in other groups). Males approached only much shorter distances, generally less than 100 m, in these cases; they sometimes performed a "leaping-about" or "branch-shaking" display, not seen in the other species, along with the rapid approach. As in *C. albigena*, dominant males did not approach calls of other males in their own groups, though they often answered them with another call.

Responses by other group members

Immediate responses by other age–sex classes are almost never observed. In *C. albigena* other vocalizations may momentarily cease when a call occurs. On two occasions estrous females (with swollen perineal skin) foraging on the periphery of the group returned to its center when one of the group males gave the call. Other responses were less immediate: An increase in the rate of staccato-bark calls and a decrease in soft grunts (quiet vocalizations used in a variety of nonagonistic contexts) follow many playbacks of calls from neighboring group males. Male calls evoke reverse trends in their own groups. *C. albigena* groups tend to approach the sites from which calls of their own RA males have been played back, ignoring the calls of other males (Waser 1977a). None of these trends follow loud-call playbacks to *C. galeritus* groups.

In both *Cercocebus* species, males other than the RA, or dominant, male also respond to loud calls in kind, particularly if they come from outside the group. The strength of the subordinate male response is, however, much less than that of the dominants; for instance, dominant male *C. albigena* are approximately 10 times more likely to answer a subordinate within their own group than vice versa (Waser, unpublished data). Other than an answering call, momentary attention is the strongest response subordinate males normally show, and many calls are not followed by any discernible reaction. Rapid approaches by more than one male were observed once in *C. galeritus*, during a period of instability in the male heirarchy.

The traditional view of male loud calls as mechanisms of intergroup spacing has been clearly borne out only in *C. albigena* (Waser 1975, 1977a); groups of this species clearly avoid both playbacks and naturally occurring calls from neighboring groups. In contrast, experiments with *C. g. galeritus* have revealed no avoidance tendencies (Waser and Homewood 1979); and none have been noted in *C. g. agilis* or baboons.

Trends in signal form

Within species, type 1 loud calls are extremely stable in form. The congruence of *C. albigena* whoopgobbles across central Africa has already been mentioned. It is perhaps conceivable that interbreeding is sufficient to maintain signal form constant throughout *C. albigena*, but this argument is difficult to extend to *C. aterrimus*, a species that shares the same call despite isolation by the Congo River. *C. galeritus* provides the most striking case of loud-call stability, as *C. g. agilis* and *C. g. galeritus* have been isolated from each other at least since the early Pleistocene and are now separated by 1,500 km. Do trends in signal form across species suggest which selective constraints act within species?

Between species, loud calls differ most clearly in the temporal structure of the gobble. The number of pulses in B1 is greater in *C. galeritus* than in *Papio* ($t(34) = 2.60, p < .05$, comparing *C. g. galeritus* with sympatric *P. cynocephalus*). In *C. albigena* it is significantly greater again ($t (105) = 15.58$, $p < .001$, *C. a. johnstoni* vs. *C. g. galeritus;* Figure 6.3, Table 6.2). Because the length of the B1 portion of the call remains similar in all species, the pulse repetition rate also increases severalfold from *Papio* to *C. galeritus* to *C. albigena*.

Fundamental frequencies are constant across species. Calls do differ in their harmonic compositions, but the rapidity with which harmonics are attenuated makes their quantitative comparison of questionable usefulness. At a distance, *C. albigena* and *C. galeritus* whoops are indistinguishable (Quris 1980). Energy in the fundamentals of parts A and B1 remain at identical low frequencies in all populations recorded (Table 6.2).

Concentration of energy at such low frequencies would seem to suggest selection favoring maximum audible range (Waser and Waser 1977). However, the lack of interspecies differences is counterintuitive, because such low frequencies are strongly absorbed by the ground (Wiley and Richards 1978) and would thus seem less than optimal for a terrestrial species.

High-frequency barks follow part B1 in terrestrial forest populations. In both *C. galeritus* subspecies, males regularly follow B1 with a series of staccato barks, B2 (Figure 6.4 B–D). These are broadband signals containing energy up to 2–4 kHz, considerably higher than the rest of the call. These may increase the locatability of nearby calls, since locatability increases with bandwidth when calls are well above background noise levels (Brown, chap. 7, this volume). This cue would be useful only over short or intermediate distances, however, as greater bandwidth is no longer advantageous for localization in the presence of strong masking noise and as B2 is not as audible as far as B1 (Quris 1980; personal observation). It is of interest in this regard that only the Gombe *Papio* population is reported to follow roar-grunts regularly with two-phase barks.

Loud calls become increasingly stereotyped and discrete in forest species, especially arboreal species. For instance, all recordings of nearby *C. albigena* males show a consistent emphasis of the third and fifth harmonics in B1 (e.g., Figures 6.2A and B). Harmonics also appear in good recordings of *C. galeritus* and *Papio* calls, but the spectral distribution of energy within them is highly variable both within and between individuals. For instance, three recordings of a single male emphasize the second through fifth harmonics; the second, third, fourth, and sixth; and the third only. As noted by Brown (chap. 7, this volume), standardization of harmonic content could provide information on vocalizer distance.

According to Ransom (1971), the "hum" initiating the *Papio* roar-grunt may be either omitted or inaudible. Neither is true in *Cercocebus* calls. In my experience the whoop is always present, and the alerting effect of the initial *Cercocebus* whoop has frequently been noted by field observers.

The presence and length of the call's B2 portion is highly variable in *Papio* and *C. g. galeritus,* relatively constant in *C. g. agilis,* and absolutely invariant (it never occurs) in *C. albigena.* Though overall variability in the temporal patterning of the B1 portion in *C. g. galeritus* and *C. albigena* is similar, calls of particular individuals are most stereotyped in *C. albigena* (Table 6.3, Figure 6.3). Coefficients of variation of the length of B1 for eight *C. albigena* males range from .02 to .12. For three *C. galeritus* males, the range is from .10 to .35. For the number of pulses in B1, the coefficients of variation are .03–.09 (*C. albigena*) and .07–.39 (*C. galeritus*) (Table 6.3).

Coefficients of variation under 10 are extremely unusual among primate vocalizations. Stereotypy per se should increase the ability of listeners to detect the loud call in noise. By eliminating transitional forms to other vocalizations, stereotypy should also increase the listener's ability to distinguish loud calls from the rest of the repertoire.

In forest and particularly in arboreal species, individual identity is more clearly coded, especially in terms of temporal patterning. C. a. johnstoni males were readily distinguishable in the field by their calls, and individual differences were almost exclusively coded by temporal parameters in the gobble (Table 6.3, Figure 6.3). (Waser, 1977a, discusses cues to individual identity in more detail.) Acoustic recognition of *C. g. galeritus* in the field was considerably more difficult; an examination of individual call characteristics (Table 6.3) shows why. Temporal patterning in B1 does not separate individuals as clearly in *C. galeritus* as in *C. albigena;* there is greater variability within individuals and less between, particularly in the number of pulses. *C. g. agilis* individuals also have calls whose temporal characteristics overlap broadly. Individuals can be identified by their calls only by using a combination of temporal and frequency characteristics (Quris 1980). In *Papio,* the call may again be recognizable individually, but again only by a combination of cues, including pitch and "quality." In forest populations, males are most easily identified by the characteristics of the following two-phase barks (C. Packer, personal communication).

These patterns suggest that all species broadcast information identifying themselves, but that *C. albigena* males do so least ambiguously by using cues that are least degraded by distance. Nuances of pitch and quality are rapidly lost through attenuation and reverberation, whereas pulse-repetition rate is not.

What can we conclude about ecological constraints on signal form?

Table 6.3. *Individual differences in call parameters: mean ± standard error (N). Second line presents coefficient of variation for seven C. a. johnstoni and three C. g. galeritus males*

Individual	A: length (sec)	A–B1: interval (sec)	B1: length (sec)	B1: number of units	A: frequency range (Hz)	B1: frequency range (Hz)
C. albigena						
hk	0.2 ± 0.01 (21) 0.16	5.7 ± 0.16 (22) 0.13	2.6 ± 0.04 (31) 0.07	16.7 ± 0.26 (31) 0.09	80–400 (28)	80–480 (31)
rf	0.3 ± 0.04 (5) 0.31	6.1 ± 0.76 (5) 0.28	1.5 ± 0.06 (10) 0.12	8.9 ± 0.18 (10) 0.06	80–560 (8)	80–400 (8)
tk	0.2 ± 0.02 (4) 0.21	6.2 ± 0.61 (4) 0.41	1.9 ± 0.08 (7) 0.11	12.8 ± 0.16 (8) 0.04	80–400 (10)	80–560 (8)
sf	0.2 ± 0.01 (7) 0.12	5.2 ± 0.94 (7) 0.18	2.2 ± 0.04 (7) 0.05	16.9 ± 0.40 (7) 0.06	80–480 (7)	80–560 (7)
ok	0.2 ± 0.01 (3) 0.13	5.4 ± 0.41 (3) 0.13	1.7 ± 0.04 (4) 0.05	11.9 ± 0.14 (7) 0.03	80–390 (3)	80–510 (4)
sh	0.2 (1)	3.8 (1)	1.3 ± 0.02 (3) 0.04	9.7 ± 0.21 (6) 0.05	160–400 (1)	80–480 (6)
ds	0.3 (1)	6.6 (1)	2.4 ± 0.03 (3) 0.02	15.3 ± 0.33 (3) 0.03	80–480 (1)	80–640 (3)
C. galeritus						
w	0.2 ± 0.01 (3) 0.25	5.1 ± 0.10 (3) 0.03	2.2 ± 0.07 (13) 0.12	5.2 ± 0.10 (13) 0.07	150–440 (3)	100–440 (13)
tr	0.2 ± 0.01 (5) 0.17	2.3 ± 0.14 (5) 0.14	2.8 ± 0.10 (8) 0.11	6.6 ± 0.26 (8) 0.11	150–440 (5)	100–440 (8)
fr	—	—	1.7 ± 0.25 (6) 0.35	4.5 ± 0.72 (6) 0.39	—	100–500 (6)

Note: The dash indicates not measurable from available recordings.

Acoustic considerations may have dictated that of all the calls in the man-gabey–baboon repertoire, the roar-grunt was best suited for elaboration as a long-distance call. However, few of the species differences in signal form reflect differences in acoustic environment directly. The spectral composition of background noise and the importance of reverberation versus amplitude fluctuations (Wiley and Richards 1978) differ between forest and open country; the importance of ground absorption differs for arboreal and terrestrial species. Yet spectral characteristics of loud calls do not obviously respond to these differences, nor is it clear that the sys-tematic interspecies differences in the timing of B1 elements are dictated by them.

Thus, males are not simply broadcasting identical information in calls differing according to the dictates of different acoustic environments. In-stead, most of the trends in loud-call structure are consistent with the sug-gestion that differences in habitat influence call structure secondarily. Males differ in the extent to which they incorporate cues to location, dis-tance, identity, and other variables in their loud calls, presumably be-cause the aspects of social structure and habitat that determine the bene-fits to males of incorporating those cues differ.

Trends in context and response

Call contexts. The contexts of type 1 loud calls, when they can be defined, are remarkably invariant across species. Calls tend to be more often given spontaneously in forest species, though even this trend is confounded by the apparent absence of the call in field populations of *C. torquatus* and *C. atys*. Since both these species can produce the call in captivity, the capa-bility is there. The simplest summary statement is perhaps that type 1 loud calls are always provoked by the same class of situations, though there may be interpopulation differences in threshold.

Influences of ecological factors on response. In contrast to contexts, re-sponses to type 1 loud calls vary considerably across species, and some of the variation may reflect habitat differences. For instance, tendencies for group members to approach calls given by their own dominant males are most prominent in *C. albigena* populations, which occupy a low-visibility habitat at low densities. Similarly, avoidance of calls by males in other groups is most pronounced in *C. albigena*. *C. galeritus* groups apparently avoid neighboring males' loud calls only during seasonal periods of food scarcity, if at all (Homewood 1976; Waser and Homewood 1979). *Papio* groups are generally believed to use visual cues in intergroup interactions.

At least two "ecological" hypotheses can be suggested to account for the response differences between *C. galeritus* and *Papio*, on the one hand, and *C. albigena*, on the other. First, although all species might share basically similar resource characteristics that dictate similar pat-terns of intergroup spacing, they might use visual or more generalized acoustic cues as proximal mechanisms of detection and interaction in populations of higher density or in higher-visibility habitats. If this is the case, *C. g. agilis* in Gabon and *C. albigena* in Uganda should show similar responses, since these populations occur at low density in surroundings with poor visibility.

Alternately, resource characteristics of the different species might differ in ways that would favor differing patterns of spacing. For example, the lack of avoidance of a neighboring group's calls by members of *C. g. galeritus* groups might reflect the food distribution, or the seasonality that characterizes the gallery forests in which they live (Waser and Homewood 1979). In this case, responses of *C. albigena* in more seasonal or vegetationally less diverse forests might be expected to resemble those of *C. g. galeritus*.

Influences of social factors on response. Baboons and mangabeys share a multi-male group structure, with female matrilines forming the nucleus of the group and males changing groups at least once during their lives (Homewood 1976; Packer 1979; Wallis 1979; Waser, unpublished data). However, the species differ substantially in group size, number of adult males and ratio of adult males to females in a group, seasonality of breeding, and rates of male turnover. Differences in the use of the male loud call may well reflect differences in the nature of male–male competition.

Dominant male mangabeys, but not baboons, leave their groups to approach loud calls from males outside the group. Even in open-country populations, in which the roar-grunt is rarely or never used, baboon males tend to "herd" females away from neighboring groups during encounters (Buskirk, Buskirk, and Hamilton 1974; Cheney and Seyfarth 1977), whereas behaviors resembling the rapid approach of the mangabey are only rarely reported (e.g., Harding 1973). Only low-ranking male mangabeys are present as potential "herders" during encounters, and such behavior has not been reported (although it might occur in attenuated form). These differences suggest that male mangabeys have less to lose or more to gain by investigating outside loud-call sources than by attempting to "protect" females in their own group from the attentions of other males.

Differences in male response to neighboring groups, and in particular to loud calls of neighboring group males, could reflect any of several social differences.

First, in larger groups, the larger number of females increases the probability that the group will contain a sexually receptive individual at any particular time; therefore, a dominant baboon might run more "risks" in leaving his group (e.g, Hausfater 1975).

Second, because of the larger number of females, variance in male reproductive success is potentially higher in a baboon group than in a mangabey group. Intrasexual selection, and selection favoring males' continued attention to particular females, would then be stronger among baboons. There are many possible variants of this hypothesis. For instance, some authors (Ransom and Ransom 1971; Hausfater 1975; Seyfarth 1978) have suggested that substructure exists within baboon groups, different females tending to associate with particular males. No such substructure occurs in the smaller mangabey groups. Relative to a mangabey male, it may be less in a baboon male's interest to "defend" the entire group. Alternately, because the mean number of males in a mangabey group is lower, a dominant mangabey male might be more likely to encounter other groups with few males, or "weak" males, in which he would have a high probability of mating.

Third, the low visibility in forest habitats and the possibly more frequent occurrence of peripheral or isolate males might increase the probability that single outside males could interact with group females without detection. Selection might therefore be greater for mangabey males to intercept such males when they are detected.

The inadequacy of existing evidence precludes a choice among these hypotheses, though the intermediate size of *C. galeritus* groups and the intermediate nature of their response is consistent with the first two. We know little about the factors that determine whether a particular male will respond with herding or rapid approach, or how extensive that response will be, and we kow almost nothing about RA males' behavior when they actually encounter outside loud-call sources. Nevertheless, the nature of this most clear-cut variation in loud-call use appears to reflect interspecies differences in social structure.

DISCUSSION: HOW DOES THIS COMMUNICATION SYSTEM RESPOND TO SELECTION?

Signal form

The increasing structural specialization of the type 1 loud call in forest species is the most striking trend within the mangabey–baboon radiation. A listener to any species is reliably apprised by this call of the presence of an adult male; in all species, a practiced listener can identify *which* adult male it is. Comparing the loud calls of baboons with those of mangabeys, and particularly with those of *C. albigena,* the progression is toward increased distinctiveness, increased stereotypy, greater audible range, increased localizability, and an increasingly clear coding of individual identity in the temporal patterning of broadband pulses. All of these trends effectively permit reliable transmission of the identity and location of particular males over longer distances.

Two factors are probably behind these trends. First, in open environments, a wide range of acoustic and visual cues is available to locate particular males; specialized calls would not markedly increase the likelihood or ability of a male's neighbors to react to his presence. In contrast, in forest habitats, with high sound attenuation, high background noise, and limited line-of-sight distances, only a small proportion of the signals in a male's repertoire would locate and identify him outside his immediate vicinity. To the extent that some responses to his presence by other individuals are advantageous to a male, selection should favor specialization of those signals that do contain this information. The low frequency and unusual temporal pattern of the roar-grunt preadapts it for such elaboration.

Second, the individuals whose actions are important to a baboon male – other males with whom he is competing for access to females, as well as the females themselves – are primarily within his group. Potential signal recipients are numerous and nearby. In contrast, potential receivers for a male mangabey's calls are frequently far away both because mangabey groups are often widely spread or split into subgroups and because a large proportion of important interactants are outside the group. Thus selection may favor signals detectable longer distances in mangabeys than in baboons.

Several of the trends apparent among these loud calls can be interpreted in these terms. For instance, male *C. galeritus* and baboons incorporate, in B2, locational cues that should work well at intermediate distances in forest environments where visual location is not possible. But if recipients whose response is important to the *C. albigena* sender are hundreds of meters away, the rapidly degraded locational cues in B2 are useless.

Likewise, that calls contain individual identifiers in all forest populations suggests that selection favors males' identifying themselves acoustically when they can't see each other; the more precise and degradation-resistant cues broadcast by *C. albigena* reflect the greater distance over which a male gains by identifying himself.

If most important recipients of the signal are relatively nearby, a male may not need to broadcast distance cues. Distance cues are indeed most striking in the *C. albigena* call.

The effects of habitat on call structure are therefore both direct (through differences in the acoustic environment) and indirect (by favoring different group sizes and social structures). The social environment determines how clearly and to whom it is in the interests of a male to broadcast information; the acoustic environment defines the range of signal forms that meet those requirements.

Signal use

The contextual similarity of the type 1 loud call across species suggests that the content of the call is also similar; only the frequency and the clarity with which that content is broadcast varies.

The substantial between-species differences in *responses* to the type 1 call, in contrast, suggest that the use signal recipients make of this information varies with social structure. Dominant males apprised of the vicinity of an unfamiliar male by hearing his loud call respond by approach if they are mangabeys and by herding females away if they are baboons; the differences are interpretable in terms of differing dynamics of male–male competition in groups of different size and dispersion. Other group members, when (as in *C. albigena*) it is to their advantage to avoid neighboring groups, respond to the information contained in the loud call by moving away. When other cues to a neighboring group's location are usually present (as in *Papio*) or where group members do not necessarily benefit by avoiding their neighbors (as in at least some *C. galeritus*), avoidance is not so clearly associated with the call.

It is not clear whether selection has favored differences in *abilities* to respond or simply differences in response *tendencies*. It is not inconceivable that a survey of perceptual abilities across species would reveal specializations paralleling those in the call; if so, one would expect such specialization to be most prominent among individuals for whom the possession of a particular sort of information would be particularly advantageous and to concern abilities to process the signal parameters coding that information (e.g., abilities to localize or decode temporal identity cues in *C. albigena* calls).

Even where interspecies differences in perceptual abilities are lacking, individuals to whom a particular set of information is available may have differing response tendencies shaped by selection. Certainly this is true *within* species. The whoopgobble and its homologs indicate the presence, identity, and location of an adult male; at least in mangabeys, their rate may indicate his dominance. If the signal recipient is a familiar animal, separated from other members of his group, the appropriate response may be approach; if he is not separated, the appropriate response is simply to continue, uninterrupted, in his ongoing activities. If the signal recipient is an estrus female and the caller is a dominant male, it may be in the female's interest to approach him (e.g., Borgia 1979:51). If the signal recipient is another male and females use the call as a basis for mate choice, it is in his interest to broadcast his location too.

Are the trends apparent in the evolution of this set of relatively simple, discrete vocalizations duplicated in more complex signal clusters? An intriguing candidate for more detailed examination is the mangabey–baboon "type 2 loud call" – the staccato bark and its variants. An individual's type 2 loud calls tend to be relatively variable in form, suggesting that they contain motivational or other cues absent from type 1 calls. But like type 1 calls, these vocalizations retain a constant basic structure while showing interspecies variation in spectral and temporal details. In all species, calls of adult males are structurally distinguishable from those of other age–sex classes; at least some species' calls contain cues to location and individual identity. And as with type 1 calls, the precision with which such information is coded appears to vary across species. The *C. torquatus* and *C. atys* are of particular interest, since their type 2 call – the two-syllable bark – appears to have become the functional equivalent of the whoopgobble in the other species, whereas the type 1 call has remained unspecialized or is perhaps even absent. In these species, the type 2 call has certainly become louder and more distinct than its counterparts in other species; does it also show increased stereotypy and the other specializations found in the *C. galeritus* or *C. albigena* type 1 call?

In all species, type 2 calls occur in the same set of contexts: They are an immediate reply to similar sets of "alarming" contexts and, in most cases, a sequel to the type 1 call. But available evidence suggests that responses to type 2 calls (as to type 1) are much more variable than the contexts that evoke them. In *C. albigena,* a staccato bark often evokes chorused answering staccato barks in a mobbing reaction; in *Papio,* the response simply may be attentiveness to the potential predator evoking the call (Rowell 1966), mobilization and herding of the group during intergroup encounters (Buskirk, Buskirk, and Hamilton 1974; Cheney and Seyfarth 1977), or reaggregation following group fragmentation (R. W. Byrne, personal communication). The responses to *C. torquatus* and *C. atys* type 2 calls remain unknown.

The data available, then, suggest that the sort of evolutionary scenario suggested for type 1 loud calls may have some generality. Similar contexts across species elicit similar basic vocal forms, thus broadcasting similar basic information. Signal recipients respond to their own advantage, the details depending on the relationship of the sender and recipient, the precision and accuracy of the information as it reaches the listener, and the social system within which they live. Where recipients' responses benefit the signaler, selection should favor elaboration of the signal form to code the relevant information more precisely and in modes more resistant to degradation. If the evolution of other primate communication systems follows this pattern, it should hint at the classes of individuals and signal types among which the most complex modes of communication must first have arisen.

SUMMARY

Within the vocal repertoires of some (but not all) species of mangabeys and baboons, there is a distinctive call type given only by adult males. Recordings from *P. anubis, P. cynocephalus, C. albigena, C. aterrimus,* and *C. galeritus* suggest that these calls are homologous. Within this species array, savanna–forest, terrestrial–arboreal, and other ecological contrasts allow examination of habitat influences on the acoustic structure of the calls, the contexts and rates of calling, and the responses by signal recipients. Observational studies, as well as sound playback experiments with *C. albigena* and *C. galeritus,* suggest: (1) The loud calls of forest *Cercocebus* males are more distinct than other calls in the repertoire, more stereotyped, and more elaborate in acoustic structure than are those of baboons; (2) loud-call contexts are similar among all species, though rates of calling may differ; and (3) between species the responses of signal recipients to male loud calls differ considerably.

Two general conclusions emerge from these results. First, differences in call structure reflect differences in the nature of male–male competition as much as the constraints of differing acoustic environments per se. The major effects of habitat differences on signal structure may be secondary, through their effects on social organization. Second, response characteristics, which are more labile than signal contexts, may determine evolutionarily the precision and the degree of elaboration of information broadcast in a signal system.

REFERENCES

Barnicot, N. A., and Hewett-Emmett, D. 1972. Red cell and serum proteins of *Cercocebus, Presbytis, Colobus,* and certain other species. *Folia Primatologica* 17:442–57.

Borgia, G. 1979. Sexual selection and the evolution of mating systems. In *Sexual selection and reproductive competition in insects,* ed. M. Blum and A. Blum, pp. 19–79. New York: Academic Press.

Buskirk, W. H., Buskirk, R. E., and Hamilton, W. J., III. 1974. Troop-mobilizing behavior of adult male chacma baboons. *Folia Primatologica* 22:9–18.

Cashner, F. M. 1972. The ecology of *Cercocebus albigena* and *Cercocebus torquatus* in Río Muni, Republic of Equatorial Guinea, West Africa. Ph.D. thesis, Tulane University.

Chalmers, N. 1968. The visual and vocal communication of free living mangabeys in Uganda. *Folia Primatologica* 9:258–80.

Cheney, D. L., and Seyfarth, R. M. 1977. Behaviour of adult and immature male baboons during inter-group encounters. *Nature* 269:404–6.

Cronin, J. E., and Sarich, V. M. 1976. Molecular evidence for dual origin of mangabeys among Old World monkeys. *Nature* 260:700–2.

Dawkins, R., and Krebs, J. R. 1978. Animal signals: Information or manipulation? In *Behavioral ecology an evolutionary approach,* ed. J. R. Krebs and N. B. Davies, pp. 282–309. Sunderland, Mass.: Sinauer.

Gautier, J.-P., and Gautier, A. 1977. Communication in Old World monkeys. In *How animals communicate,* ed. T. Sebeok, pp. 890–964. Bloomington: Indiana University Press.

Groves, C. P. 1978. Phylogenetic and population systematics of the mangabeys (Primates: Cercopithecoidea). *Primates* 19(1):1–34.

Hall, K. R. L., and DeVore, I. 1965. Baboon social behavior. In *Primate behavior: Field studies of monkeys and apes,* ed. I. DeVore, pp. 53–110. New York: Holt, Rinehart and Winston.

Harding, R. S. O. 1973. Range utilization by a troop of olive baboons (*Papio anubis*). Ph.D. dissertation, University of California, Berkeley.

Hausfater, G. 1975. *Dominance and reproduction in baboons (Papio cynocephalus).* Basel: Karger.

Homewood, K. 1976. Ecology and behaviour of the Tana mangabey. Ph.D. thesis, University College, London.

1978. Feeding strategy of the Tana mangabey (*Cercocebus galeritus galeritus*). *Journal of Zoology* 186:375–92.

Jolly, C. J. 1967. The evolution of the baboons. In *The baboon in medical research,* Vol. 2., ed. H. Vagtborg, pp. 323–38. Austin: University of Texas Press.

Kummer, H. 1970. Behavioral characters in primate taxonomy. In *Old World monkeys: evolution, systematics and behavior,* ed. J. R. Napier and P. H. Napier, pp. 25–37. New York: Academic Press.

Michelson, A. 1978. Sound reception in different environments. In *Sensory ecology,* ed. M. Ali, pp. 345–73. New York: Plenum.

Packer, C. 1979. Inter-troop transfer and inbreeding avoidance in *Papio anubis. Animal Behaviour* 27:1–36.

Quris, R. 1973. Emissions sonores servant au mantien du groupe social chez *Cercocebus galeritus agilis. Terre et la vie* 27:232–67.

1975. Ecologie et organisation sociale de *Cercocebus galeritus agilis* dans le nord-est du Gabon. *Terre et la vie* 29:337–98.

1976. Données comparatives sur la socio-ecologie de huit especes de cercopithicidae vivant dans une meme zone de foret primitive periodiquement inondée (nord-est du Gabon). *Terre et la vie* 30:193–209.

1980. Emission de forte intensité chez *Cercocebus galeritus agilis:* structure, characteristiques spécifiques et individuelles, modes d'émission. *Mammalia* 44:35–50.

Ransom, T. W. 1971. Ecology and social behavior of baboons (*Papio anubis*) at the Gombe National Park. Ph.D. thesis, University of California, Berkeley.

Ransom, T. W., and Ransom, B. S. 1971. Adult male–infant relations among baboons (*Papio anubis*). *Folia Primatologica* 16:179–95.

Rowell, T. 1966. Forest living baboons in Uganda. *Journal of Zoology (London)* 149:344–64.

Seyfarth, R. M. 1978. Social relationships among adult male and female baboons. II. Behaviour throughout the female reproductive cycle. *Behaviour* 64:227–47.

Struhsaker, T. T. 1975. *The red colobus monkey.* Chicago: University of Chicago Press.

Wallis, S. 1979. The sociology of *Cercocebus albigena johnstoni* (Lyddeker): an arboreal rain forest monkey. Ph.D. thesis, University College, London.

Waser, P. M. 1975. Experimental playbacks show vocal mediation of intergroup avoidance in a forest monkey. *Nature* 255:56–8.

1976. *Cercocebus albigena:* site attachment, avoidance, and intergroup spacing. *American Naturalist* 110:911–35.

1977a. Individual recognition, intragroup cohesion, and intergroup spacing: evidence from sound playback to forest monkeys. *Behaviour* 60:28–74.

1977b. Sound localization by monkeys: a field experiment. *Behavioral Ecology and Sociobiology* 2:427–31.

Waser, P. M., and Homewood, K. 1979. Cost-benefit approaches to territoriality: a test with forest primates. *Behavioral Ecology and Sociobiology* 6:115–20.

Waser, P. M., and Waser, M. S. 1977. Experimental studies of primate vocalization: specializations for long-distance propagation. *Zeitschrift für Tierpsychologie* 43:239–63.

Wiley, R. H., and Richards, D. G. 1978. Physical constraints on acoustic communication in the atmosphere: implications for the evolution of animal vocalizations. *Behavioral Ecology and Sociobiology* 3:69–94.

Forthcoming. Adaptations for acoustic communication in birds. In *Ecology and evolution of acoustic communication in birds,* ed. D. Kroodsma and E. H. Miller. New York: Academic Press.

7 · Auditory localization and primate vocal behavior

CHARLES H. BROWN

One emerging theme in the present volume is the interplay between laboratory- and field-centered inquiries into the nature of primate languages and communication. This is a new development. Until recently, communication research in the primate laboratory was confined to the study of the acquisition and use of synthetic languages by apes. This research, consonant with the traditions of comparative psychology, infrequently addressed the natural history, ecology, or social organization of the species under study. As a consequence, the behavior of animals in the wild and the social and ecological forces pertinent to the development of sophisticated communicative abilities were rarely considered. In contrast, these neglected considerations figured prominently in the research objectives of field biologists schooled in the European ethological tradition. Teams of investigators developed detailed descriptions and catalogs of the communicative displays of a variety of primates in their natural habitats. Yet these descriptive accounts rarely permitted sophisticated comparisons between species (see Gautier and Gautier-Hion, chap. 1, this volume), and the cognitive and communicative capacities of nonhuman primates remained obscured. Recent work by a number of cross-disciplinary research teams has made considerable progress (see Petersen, chap. 8, this volume), and laboratory–field collaborations hold considerable promise for the future. My objective here is to focus on a superficially simple facet of vocal communication – sound localization – to examine the interplay between field and laboratory approaches to the problem of sound localization and to explore its role in primate communication.

SOUND LOCALIZATION AND VOCAL SIGNALS

Evidence from a variety of sources suggests that an evolutionary premium has been placed on the ability of animals to resolve acoustically the site or origin of sounds in space (Harrison and Irving 1966; Masterton,

I thank H. C. Gerhardt, M. R. Petersen, and C. T. Snowdon for commenting on an earlier draft of this manuscript. The preparation of this chapter was partially supported by NIH Grant R 01 NS16632-01.

Heffner, and Ravizza 1969; Konishi 1973; Masterton and Diamond 1973; Masterton 1974; Knudsen, Blasdel, and Konishi 1979). Since one's location may be revealed by acoustic signals, animals may emphasize this attribute of vocal signals to favor the delineation of territories, the attraction of mates, and a variety of other biological pursuits. In many instances, the determination of the origin of the signal may be as important as its recognition (Konishi 1977). This is particularly true for predators and prey. The marsh hawk and the barn owl go hungry if the location of the rustling leaves or the calls of prey are unknown. Because of the evolutionary significance of sound localization, it may be practically impossible to signal vocally without some degree of disclosure of one's location. In the final analysis, the initial processes of communication – detection, discrimination, and recognition – are inextricably linked to the processes of localization.

All sounds, however, are not equally locatable. Some sounds virtually command one to orient to their spatial address, whereas the location of other sounds is uncertain. At least four factors influence the acuity of sound localization in natural environments:

1. The physical degradation of the signal, particularly the processes of refraction and reflection of the sound wave with environmental surfaces
2. The design of the auditory and perceptual system, notably the organization of the auditory brain, the size of the head, the biophysical transfer functions of the external ears, and so forth
3. The signal's acoustic context, including the amplitude, spectra, and location of the masking noises in the habitat
4. The physical structure of the signal: its amplitude, duration, and frequency.

Whereas the degradation of vocal signals by environmental factors has received considerable attention (Chappius 1971; Morton 1975; Marten and Marler 1977; Marten, Quine, and Marler 1977; Waser and Waser 1977; Michelson 1978; Wiley and Richards 1978; Richards and Wiley 1980), the perceptual consequences of these various types of environmental degradation have yet to be directly studied. In contrast, anatomical, physiological, and psychophysical observations pertinent to the design of the auditory system (factor 2) have been examined by a number of researchers (Neff 1968; Erulkar 1972; Konishi 1977; Gourevitch 1980; Knudsen 1980; Brown, in press a) and will not be addressed here. This chapter is focused on the third and fourth factors, the design features of signals and their acoustic context, with regard to their significance for sound localization and communication.

LOCALIZATION IN THE NATURAL HABITAT

Animals in a diversity of habitats exhibit phonotaxis and orient and move toward the location of species-specific vocal signals. Acoustically evoked orientation and phonotaxis have been examined in insects (Busnel, Du-

100 m

Figure 7.1. Phonotaxis in grey-cheeked mangabeys evoked by playback of a whoopgobble vocalization. The playback was conducted with habituated native populations in the Kibale forest, Uganda. (From Waser 1977. Reprinted with permission.)

mortier, and Pasquinelly 1955; Morris 1972), frogs (Feng, Gerhardt, and Capranica 1976; Rheinlaender, Gerhardt, Yager, and Capranica 1979), birds (Konishi 1973; Knudsen, Blasdel, and Konishi 1979; Brown, in press b), and mammals (Waser 1977; Clements and Kelly 1978) and may be judiciously employed to study the acuity of sound localization. Waser (1977) broadcast the loud whoopgobble call of gray-cheeked mangabeys and was able to measure the acuity of sound localization under field conditions. The whoopgobble vocalization probably mediates both inter-group spacing and intragroup cohesion (Chalmers 1968; Waser 1975a, 1976, 1977), and it typically elicits rapid approach by specific dominant adult males. Waser was able to observe and track the course of phonotaxis over playback distances as great as 575 m. Figure 7.1 schematizes the search of one mangabey for the location of a presumed intruder.

Playbacks of 10 trials at distant sites (275–575 m) yielded an average error of localization of only 6° (range 0° to 21°). The playback-phonotaxis paradigm convincingly demonstrates the significance and potency of the spatial information inherent in acoustic displays. However, playback experiments are limited in their ability to explore the signal features critical to localization. First, phonotaxis is dependent upon the motivational state of the subject. In Waser's experiment, only the dominant male responded, and in other species (notably, some of the frogs) testing may be restricted to an interval of a day or less within the yearly reproductive cycle (H. C. Gerhardt, personal communication). Second, only the limited class of signals that naturally elicit approach may be tested. Although a variety of signals may have evolved features that enhance or impair sound localization (as has been attributed to some of the mobbing and alarm calls, respectively), their presentation may fail to evoke phonotaxis. Third, the phenomenon of habituation may severely constrain the number, spacing, and conduction of test trials. Fourth, in the field the investigator is often unable to control or to specify either the acoustics or geography of the habitat. Thus, variation in performance both between subjects and across trials may be due, in part, to differences in masking, reflection, refraction, and absorption of the signal. Fifth, the local geography of the habitat may dictate the prudent route the approaching organism should take, and this route may deviate from the perceived acoustic location (Waser 1977). Thus, the perceived location may not correspond with the course of phonotaxis. Sixth, the playback presentation and/or the presence of the experimenter and playback apparatus may alter the behavior of the local fauna and potentially cue the approaching organism to the site of the playback. This difficulty has been repeatedly brought to my attention as my stealthiest sorties in the Missouri woodlot have been foiled by the raucous alarms of blue jays. The significance of the "blue jay factor" for playback research is difficult to assess and to control.

At a cost of simplicity, a number of these difficulties may be overcome through some laboratory approaches.

LABORATORY LOCALIZATION EXPERIMENTS

Monkeys have been trained to localize conspecific vocalizations (Brown, Beecher, Moody, and Stebbins 1978b, 1979) in addition to a variety of biologically arbitrary tones, clicks, and noise bands (Wegener 1974; Heffner and Masterton 1975; Brown, Beecher, Moody, and Stebbins 1978a, 1980; Houben and Gourevitch 1979). In the laboratory, the acoustics of the test situation can be easily specified and controlled. As shown in Figure 7.2, testing can be conducted in specially designed anechoic chambers, which are very quiet (ambient noise levels of 25 dBA or less) and free of echoes and reflected sounds. My associates and I have trained monkeys by operant conditioning procedures (Moody, Beecher, and Stebbins 1975; Stebbins, Brown, and Petersen, in press) to report changes in the spatial coordinates of acoustic signals. From the monkey's performance, an estimate may be made of the minimum audible angle, a measurement of the spatial acuity of the auditory system that is analogous to the minimum visual angle. Monkeys earned a food reward for correctly reporting when a pulsed test signal changed location from a reference position directly in front of them (0° azimuth, 0° elevation) to some other comparison location. The acuity of localization of the test signal is expressed by the magnitude of displacement of the sound required for the monkey to report that the origin of the signal has changed. One advantage of this approach is that the resolution of the auditory system may be determined for any arbitrary sound. Consequently, the design features of signals that promote or hinder sound localization are accessible to study.

Figure 7.2. Sound localization in the laboratory. The monkey earns a food reward for detecting changes in the location of sounds. See text. (From Brown et al. 1978a. Reprinted with permission.)

Horizontal sound localization

Figure 7.3 presents individual psychophysical sound localization functions for three macaque monkeys (one rhesus macaque, *Macaca mulatta;* and two pig-tailed macaques, *M. nemestrina*). The test signal was the macaque coo vocalization depicted in the sound spectrogram in the right panel of Figure 7.3. Vocalizations from both rhesus and pig-tailed macaques have been used, and no acoustical differences have been noted between the calls of these two species. The psychophysical functions in Figure 7.3 show the monkey's rate of guessing, or its catch-trial rate over the 0° point, and the monkey's detection of the change in the position of the sound at four comparison positions. The catch-trial rate for each monkey is very low (less than 10%); it remains stable over time and is independent of the signal being tested. The percentage of test trials detected correctly increases with angle and usually reaches 100% correct by 30°. The psychophysical functions exhibit the classic ogive shape, and the minimum

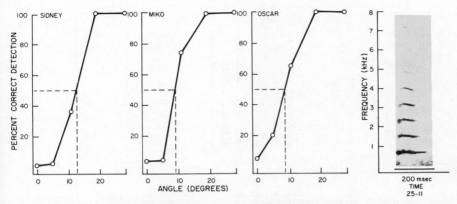

Figure 7.3. Psychophysical functions for the localization of a macaque clear call, signal 25-11. The sonagram of the signal is displayed in the right panel. Behavioral functions are shown for three individual monkeys, Sidney, Miko, and Oscar. The monkey's catch-trial rate was displayed over the zero-degree point, and the monkey's performance trials presented at the four comparison locations increased with angle. See text. The calculation of the minimum audible angle is expressed by the dashed line. (From Brown et al. 1979. Reprinted with permission.)

audible angle, the 50% correct detection point, is measured in degrees and calculated from the curves.

Macaque vocalizations are produced by laryngeal structures that in gross morphology resemble those structures engaged by humans during speech production (Negus 1949); however, some other primates such as the howler monkey (*Alouatta villosa*) and the spot-nosed guenon (*Cercopithecus nictitans*) possess accessory phonatory organs that may act to selectively amplify certain frequency regions in their vocal range (Gautier 1971). The macaque vocal repertoire is quite extensive, and subtle acoustical distinctions may be communicatively significant (Rowell 1962; Rowell and Hinde 1962; Grimm 1967; Green 1975). According to gross acoustic considerations, the macaque repertoire may be partitioned into two general categories: clear calls, and harsh calls (Rowell and Hinde 1962). The clear–harsh categorization is arbitrary and reflects qualitative differences in the perception of macaque calls by human listeners. The clear calls are tonal and often, though not exclusively, harmonically structured vocalizations; these calls, like many bird calls, are somewhat musical in character. Harsh calls, akin to the barks and roars of other mammals, are noisy atonal sounds. Although the structure of harsh calls may be harmonic, it is usually blurred by a wide-band noisy overlay. The principal dimension that distinguishes these two categories is the relative bandwidth, or the range of the call's spectrum, over brief intervals. Representative sound spectrograms of macaque clear and harsh calls are displayed in the bottom panel of Figure 7.4. Calls of both types are locatable, and minimum audible angles for two clear calls and two harsh calls are presented in the top panel of Figure 7.4. The acuity of localization was

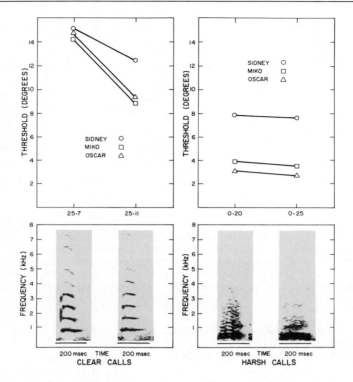

Figure 7.4. Minimum audible angles for two representative clear calls and two representative harsh calls. The bottom panel presents sonagrams corresponding to each vocalization. (From Brown et al. 1979. Reprinted with permission.)

dependent upon the acoustical structure of each signal. In general, the harsh calls were readily localized, the acuity of localization was quite good, approximately 5° or less, and although differences in the spectrograms are apparent between the two representative harsh calls, they were equally locatable. Subsequent research has shown that synthetic signals one octave or greater in bandwidth in this region of the spectrum are all highly locatable (Brown et al. 1980). This suggests that harsh calls should vary little from one type to another in their suitability for sound localization.

In contrast with the harsh calls, rather subtle phonatory differences between clear calls are critical for sound localization. The two clear calls presented in Figure 7.4 differed markedly in their acuity of localization. The acoustical features of clear calls relevant to localization have been experimentally investigated (Brown et al. 1979). A comparison of the acuity of localization of a natural coo (harmonics present) with a low-pass filtered coo (harmonics absent) is shown in Figure 7.5. Localization acuity was only slightly degraded by the elimination of the signal's harmonics. This finding suggests that the acoustic structure of coo vocalizations im-

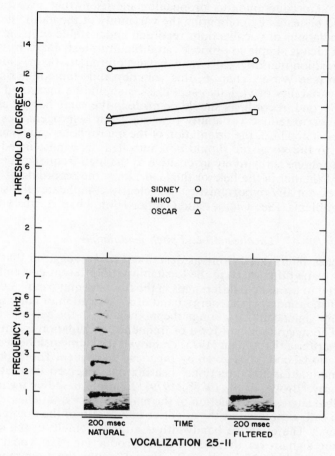

Figure 7.5. Minimal audible angles for natural and low-pass-filtered versions of call 25-11. Sonagrams of the signals are presented in the bottom panel. (From Brown et al. 1979. Reprinted with permission.)

portant for localization is contained in the fundamental, or dominant, frequency band of the call. Although the harmonic structure of macaque coo calls is not critical for localization in the horizontal plane, it may play a significant role for coding distance. In a number of habitats the overall amplitude of the signal is a poor predictor of transmission distance. In environments subject to wind and thermal drafts the peak-to-peak amplitude of a signal may vary over a range of 40 db or more in an interval of less than 1 min (Richards and Wiley 1980). On the other hand, the relative attenuation of the various frequency bands in a complex signal appears to vary rather systematically with transmission distance in homogeneous habitats (Wiley and Richards 1978). These observations suggest that position marking calls should be rigidly stereotyped in the relative level of the harmonics at production; this proposition merits further investigation, and it could be tested by comparing the variability of the harmonic structure of populations of vocalizations recorded under stable acoustic conditions. This idea is simple to propose but difficult to test. First, it is possible that position-marking calls may be structured to favor individual recognition (see Waser, chap. 6, this volume); thus the comparisons of harmonic variability of various vocal classes should be measured within subjects. Second, recordings should be made in the same habitat and at a fixed distance to reduce variability resulting from heterogeneities in the acoustic habitat. Third, the orientation of the microphone and its azimuth in respect to the vocalizer should be controlled, as a misaligned directional microphone is relatively insensitive to the high-frequency elements of signals. Whether in the field or the laboratory, the recording of vocalizations is essentially opportunistic. This clearly complicates the solution to these problems. (See Gautier and Gautier-Hion, chap. 1, this volume.)

Localization and pitch modulation

The low-pass filtering experiment described above suggests that in the horizontal plane differences in the locatability of different coo calls must be attributed to acoustical differences in the fundamental band. A detailed analysis of the fundamental energy band of a population of coo vocalizations reveals significant variation in the modulation of the call's frequency or pitch. The magnitude and form of frequency modulation is one of the acoustical criteria that Green (1975) employed to distinguish between different functional coo types given by Japanese macaques. The importance of pitch modulation for directional hearing has received little attention until recently (Brown et al. 1978b, 1979). Figure 7.6 shows mean minimum audible angles, as a function of the magnitude of frequency modulation of the call, expressed as the effective bandwidth of the dominant frequency band. The effective bandwidth is a measurement of the spectral range of the signal over the course of its production. The vocal signals used were only about 200 msec in duration. Though the most modulated of these calls had an effective bandwidth of more than 200 Hz, at any brief interval through the course of the call the spectrum would be very narrow, only a few Hertz in bandwidth. (Because the frequency of sound is a time-dependent dimension, the measurement of a signal's spectrum, as performed by a real-time analyzer, is conducted in brief, fixed intervals. If the measurement interval was reduced to zero, the signal would have no frequency or spectrum.) The minimum audible angle of some frequency modulated coos (approximately 4°) approached that obtained for some noisy harsh calls (Figure 7.4) and broad bands of noise (Brown et al. 1980). Thus monkeys may alter the acuity of sound localization over at least a fourfold range by remarkably subtle adjustments in the acoustical structure of their vocalizations. It is quite possible that other calls are either more or less locatable than those tested to date; thus the finding of a fourfold range in acuity is probably conservative.

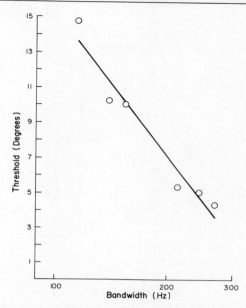

Figure 7.6. Minimum audible angles as a function of the effective bandwidth (or magnitude of frequency modulation) of the dominant frequency band of six macaque vocalizations. The data are averaged for three monkeys. The correlation coefficient is − .98. (From Brown et al. 1978b. Reprinted with permission.)

Vertical sound localization

Most research on sound localization reflects a terrestrial perspective, and, consequently, relatively few studies have addressed localization in the vertical plane (Grinnell 1963; Butler, 1969; Knudsen, Konishi, and Pettigrew 1977). Primates, of course, either occupy, or have radiated from, an arboreal niche. In visually restrictive arboreal habitats, selection may have favored the design of both signals and auditory systems suitable for vertical localization. Horizontal and vertical minimum audible angles for two macaque vocalizations are shown in Figure 7.7. The left panel is for a harsh call and the right panel is for a clear call (representative spectrograms are displayed in Figure 7.4). The data show that the acuity of localization in the vertical plane is less than that observed for these signals in the horizontal plane. Nevertheless, the macaques do produce signals that permit conspecifics to detect differences in the elevation of the source. The results also show that the noisy, broader bandwidth harsh call is more amenable to vertical localization than is the harmonically structured clear call. This is consistent with a number of studies that indicate the importance of signal bandwidth for vertical localization in human listeners (Butler 1969; Gardner 1973; Hebrank and Wright 1974; Shaw 1974; Kuhn 1977). Some recent studies have suggested that reflections from an orga-

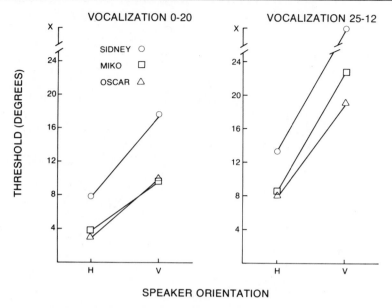

Figure 7.7. Horizontal (H) and vertical (V) minimum audible angles for a macaque harsh call (0-20) and for a macaque clear call (25-12). Representative sonagrams of these classes of signals are presented in Figure 7.4. (From Brown et al., in press. Reprinted with permission.)

nism's torso may be important for vertical localization (Kuhn 1979; Brown in press; Brown, Schessler, Moody, and Stebbins in press), and the data in Figure 7.7 are probably in part dependent upon torso reflections. However, it is likely that the biophysical transformation characteristics of the pinna, or the external ear, and that of the ear canal itself are of central importance for resolving sound elevation. In humans, elevation-dependent transformations are apparent in the two-octave span between 4 and 16 kHz (Butler 1969; Gardner 1973; Hebrank and Wright 1974; Shaw 1974; Kuhn 1977, 1979). The relative small size of the monkey pinnae compared to those of humans should require somewhat higher frequencies for the full realization of elevation-dependent transformations. However, the transformation characteristics of nonhuman primate pinnae have not been carefully studied; indeed, some primates such as the bonnet macaque (*M. radiata*) may make rather dramatic changes in the shape and orientation of their pinnae, and these changes may markedly influence their transformation characteristics. Furthermore, many monkeys may rotate the head and employ binaural cues while making elevation judgments or when hearing signals long enough or repetitive enough to permit this strategy (see Menzel 1980). Hence, the question of vertical localization requires further research. It is quite possible that some harmonically structured, frequency modulated calls such as the screams of colobus (Marler 1972) or type 1 talapoin calls as well as many broad-band harsh calls (Gautier 1974) would be highly appropriate for vertical localization. The structural features of the vocalizations of related arboreal and terrestrial species have not been compared. Although it is difficult to study communication in forest monkeys, and particularly so for species ranging through the forest canopy, it would be interesting to examine the structure of signals designed to advertise elevation-critical events. Calls that act to recruit conspecifics dispersed over various elevations or strata should be carefully examined.

Localization in noisy habitats

Many primates, particularly those occupying forest habitats, must communicate under extremely noisy conditions. The noise may include remarkably incessant and ubiquitous insect sound, wind noise, the vocal displays of a variety of sympatric species, as well as sounds produced by conspecifics. Collectively, these sources may sum to ambient levels in excess of 70 dBA with peak levels much higher (Waser and Waser 1977). Preliminary measurements of the impact of masking noise on sound localization have been made (Brown, in preparation).

Minimum audible angles for simulated macaque vocalizations of various bandwidths (1 Hz, 125 Hz, 1,000 Hz; centered at 800 Hz) in the presence or absence of a white-noise masking stimulus are shown in Figure 7.8. In the quiet, a 1,000-Hz bandwidth signal (simulating a harsh call) was more accurately localized than were the two narrower bandwidth signals (simulating clear calls). In the presence of masking noise, localization of all signals was impaired; however, the impairment was much less for the narrow band signals.

Although the present data do not show that the impairment in localization was independent of a decrease in audibility, the results nevertheless emphasize that the concentration of the available acoustic energy into a narrow band may be strategic for communication under noisy conditions or over long distances. Given the simplifying assumption that primate vocalizations are rectangular in their power density spectrum (that is, the call's spectral structure resembles band-limited white noise), the relationship between the sound level of the call (P), its spectrum level (S), and its bandwidth (W) is given by:

$$P = S + 10 \log W$$

Where P is measured in dB (sound pressure level), S is measured in dB/Hz, and W is measured in Hz.

Accordingly, if the energy of the call may be concentrated into one-fourth of its initial bandwidth, the spectrum level of the signal and the call's signal-to-noise ratio would be elevated 6 dB. This would increase the call's effective transmission distance and the potential number of recipients. If the call is propagated spherically, a 6-dB increase in its spectrum level would double its effective distance (ignoring excess attenuation as a complication; see Wiley and Richards 1978). However, some specialized loud calls, such as the boom of blue monkeys (*Cercopithecus*

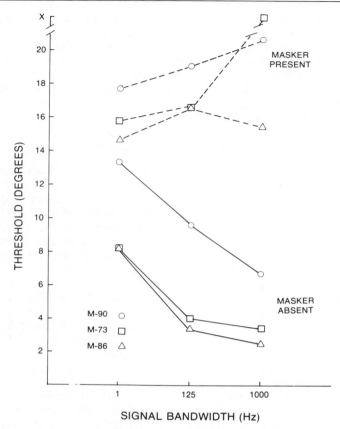

Figure 7.8. Minimum audible angles in the presence and absence of a white-noise masker. The stimuli were geometrically centered at 800 Hz and were presented at 30 dB SPL; the masking level was 55 dBA SPL. (From Brown, in preparation.)

mitis), are propagated even more favorably (Waser and Waser 1977). A 6-dB increase in the spectrum level of a boom-type call may quadruple its transmission distance, and a fourfold increase in range would, on the average, increase the potential number of recipients by a factor of 16 (assuming that the recipients are randomly distributed on the ground around the vocalizer). Hence, the bandwidth of a call may be of considerable significance for communication. In addition, the reduced locatability of the narrow bandwidth call could be overcome by frequency modulation of the signal. The success of this strategy is exemplified by the narrow-bandwidth, frequency-modulated whoopgobble call that mangabeys (*C. albigena*) easily localize and detect over great distances in the wild (Waser 1977).

SOUND LOCALIZATION AND DESIGN FEATURES
FOR COMMUNICATION

Macaques produce clear calls, as well as harsh calls, suitable for sound localization. However, these different signal types tend to be given in different contexts. In semiterrestrial and terrestrial primates, such as the macaques, harsh calls are often emitted at short range and are typically accompanied by facial displays. Van Hooff (1967) described 11 situationally specific facial signals widely observed in the catarrhine monkeys and apes. Of these displays, only 1 is associated with clear calls. The other 10 are either accompanied by harsh vocalizations or are emitted silently, presumably after the establishment of visual contact. In the composite visual–acoustic signals, highly locatable harsh vocalizations may have evolved to facilitate the establishment of visual contact. With the evolution of the fovea and binocular vision, specializations for high-acuity frontal vision may have increased selection for composite signals with vocal elements capable of acoustically redirecting the site of gaze. Compared to some more specialized narrow-bandwidth clear calls, harsh calls may be less suitable for long-distance signaling. The localization and masking data described earlier (Brown, in press) clearly reveal the superiority of narrow-bandwidth signals in noisy environments. However, in dense foliage habitats, visual signals are only possible at short range, and low-amplitude harsh calls may be appropriate for restricting the signal to the intended recipients.

The members of macaque troops commonly exchange coo-type clear calls when out of visual contact. For example, solitary Japanese macaques may emit a distinctive coo call, dubbed the "double," when isolated 50 m or more from the core troop. Other clear-call variants, including the "smooth early high" and "dip late high," may be produced by scattered subgroups and individuals isolated from the main troop (Green 1975). Although clear calls are often given at short and intermediate distances as well, these calls share some general acoustical properties with the highly specialized long-range loud calls of some other primates (MacKinnon 1974; Marler and Hobbett 1975; Marshall and Marshall 1976; Tenaza 1976; Gautier and Gautier 1977; Waser 1977 and chap. 6, this volume; Waser and Waser 1977).

In general, at least 10 factors determine the maximum distance over which a signal may be audible in nature:

1. the signal amplitude at the source
2. the noise amplitude at the receiver
3. the critical bandwidth of the receiver's auditory system, the bandwidth of the signal, and the signal-to-noise ratio within that band
4. the spectrum of the signal relative to that of the noise
5. the direction of the signal relative to that of the noise
6. the auditory sensitivity of the receiver
7. the directional radiation pattern of the signal

158 *Charles H. Brown*

8. the susceptibility of the signal to degradation via refraction and reverberation
9. the rate at which the signal attenuates with propagation distance
10. the duration and repetition of the signal, the auditory time constant of the receiver's auditory system, and the manner by which it integrates transient fluctuations in the level of the signal or noise

These factors implicate a critical interplay between the organization of the auditory and perceptual systems, on one hand, and the structure of the signal and the organism's ecology, on the other. Only a few of these factors have received significant study; nevertheless, several observations are pertinent to the structure of macaque clear calls. The low-frequency dominant frequency band of macaque clear calls (centered near 1,000 Hz) matches a region of optimal acoustic sensitivity in the macaque (threshold of audibility is about 0 dB, re 20 μPa at 1,000 Hz; Stebbins 1973). In addition, low-frequency signals (in the absence of certain acoustical complications) are propagated more favorably over long distances (Konishi 1970; Morton 1975; Marten and Marler 1977; Marten, Quine, and Marler 1977; Michelson 1978; Wiley and Richards 1978; Richards and Wiley 1980). Furthermore, as already discussed, if the maximal acoustic energy an organism can generate may be restricted to a narrow bandwidth, the resulting signal will have a higher amplitude per cycle than will a wider bandwidth signal and, consequently, an advantageous signal-to-noise ratio. Thus, the general features of macaque clear calls exhibit some of the parameters of the specialized long-range signals of some other primates. The laboratory experiments show that the location of the caller may be revealed by modulating the frequency of the dominant frequency band of the signal. And field observations (Waser 1977; Snowdon and Hodun 1981) reveal the potency of this strategy.

SOUND LOCALIZATION AND THE EVOLUTION OF LANGUAGE

Variations in the coo calls of pig-tailed macaques resemble in both form and magnitude the phonatory features that are communicatively significant for Japanese macaques (Green 1975). Green has argued that different coo subtypes may communicate differences in the vocalizer's level of arousal. Signal locatability may covary positively with the level of arousal. Clear calls that are not modulated in pitch, such as the Japanese macaque's long low coo, are restricted to social contexts at the low end of the arousal continuum. For example, in the breeding season, the long low coo may be emitted by a female when she is least receptive to the establishment of a consort relationship. Frequency-modulated coo vocalizations, such as the Japanese macaque's smooth late high, tonal calls with noisy overlap (Japanese macaque class 4 and class 5 vocalizations; talapoin type 1-4 and 4-1 flight and aggression calls), as well as many harsh calls of other primates scale the upper end of the arousal continuum (Rowell 1962; Grimm 1967; Bertrand 1969; Gautier 1974; Green 1975; Deputte, chap. 4, this volume). These broader bandwidth signals appear to be optimal for perceptually focusing attention on the vocalizer. Complementary observations have been provided by other investigators as well. Newman (personal communication; also Newman and Symmes, chap. 11, this volume) has noted that the isolation peep of some infant squirrel monkeys acquires acoustical features (broadband noisy overlay) that should favor localization at a time coincident with the developmental stage when the infant is rather vulnerable, having initiated exploration away from its mother. During this time the call acts to elicit search and retrieval of the infant by its mother. Similarly, Snowdon and Hodun (1981) have observed that pygmy marmoset contact calls increase in effective bandwidth (via the magnitude of frequency modulation) and presumably increase in locatability as the distance between vocalizer and intended recipient (and the uncertainty in the intended recipient's location) increase (see Snowdon, chap. 9, this volume). Collectively, these observations are consistent with the proposition that the level of arousal and the level of contact seeking registered by different calls may be communicatively reemphasized by the signal's relative locatability. The ease of localization may be one of a number of prelinguistic codes in which a facet of the affective state of the vocalizer covaries with a perceptual dimension. As such, sound localization may prove as central to cognitive and symbolic processes as it is to spatial orientation.

One unsolved problem concerning the origin of language is the presumed transition from emotionally labelled signals into articulated utterances freely composed of emotionally vacant and symbolically arbitrary phonemelike elements (Marler 1975). Although the human voice may communicate emotion, the quality of any emotion is not confined to any subset of utterances, words, or phonemes. Consider the stop-consonant phoneme "bə." It is emotionally neutral, and it becomes semantically active only within the context of other phonemes according to a particular linguistic tradition as exemplified by the French word "*ba*rouche" and the English word "hali*but*." The versatility and power of human speech is attributed to its phonetic structure (Liberman, Cooper, Shankweiler, and Studdert-Kennedy 1967; Studdert-Kennedy 1975, 1976; Liberman 1982). Lacking a phonetic basis, the voices of nonhuman primates may be restricted to emotive expression. Recent field and laboratory investigations have begun to challenge this view. Primate vocal signals appear to be syntactically structured (Robinson 1979; Smith, Newman, and Symmes, chap. 2, and Snowdon, chap. 9, this volume); they have been shown to convey semantic, as well as emotional, information (Seyfarth and Cheney, chap. 10, this volume); and they engage special processing mechanisms akin to those employed during speech perception (Petersen, Beecher, Zoloth, Moody, and Stebbins 1978; Snowdon and Pola 1978; Zoloth, Petersen, Beecher, Green, Marler, Moody, and Stebbins 1979; Petersen, chap. 8, and Snowdon, chap. 9, this volume). Nevertheless, the evolution of prephonetic structures remains a puzzle. The "acoustical optimization" perspective advocated here may offer a possible solution. The structure of every vocalization is likely to constitute an acoustic compromise, blending a number of complementary and potentially contradictory specifications pertinent to its transmission, detection, discrimination, localization, and recognition. A radical shift in an organism's acoustic ecology may strongly favor alterations in signal specifications that contradict the previously established code. It follows then that selection for heightened audibility or discriminability may favor the emergence of signals that were emancipated from the conventional emotive code, and this may set the stage for the emergence of emotionally unlabeled vocal phonemelike elements. The likelihood of this scenario is a matter of speculation. Waser (chap. 6, this volume) elegantly documents the fundamental conservatism of the loud calls of some *Cercopithecine* primates. This conservatism, seen in the regularity of vocal structure, may suggest that selection for acoustic optimization has not been very intense or at least has been achieved infrequently. Alternatively, the stereotypy of many of these signals may hint at their resistance to modification compared with some of the softer calls. In addition, the optimum signal specifications for the range of habitats in the *Cercopithecine* radiation is a matter requiring considerable experimental attention (Waser and Waser 1977). Much more research is required to evaluate these possibilities.

The organization and evolution of speech is central to our understanding of man. Although only man has speech, after listening to the calls of many primates, few scientists would argue that meaning is absent from these calls. Yet the richness, organization, and complexity of the referents of these signals and the actual intentions of the vocalizer remain unknown. Perhaps the blending of field and laboratory inquiries, as advocated here, will bring us somewhat closer to a more accurate appreciation of speech and human origins.

REFERENCES

Bertrand, M. 1969. The behavioral repertoire of the stumptail macaque. *Bibliotheca Primatologica* 11:1–273.
Brown, C. H. In press a. Primate auditory localization. In *Sound localization: theory and practice,* ed. R. W. Gatehouse (*Journal of Auditory Research*). Groton, Conn.: Amphora Press.
    In press b. Ventriloquial and locatable vocalizations in birds. *Zeitschrift für Tierpsychologie.*
    In preparation. Primate sound localization in quiet and in noise.
Brown, C. H., Beecher, M. D., Moody, D. B., and Stebbins, W. C. 1978a. Localization of pure tones by Old World monkeys. *Journal of the Acoustical Society of America* 63:1484–92.
    1978b. Localization of primate calls by Old World monkeys. *Science* 201:753–54.
    1979. Locatability of vocal signals in Old World monkeys: design feature for the communication of position. *Journal of Comparative and Physiological Psychology* 93:806–19.
    1980. Localization of noise bands by Old World monkeys. *Journal of the Acoustical Society of America* 68:127–32.

Brown, C. H., Schessler, T., Moody, D. B., and Stebbins, W. C. In press. Vertical and horizontal sound localization in primates. *Journal of the Acoustical Society of America.*

Busnel, R. G., Dumortier, B., and Pasquinelly, F. 1955. Phonotaxie de ♀ d'Ephippiger (orthoptere a des signaux acoustiques). *Comptes rendus des séances Société de Biologie et de ses Filiales* (Paris) 149:11–13.

Butler, R. A. 1969. Monaural and binaural localization of noise bursts vertically in the median sagittal plane. *Journal of Auditory Research* 3:230–5.

Chalmers, N. 1968. The visual and vocal communication of free living mangabeys in Uganda. *Folia Primatologica* 9:258–80.

Chappius, C. 1971. Un exemple de l'influence du milieu sur les emissions vocales des oiseaux: l'evolution des chant en foret equatoriale. *Terre et la vie* 118:183–202.

Clements, M., and Kelly, J. B. 1978. Auditory spatial responses of young guinea pigs (*Cavia porcellus*) during and after ear blocking. *Journal of Comparative Physiological Psychology* 92:34–44.

Erulkar, S. D. 1972. Comparative aspects of spatial localization of sound. *Physiological Reviews* 52:237–60.

Feng, A. S., Gerhardt, H. C., and Capranica, R. R. 1976. Sound localization behavior of the green tree frog (*Hyla cinerea*) and the barking tree frog (*H. gratiosa*). *Journal of Comparative Physiology* 107:241–52.

Gardner, M. B. 1973. Some monaural and binaural facets of median plane localization. *Journal of the Acoustical Society of America* 54:1489–95.

Gautier, J.-P. 1971. Étude morphologique et fonctionnelle des annexes extralaryngées des Cercopithecinae; liaison avec les cris d'éspacement. *Biologia Gabonica* 7:229–67.

1974. Field and laboratory studies of the vocalizations of talapoin monkeys (*Miopithecus talapoin*); structure, function, ontogenesis. *Behaviour* 49:1–64.

Gautier, J.-P., and Gautier, A. 1977. Communication in Old World monkeys. In *How animals communicate,* ed. T. A. Sebeok. Bloomington: Indiana University Press.

Gourevitch, G. 1980. Directional hearing in terrestrial mammals. In *Comparative studies of hearing in vertebrates,* ed. A. N. Popper and R. R. Fay. Berlin, Heidelberg, New York: Springer.

Green, S. 1975. Variation of vocal pattern with social situation in the Japanese monkey (*Macaca fuscata*): a field study. In *Primate behavior,* Vol. 4: *Developments in field and laboratory research,* ed. L. A. Rosenblum, pp. 1–102. New York: Academic Press.

Grimm, R. J. 1967. Catalogue of sounds of the pig-tailed macaque (*Macaca nemestrina*). *Journal of Zoology* 152:361–73.

Grinnell, A. D. 1963. The neurophysiology of audition in bats: directional localization and binaural interaction. *Journal of Physiology* 167:97–113.

Harrison, J. M., and Irving, R. 1966. Visual and non-visual auditory systems in mammals. *Science* 154:738–43.

Hebrank, J., and Wright, D. 1974. Spectral cues used in the localization of sound sources in the median plane. *Journal of the Acoustical Society of America* 65:1829–34.

Heffner, H., and Masterton, B. 1975. Contribution of auditory cortex to sound localization in the monkey (*Macaca mulatta*). *Journal of Neurophysiology* 38:1340–58.

Houben, D., and Gourevitch, G. 1979. Auditory lateralization in monkeys: an examination of two cues serving directional hearing. *Journal of the Acoustical Society of America* 66:1057–63.

Knudsen, E. I. 1980. Sound localization in birds. In *Comparative studies of hearing in vertebrates*, ed. A. N. Popper and R. R. Fay. Berlin, Heidelberg, New York: Springer.

Knudsen, E. I., Blasdel, G. G., and Konishi, M. 1979. Sound localization in the barn owl (*Tyto alba*) measured with the search coil technique. *Journal of Comparative Physiology: Sensory, Neural, and Behavioral Physiology* 133:1–11.

Knudsen, E. I., Konishi, M., and Pettigrew, J. D. 1977. Receptive fields of auditory neurons in the owl. *Science* 198:1278–80.

Konishi, M. 1970. Evolution of design features in the coding of species specificity. *American Zoologist* 10:67–72.

    1973. Locatable and nonlocatable acoustic signals for barn owls. *American Naturalist* 107:775–85.

    1977. Spatial localization of sound. In *Recognition of complex acoustic signals: Report of the Dahlem Workshop*, ed. T. H. Bullock, pp. 127–43. Berlin: Dahlem Konferenzen.

Kuhn, G. F. 1977. Model for the interaural time differences in the azimuthal plane. *Journal of the Acoustical Society of America* 62:157–67.

    1979. The effect of the human torso, head and pinna on the azimuthal directivity and on the median plane vertical directivity. *Journal of the Acoustical Society of America* 65:58.

Liberman, A. 1982. On finding that speech is special. *American Psychologist* 37:148–67.

Liberman, A. M., Cooper, F. S., Shankweiler, D. P., and Studdert-Kennedy, M. 1967. Perception of the speech code. *Psychological Review* 74:431–61.

MacKinnon, J. 1974. The behaviour and ecology of wild orangutans (*Pongo pygmaeus*). *Animal Behaviour* 22:3–74.

Marler, P. 1972. Vocalizations of East African monkeys. II. Black and white colobus. *Behaviour* 42:175–97.

    1975. On the origins of speech from animal sounds. In *The role of speech in language*, ed. J. Kavanagh and J. Cutting, pp. 11–40. Cambridge, Mass.: M.I.T. Press.

Marler, P., and Hobbett, L. 1975. Individuality in a long-range vocalization of wild chimpanzees. *Zeitschrift für Tierpsychologie* 38:97–109.

Marshall, J. T., and Marshall, E. R. 1976. Gibbons and their territorial songs. *Science* 193:235–7.

Marten, K., and Marler, P. 1977. Sound transmission and its significance for animal vocalizations. I. Temperate habitats. *Behavioral Ecology and Sociobiology* 3:271–90.

Marten, K., Quine, D., and Marler, P. 1977. Sound transmission and its significance for animal vocalizations. II. Tropical forest habitats. *Behavioral Ecology and Sociobiology* 3:291–302.

Masterton, R. B. 1974. Adaptation for sound localization in the ear and brainstem of mammals. *Proceedings of the Federation of American Societies for Experimental Biology* 33:1904–10.

Masterton, R., and Diamond, I. T. 1973. Hearing: central neural mechanisms. In *Handbook of perception*, Vol. 3, ed. E. C. Carterette and M. P. Friedman, pp. 408–48. New York: Academic Press.

Masterton, R., Heffner, H., and Ravizze, R. 1969. The evolution of human hearing. *Journal of the Acoustical Society of America* 45:966–85.

Menzel, C. R. 1980. Head cocking and visual perception in primates. *Animal Behaviour* 28:151–9.

Michelson, A. 1978. Sound reception in different environments. In *Sensory ecology*, ed. M. A. Ali, pp. 345–73. New York: Plenum.

Moody, D. B., Beecher, M., and Stebbins, W. C. 1975. Behavioral methods in auditory research. In *Handbook of auditory and vestibular research methods,* ed. C. Smith and J. Vernon, pp. 439–97. Springfield, Ill.: Thomas.

Morris, G. K. 1972. Phonotaxis of mole meadow grasshoppers (*Orthoptera tettigoniidae*). *Journal of the New York Entomological Society* 80:5–6.

Morton, E. S. 1975. Ecological sources of selection on avian sounds. *American Naturalist* 108:17–34.

Neff, W. E. 1968. Localization and lateralization of sound in space. In *Ciba Foundation Symposium on Hearing Mechanisms in Vertebrates,* ed. A. V. S. DeReuck and J. Knight, pp. 207–33. London: Churchill.

Negus, V. E. 1949. *The comparative anatomy and physiology of the larynx.* London: Heinemann.

Petersen, M. R., Beecher, M. D., Zoloth, S. R., Moody, D. B., and Stebbins, W. C. 1978. Neural lateralization of species-specific vocalizations by Japanese macaques (*Macaca fuscata*). *Science* 202:324–6.

Rheinlaender, J., Gerhardt, H. C., Yager, D. D., and Capranica, R. R. 1979. Accuracy of phonotaxis by the green treefrog (*Hyla cinerea*). *Journal of Comparative Physiology: Sensory, Neural, and Behavioral Physiology.* 133: 247–55.

Richards, D. G., and Wiley, R. H. 1980. Reverberations and amplitude fluctuations in the propagation of sound in a forest: implications for animal communication. *American Naturalist* 115:381–99.

Robinson, J. G. 1979. An analysis of the organization of vocal communication in the titi monkey, *Callicebus moloch. Zeitschrift für Tierpsychologie* 49:381–405.

Rowell, T. E. 1962. Agonstic noises of the rhesus monkey. *Symposium of the Zoological Society of London* 8:91–6.

Rowell, T. E., and Hinde, R. A. 1962. Vocal communication by the rhesus monkey (*Macaca mulatta*). *Symposium of the Zoological Society of London* 138:279–94.

Shaw, E. A. G. 1974. The external ear. In *Handbook of sensory physiology,* ed. W. Kleidel and W. D. Neff, 5, pt. 1, pp. 455–90. Berlin: Springer.

Snowdon, C. T., and Hodun, A. 1981. Acoustic adaptations in pygmy marmoset contact calls: locational cues vary with distances between conspecifics. *Behavioral Ecology and Sociobiology* 9:295–300.

Snowdon, C. T., and Pola, Y. V. 1978. Interspecific and intraspecific responses to synthesized marmoset vocalizations. *Animal Behaviour* 26:192–206.

Stebbins, W. C., 1973. Hearing of Old World monkeys (Cercopithecine). *American Journal of Physical Anthropology* 38:357–64.

Stebbins, W. C., Brown, C. H., and Petersen, M. 1982. Measurement of sensory function in non-human species. In *Handbook of physiology,* vol. 2, *Sensory Processes,* ed. I. Darian-Smith. Washington, D.C.: American Physiological Society.

Studdert-Kennedy, M. 1975. The perception of speech. In *Current trends in linguistics,* vol. 12, ed. T. Sebeok, pp. 2349–85. The Hague: Mouton.

1976. Speech perception. In *Contemporary issues in experimental phonetics,* ed. N. J. Lass, pp. 243–93. New York: Academic Press.

Tenaza, R. C. 1976. Songs, choruses, and countersinging of Kloss' gibbons (*Hylobates klossii*) in Siberut Island, Indonesia. *Zeitschrift für Tierpsychologie* 40:37–52.

Van Hooff, J. A. R. A. M. 1967. The facial displays of the catarrhine monkeys and apes. In *Primate ethology,* ed. D. Morris, pp. 7–68. Chicago: Aldine.

Waser, P. M. 1975a. Experimental playbacks show vocal mediation of intergroup avoidance in a forest monkey. *Nature* 255:56–8.

1976. *Cercocebus albigena:* site attachment, avoidance, and intergroup spacing. *American Naturalist* 110:911–35.

1977. Sound localization by monkeys: a field experiment. *Behavioral Ecology and Sociobiology* 2:427–31.

Waser, P. M., and Waser, M. S. 1977. Experimental studies of primate vocalization: specializations for long-distance propagation. *Zeitschrift für Tierpsychologie* 43:239–63.

Wegener, J. G. 1974. Interaural intensity and phase angle discrimination by the rhesus monkey. *Journal of Speech and Hearing Research* 17:638–55.

Wiley, R. H., and Richards, D. G. 1978. Physical constraints on acoustic communication in the atmosphere: implications for the evolution of animal vocalizations. *Behavioral Ecology and Sociobiology* 3:69–94.

Zoloth, S. R., Petersen, M. R., Beecher, M. D., Green, S., Marler, P., Moody, D. B., and Stebbins, W. C. 1979. Species-specific perceptual processing of vocal sounds by monkeys. *Science* 204:870–73.

PART III

Perceptual and psycholinguistic approaches to primate communication

In recent years students of communication have become increasingly aware of the anthropocentric error in studying animal communication. The animal signaling capacities existing outside human sensory modalities, such as the ultrasounds of bats, the pheromones of insects, and the electric signals of some fishes, have emphasized the importance of exploring the perception and evaluation of signals from the animal's, rather than the observer's, point of view. Each chapter in Part III is fundamentally concerned with the determination of how animals perceive and respond to their own signals.

Each investigation uses techniques borrowed from the fields of linguistics, psycholinguistics, and perception. The choice of linguistic and psycholinguistic techniques makes two assumptions: first, that the proper study of the highly complex system of primate communication requires the use of techniques derived from the study of human language and, second, that the use of these techniques allows one to demonstrate parallels (or the lack thereof) between primate communication and human speech and language.

It has become clear from studies of human speech that an utterance contains two components: the linguistic component, which is, in human terms, the phonemes and sequences of phonemes from which words and sentences are constructed, and the paralinguistic component, which is, in human terms, the emotional tone, individual identity, gender identity, and dialect. In any analysis of primate vocal systems a similar distinction must be made – what are the aspects of the signals that convey information about state, behavioral activities or future activities, and what are the aspects of the signals that convey information about species, individual, populational, and gender identity? The chapters of this section present some analytical solutions to the separation of components within primate signals. As Petersen's chapter makes clear, one must distinguish between the "linguistic" and the "paralinguistic" components of signals to develop adequate tests and predictions about how animals perceive their species' specific calls.

Despite the sharing of general goals, each chapter presents a different set of methods and techniques for studying the perceptual and psycholinguistic aspects of primate vocalizations. Petersen's work, in the tradition of behavioral psychology, involves the testing of animals in controlled environments with precisely definable stimuli and recordable response measures. His chapter is a model of the application of laboratory techniques to the understanding of the perceptual processes of species-specific sounds.

The chapter by Seyfarth and Cheney represents the other end of the continuum. In their study of wild animals in their natural habitats, they have adopted a set of elegant techniques. Prior to their work, most primatologists had despaired of using playback techniques with wild primates. Seyfarth and Cheney have shown, however, that,with sufficient attention to detail and sufficient patience, the testing of wild animals in the field can be accomplished. Although the playback technique does not allow the degree of control possible in a laboratory setting, it does have an ecological validity. When animals hear sounds played back from hidden but appropriate locations, their responses are within their natural behavioral repertoire, not arbitrary responses derived from extensive conditioning.

Snowdon's chapter represents an intermediate form of methodology. Although he has studied wild animals in natural habitats, his investigation here is concerned with captive animals in natural social groupings. Like Petersen, he maintains a degree of control that is possible only in the laboratory, but like Seyfarth and Cheney, he observes natural, spontaneous responses elicited through the use of playbacks. Despite the diversity in methodology, each chapter converges toward a similar set of problems and a closely related set of findings.

Petersen's chapter deals with three major issues in the study of animal perception: the processing of species-specific calls in the left cerebral hemisphere, the categorical perception of calls, and perceptual constancy. In addition he addresses the stages of signal processing and the animal's ability to retain or not retain in its memory the linguistically "relevant" and "nonrelevant" aspects of signals. He first argues that the techniques and paradigms developed by psychologists to study the perception and cognitive processing of speech in human beings provide a valid framework for similar research on the perception and processing of species-specific sounds by nonhuman animals. He argues convincingly that the use of parallel techniques has been quite productive to date, both in elucidating the mechanisms of perception in monkeys and in providing clear demonstrations of parallels between nonhuman and human primates.

Although considerable evidence exists that human speech is produced and perceived by one cerebral hemisphere (generally the left), little comparable information is available for other species. Nottebohm (1971) has shown the importance of left-hemisphere control over song production in certain bird species, but no demonstration of left-hemisphere involvement in the perception of species-specific sounds in nonhuman animals has been reported. Petersen documents that Japanese macaques are better able to discriminate between two "linguistic" elements when the elements are presented to the right ear (and hence to the left cerebral hemisphere) than when they are presented to the left ear. He further shows that "nonlinguistic" discriminations do not display this right-ear advantage, nor do other species lacking the same "linguistic" discriminations in their vocal communication show a right-ear advantage in processing Japanese macaque vocalizations. Thus, the left hemisphere seems to be critical in processing the "linguistic" aspects of species-specific vocalizations in monkeys.

Another parallel with human speech occurs in the phenomenon of categorical perception. When phonetic elements are varied along a continuum, human observers do not report the existence of several different sounds. Rather, the continuum is divided into two or three categories, with each position on the continuum assigned unambiguously to one category. Petersen shows that monkeys trained to respond positively to "coo" calls with a peak frequency early in the call and negatively to a call with the peak frequency occurring late in the call (a "linguistic" contrast) show a categorical response to synthesized intermediate calls. Thus several calls with progressively later frequency peaks are treated positively, as equivalent to the early-peak call, but then a sudden shift in response occurs and progressively later peaks are treated negatively, the normal response to late-peak calls.

Petersen also demonstrates that perceptual constancy occurs. Animals were trained to discriminate between individual exemplars of early-peak and late-peak coos. Subsequent exemplars of each type were added until the animals were easily discriminating among a stimulus set of 15 different calls. Individual differences and other nonlinguistic aspects of the signal were ignored as the animals responded to the consistent linguistic differences in the call set. Thus, Petersen argues, the animals' retention of the "phoneticlike" features of the call remains constant over a number of other variations in call structure.

Finally, Petersen suggests several paradigms from human information processing that will, when applied to monkeys, serve to discriminate between different types of information processing, such as between an auditory analyzer and a phonetic analyzer. He argues that application of such paradigms will allow investigators to differentiate with greater accuracy between the linguistic and the paralinguistic aspects of animal signals. He concludes with a brief discussion of the role of memory in the perception and processing of speech versus nonspeech stimuli in human beings and the application of methods used in the study of memory to the study of animal vocalizations.

Snowdon's chapter begins with an argument that psycholinguistic and linguistic methods are necessary to deal adequately with the variability and complexity found in primate communication. He then develops a set of quantitative techniques by which he has measured the components of variability within animal vocalizations and separated them into their linguistic and paralinguistic components.

He presents as case studies of phoneticlike variability the trills of pygmy marmosets, the eight chirp variants of cotton-top tamarins, and the long calls of cotton-top tamarins. In each case there is evidence of discretely different variants within a more general class of call structures. In many cases these variants are highly specific to particular behavioral or contextual situations. For example, long calls, which are equivalent to the song of gibbons (see Deputte, chap. 4, this volume) and the loud calls of cercopithecine monkeys (see Waser, chap. 6, this volume), occur in three different contexts: intergroup interactions, intragroup cohesion, and isolation. Each context is associated with a different form of long-call variant. Thus the vocabulary of monkeys is much larger than it was once thought to be.

Other acoustic parameters are different from those used to distinguish call variations in behavioral situations. The paralinguistic variables are: locational variability, in which the call structure is varied to increase the localizability of calls as the distance from caller to recipient increases; individual variability, in which identification of each individual can be made on the basis of vocal cues alone; populational variability, in which subspecies have different call parameters; and ontogenetic variability, in which changes in call structure occur as a function of development. It is possible to assign one set of acoustic parameters to determine the variants of calls used in linguistic ways, for example, in different situations or contexts, and assign another set of acoustic parameters to each of the paralinguistic variations, for example, individual differences, populational differences. Thus, each component of variance in primate vocalizations can be identified and described.

Like Petersen, Snowdon has also shown evidence of a categorical response to synthesized variants of calls in monkeys. Using pygmy marmosets, which have two semantically different forms of trills that differ only in duration, it was possible to show that as duration was increased up to a point there was no differentiation of call type by the monkeys but that a sudden, sharp drop appeared in response to the positive trill at exactly that point where the distributions of the two trill types abutted. Thus, a perceptual distinction between the two call types that exactly matches the production distinction seems to exist. As Petersen notes, neither of these primate demonstrations have been followed up with discrimination testing, as has been done with human speech categorization studies. Snowdon suggests that detection of a categorical discrimination function is dependent upon how the question is asked. When synthetic calls are used and when monkeys are measured in terms of appropriate linguistic responses, the function is likely to be categorical. However, when additional information, such as individual identity, is present within the category, and when the animals are measured by responses appropriate to individual discrimination, continuous, rather than categorical, response is likely. Studies of human speech perception indicate parallel results.

Snowdon also describes two examples of syntax in the calls of monkeys. In the first, a type of conversational syntax, a group of monkeys develops fixed sequences for the emission of contact calls. Not only is an animal likely to wait until all other animals have called before repeating his own call, but the sequence of which animals call when is also determined. A second example of syntax, which is provided by the cotton-top tamarin, is described as the stringing together of several individual call types to form sequences. In many cases these sequences are simple repetitions of elements, which serve as intensifications of the single unit. But there are also examples of different units being combined, in some cases to form sequences that have a semantic value different from the sum of its component semantic values and in other cases sequences that have a semantic value equivalent to the multiplexing of the values of the individual components. The formation of all sequences is governed by strict rule systems, which can easily be derived as a grammar for sequence formation.

Finally, Snowdon proposes the ontogeny of vocal behavior as a critical area for future work in communication. He notes that marmosets and tamarins exhibit an early "babbling" stage, much like that of the human infant, during which imperfect versions of adult calls are given in inappropriate contexts and sequences. Perhaps this serves as a practice phase, during which an infant learns through its own experience and feedback from others how to structure its calls. The marmosets and tamarins may prove to be promising subjects for future work on vocal ontogeny.

Seyfarth and Cheney provide the first convincing demonstration that monkeys are able to refer to external objects in their communication signals. Previously, all animal communication had been considered egocentric, that is, that animals communicated only about their internal states or the probabilities of their future actions, that none of their communication was referential. Seyfarth and Cheney used the three alarm calls of the vervet monkey to demonstrate that the monkeys use a call specific to each major type of predator – pythons, leopards, and martial eagles. When they played back each call type, the recipient monkeys responded appropriately to the predator being referenced. Thus, when an eagle alarm was given, the animals looked up in the sky, and if in a tree, they took cover under brush. When a snake alarm was played back, they looked down, and if on the ground, they climbed a tree. That is, in the absence of an actual predator, the appropriate alarm call elicited behavior appropriate to the presence of an actual predator. That the monkeys understood the referent of the call seems incontrovertible.

Seyfarth and Cheney also describe the ontogeny of alarm calls. From the earliest ages, the monkeys gave alarm calls appropriate to the general class of predators. Thus, an infant might give an eagle alarm to any large bird, eagle or not, or even to a falling leaf but never to an object or activity on the ground. The class of objects to which they gave alarm calls became more restricted as they grew older, and finally the adults restricted their alarms to the one species of bird that was an actual predator. The authors argue that the ontogenetic process of responding to broad, general categories that are progressively refined and sharpened through observational learning, imitation, and reinforcement is a general model for the acquisition of communication signals and the development of cognitive categories. Snowdon's preliminary data on the "babbling" of young marmosets and tamarins provides parallel data of a progressive ontogenic tuning of communication signals to highly specific situations.

Finally, Seyfarth and Cheney deal with individual recognition as another type of classification. Although the playback of an infant distress call arouses response in only the mother of that infant, other females hearing the call indicate by looking at the mother of the infant that they "know" which infants belong to which mothers. Thus, there is an identification system by which mothers recognize not only their own offspring but the offspring of other mothers as well. It is a recognition based not simply on familiarity but on the knowledge of the relationships of other members of the group.

Taken together, these chapters provide considerable evidence of the complexity of the communication system of monkeys – more complex than we might have expected a decade ago. And by the use of techniques used to analyze human speech and language, they also demonstrate a large number of parallels between human and nonhuman communication.

REFERENCE

Nottebohm, F. 1971. Neual lateralization of vocal control in a passerine bird. I. Song. *Journal of experimental zoology* 177:229–62.

8 · The perception of species-specific vocalizations by primates: A conceptual framework

MICHAEL R. PETERSEN

Several primatologists have succeeded in describing the vocal lexicons of a variety of primate species in considerable detail (Green 1975a; Richman 1976; Gautier and Gautier 1977; Robinson 1979). Consequently, we now have a much greater appreciation of the richness and complexity of primate vocal behavior. Unfortunately, the elucidation of the perceptual processes and mechanisms used by animals to comprehend their vocal messages has lagged behind efforts to develop descriptive catalogs of the vocalizations, largely because a frame of reference that could provide some direction for such perceptual studies has been lacking. The problem is akin to that facing visitors from another planet saddled with an assignment to describe the rules that govern the perception and production of human language: Where and how should they begin? The observation that human speech and primate vocalizations share certain morphological complexities and organization properties (Marler 1975; Zoloth and Green 1979; Snowdon, chap. 9, and Seyfarth and Cheney, chap. 10, this volume) suggests that a potentially useful approach in studying the perception of primate vocalizations might be to search for parallels with human speech perception.

The principal objective of this chapter is to outline a conceptual framework, derived from research on the perception of speech by human beings, that might guide explorations of the perceptual processes employed by nonhuman primates to comprehend species-specific vocalizations. As an illustration of the utility of this approach I shall review in some detail the results of a long-term project in which I have been involved that aims to characterize the perception of vocalizations by the Japanese macaque (*Macaca fuscata*). Since the framework itself leans heavily on phenomena that are characteristic of human speech percep-

The author is an Alfred P. Sloan Research Fellow in neuroscience. Preparation of this chapter was supported by NSF Grant BNS 79-24477 and a NIH Biomedical Sciences Research Grant RR7031. The research was funded by these grants as well as NSF Grants BNS 77-19254 to M. Beecher, W. Stebbins, and D. Moody and BNS 75-19431 to P. Marler. I gratefully acknowledge the advice and support of the above individuals as well as S. Zoloth and S. Green. N. Layman provided excellent secretarial assistance and C. Snowdon and C. Brown provided a critique of the manuscript.

tion, it allows us to ask the simple question: To what extent are the perceptual processes enlisted by nonhuman primates to extract information from their vocalizations isomorphic with those used by humans to perceive speech? Although such a query might seem to disclose a distinct anthropocentric bias, the intent is not to show that monkeys are furry little humans (or that humans are large, hairless monkeys) with regard to their communicative abilities. Rather, I acknowledge that human speech is probably the most sophisticated system of communication and that it is highly unlikely that any animal will ever be shown to possess a system capable of transmitting the variety of messages that can be encoded through speech. But, given that we now know so much about certain key aspects of speech production and perception, it seems prudent to use that information as a frame of reference for the study of other vocally prolific species.

This approach is identical to that taken by any comparative psychologist or biologist interested ultimately in tracing the evolution of certain anatomical structures, physiological processes, or behavioral traits. One compiles a list of distinctive features or properties that characterize a particular target species and uses it as a yardstick to determine the qualitative and/or quantitative degree of similarity between that species and others with regard to those characteristics.

For example, Hockett (1960) offered a list of the fundamental properties of human language and then proceeded to determine how other animal communication systems measured against it. Using the small body of literature then available, Hockett concluded that the most sophisticated of animal communication systems, that of the gibbon, exhibited 9 of the 13 fundamental design features of human language. The veritable explosion of interest and active research in animal communication, especially primate communication, in the last 20 years has uncovered details of other primate vocal systems that are considered even more complex than that of the gibbon (Green 1975a; Seyfarth, Cheney, and Marler 1980a). In addition, although Hockett reported no examples of the cultural transmission of vocal traditions and cited only the honeybee's dance and gibbon vocalizations as having semantic properties, recent evidence suggests that many species acquire certain vocalizations and vocal traits via protocultural influences (e.g., vocal dialects in Japanese monkeys, Green 1975b; the acquisition of predator alarm calls by vervets, Seyfarth, Cheney, and Marler 1980b; Seyfarth and Cheney, chap. 10, this volume), and some use vocalizations that are representational or semantic in character (Seyfarth, Cheney, and Marler 1980a, b). (See also Newman and Symmes, chap. 11, this volume.)

The essential point is that Hockett's scheme brings a certain degree of systematization to the description of the properties of any species' communication system. Moreover, whereas Hockett's design features find their origins as descriptors of human language and thereby permit comparisons between animals and humans, they also provide a common ground for making useful comparisons among different animals. Thus, his scheme might be viewed as an important heuristic, pointing to gaps in our knowledge of the communicative capacities of different species and providing a basis for the comparison of these species with one another.

I offer here a set of processing phenomena characteristic of human speech perception for use as a checklist in the comparative study of the perceptual facts of primate vocal communication. Once we have sufficiently characterized the perceptual processes used by different species to analyze their vocalizations, it ought to be possible to begin tracing the evolution of such capacities by correlating ecological demands and social organization with the presence or absence of particular perceptual traits among different species.

In addition to providing a means of describing a species' perceptual capacities and then placing them in a comparative context, I expect that a perceptual assay will also yield information about the way in which primates categorize and apprehend the different classes of information encoded in their vocalizations. A most intriguing property of human speech is that a single utterance is capable of encoding simultaneously several related, but essentially independent, bits of information. Thus, within a single vocalization or series of vocalizations a speaker may transmit information from which a listener could conceivably extract the central message intended as well as the speaker's sex, relative age, geographic origin, emotional tone or intent, and individual identity. Most of these so-called paralinguistic features are realized acoustically by a speaker through manipulation of the speech signal's fundamental frequency, amplitude, overall timbre, and subtle frequency and amplitude modulations over time. This information is then multiplexed with the acoustic cues underlying the linguistically relevant unit. The listener's task is to analyze the linguistic and paralinguistic aspects of the signal to comprehend the meaning of the message and, as needed, to extract other pieces of information about the vocalizer. The high information-bearing capacity of speech thereby allows a listener to distinguish between the affective states of a child uttering "father" in a moment of exasperation and crying "father" when in danger.

Recent research in animal acoustic communication provides evidence for a duality of function in animal vocalizations (Smith 1968, 1977; Green 1975a, 1981; Beer 1976; Snowdon 1979 and chap. 9, this volume; Seyfarth, Cheney, and Marler 1980a, b) paralleling that seen in human speech. This work suggests that animals are transmitting specific messages (in analogy to the linguistic dimensions of speech) interdigitated with auxiliary information about the animal's identity, emotional state, sex, age, and species or group membership. In an effort to simplify the task of cataloging a species' repertoire, animal communication scientists generally tend to concentrate on only one or two of the pieces of information that a signal comprises. Thus, one investigator focuses on the signal parameters that might code for species-identity, whereas another is interested in aspects of the signal pertinent to individual identification, and still another attends to the nature of the central message being transmitted. This approach is especially valuable and appropriate when each of these functions is subserved by a different signal type. For example,

Beecher, Beecher, and Nichols (1981) have shown that each young bank swallow has a unique signature call that seems to have no function beyond allowing adults to identify their offspring. But what of species with more complex repertoires in which each signal encodes multiple pieces of information? One might reasonably expect that an investigator would have considerable difficulty teasing the signal apart to determine the specific function of its acoustic components. Add to this the fact that the value of each component is likely to vary as a function of the age, sex, and acoustic surroundings of the vocalizer and it might be easier to appreciate the monumental task a field investigator faces in attempting to assign vocalizations to distinct functional classes on the basis of particular acoustic features. Indeed, this raises the question of whether the traditional distinction in ethology (e.g., Marler 1975) between discrete and graded signal repertoires might simply be a reflection of the relative ease with which one can assign to categories signals that seem to exhibit little acoustic or functional overlap because they contain fewer pieces of information (discrete?) versus the difficulty of unambiguously categorizing signals that vary along multiple dimensions, carrying information simultaneously along several channels (graded?). Perhaps if we better understood the relation between the acoustic features that constitute a graded signal and their putative individual functions we would be less likely to label them as graded. Each feature might have a categorically distinct function but when multiplexed with several other features and the inherent, perhaps orthogonal, variability of each component is added, the human observer is likely to see intergradations that may or may not be functionally (perceptually) important.

One contribution to the resolution of such problems in human speech has been the identification of distinct processing modes for different aspects of the speech signal. Granted, the linguist has a special intuition about what constitutes a vowel, a consonant, or a paralinguistic feature simply because his own acquired knowledge of the language through daily usage provides him with the requisite skills and insights. Beyond this fact though, there is mounting evidence that it may be possible to distinguish among these different facets of speech by examining the fine-grain details of the human being's perceptual response to such features. This is especially true at the phonetic level of analysis where converging lines of evidence suggest that stop consonants (e.g., /b/, /d/, and /g/) are processed differently than vowels (e.g., /a/, /i/, /u/), which seem, in turn, to be processed differently than such paralinguistic dimensions as pitch and intensity (Pisoni 1973, 1975; Wood 1975). We shall defer discussion of the specific details of these processing differences to later sections of this chapter where we describe some of the perceptual tests that reveal them. For now, the important points are that the perceptual tests used in studies of parallel information transfer in human beings may: (1) illuminate the character of the modes of processing that animals enlist when analyzing different aspects of their vocalizations and (2) provide evidence of the functional separability of the different classes of information within the calls.

PERCEPTUAL PHENOMENA CHARACTERISTIC OF
SPEECH PERCEPTION

Much research on the perception of speech signals has evolved from information-processing theories of human cognition (Neisser 1967; Haber 1969; Massaro 1972; Garner 1974; Pisoni 1978; Posner 1978). Perception is assumed to be an end result of processing at several interdependent stages of analysis. Sensation, memory, attention, and thought are viewed as separate but highly interactive components responsible for all perceptual activity. As applied to speech perception (see Studdert-Kennedy 1974, 1976; Pisoni 1975, 1978), the information-processing perspective holds that the recognition and identification of a speech sound results from processing at several different stages of analysis beginning with a packet of acoustic energy impinging on the auditory sensorium. From the information received at the ear, a collection of individual acoustic features are extracted and transformed (in some as yet unknown way) into a set of phonetic features that can be combined to yield an integrated speech percept. Because the results of analysis at each processing stage must be retained for subsequent stages and because the processing takes a finite amount of time, two types of memory device are also invoked: one is a short-term, temporary store for information extracted from the signal under analysis and the other is a more permanent long-term store used to compare information in the current signal with that acquired in the past so that an identity or recognition decision can be made. Thus, this view of speech perception has led to attempts to characterize the independent and combined contributions of sensory, attentional, and memorial processes to speech analysis and to uncover the neural mechanisms underlying such processes. We shall review this work in the context of five speech-processing characteristics:

1. neural asymmetries in speech-perception processes
2. categorical versus continuous perception of acoustic dimensions cueing distinctions among speech sounds
3. selective attention to linguistically distinctive acoustic features
4. interactions among different processing stages in the speech perception network
5. short-term memory involvement in processing different aspects of speech

The ability of an information-processing model to accommodate such a wide range of issues speaks to its potential as a theoretical basis for studies of nearly any perceptual system. Important insights into the perception of species-specific communication sounds can be gained by taking such an approach. Some precedent has been established by other investigators who have successfully adopted such a perspective in their studies of various aspects of cognitive function in animals (Riley and Leith 1976; Blough 1977, 1979; Hulse, Fowler, and Honig 1978). I extend it to the analysis of primate communication.

Neural lateralization

A large literature now suggests that the neural mechanisms governing a wide range of behaviors, including the perception and production of speech, are confined primarily to one of the two cerebral hemispheres. A confluence of evidence from neuroanatomy, electrophysiology, and psychology points to a predominant role for the left cerebral hemisphere in managing human speech. P. Broca and C. Wernicke (see Benton 1965) first reported that damage to temporal regions of the left cerebral hemisphere more severely disrupted the comprehension and production of language than did damage to corresponding regions of the right hemisphere. Their initial conclusions have withstood more than 100 years of research aimed at identifying the anatomical substrata responsible for this correlation (Geschwind 1970). A number of studies (Geschwind and Levitsky 1968; Teszner, Tzavaras, Gruner, and Hécaen 1972; Witelson and Pallie 1973; Wada, Clarke, and Hamm 1975; Galaburda, LeMay, Kemper, and Geschwind 1978; see Witelson 1977 for a recent and thorough review of this area) have reported gross anatomical asymmetries between those regions of the left and right temporal lobes considered responsible for language behavior. Interestingly, Wada, Clarke, and Hamm (1975) and Witelson and Pallie (1973) have found that this asymmetry is present in newborns, suggesting that the left hemisphere is predisposed for the mediation of language-related behaviors.

Noninvasive electrophysiological and behavioral assessment procedures provide evidence bearing on the functional significance of these cerebral asymmetries. Much direct evidence for the role of the left-cerebral-hemisphere regions in speech is provided by recordings of surface electroencephalographic (EEG) activity over the temporal lobes of intact humans (Cohn 1971; McAdam and Whitaker 1971; Morrell and Salamy 1971; Wood, Goff, and Day 1971; Matsumiya, Tagliasco, Lambroso, and Goodglass 1972; Desemedt 1977; Molfese 1978, 1979). Wood, Goff, and Day (1971) conducted comprehensive examinations of hemispheric asymmetry in electrical activity, comparing the average auditory evoked potentials over the left and right hemispheres of subjects performing two different tasks: speech discrimination (/ba/ vs. /da/) and nonspeech discrimination (high-pitched /ba/ vs. low-pitched /ba/). They found significant differences between the EEG waveforms recorded from the left temporal lobe during performance of the two discriminations, whereas the responses measured over the right hemisphere during both tasks were statistically and visually indistinguishable. Molfese and his colleagues have recently reported evidence of electrophysiological asymmetries in infants listening to speech (Molfese, Freeman, and Palermo 1975; Molfese and Molfese 1979). Coupled with reports of anatomical asymmetries in the speech areas of infant brains, this work suggests an ontogenetic predisposition for speech perception among infants.

The central behavioral finding of neural lateralization is that listeners performing discrimination and identification tasks requiring analysis of the linguistically relevant features of speech stimuli respond more accurately and more rapidly to speech signals presented to the right ear, which transmits information primarily to the left hemisphere, than to signals presented to the left ear, which communicates primarily with the right hemisphere. The effect has proved reliable: Numerous investigators, studying subjects listening to a variety of different speech sounds while performing in various kinds of experimental situations, have consistently reported the right-ear advantage in the perception of speech (Kimura 1961a, b, 1964; Studdert-Kennedy and Shankweiler 1970; Springer 1973a; Catlin, Van Derveer, and Teicher 1976).

The standard technique for the study of ear performance advantages has been the dichotic listening procedure, which involves the simultaneous presentation of competing auditory stimuli to the two ears, one stimulus to each ear, and then a test of the subject's ability to report the information presented to the individual ears. The right ear typically proves superior to the left: The percent identification score is higher (Kimura 1961a, b; Studdert-Kennedy and Shankweiler 1970), and average reaction time is faster (Springer 1971). Kimura (1964) also reported a dissociation in auditory asymmetries; when melodies served as the dichotic stimulus material, the left ear outperformed the right.

The neural model proposed by Kimura (1980) to explain these findings is derived from anatomical, physiological, and clinical considerations. Each cerebral hemisphere receives a larger proportion of projections from the contralateral ear than it does from the ipsilateral. The functional advantage of this disproportionate distribution of projections is borne out by physiological recordings, which show that in most areas of auditory cortex the contralateral gross neural response is larger than the ipsilateral response and that, under binaural stimulation, the contralateral pathway can occlude the responses from the ipsilateral pathway (Rosenzweig 1951; Hall and Goldstein 1968). Hence, the auditory system is, in a functional sense, predominantly crossed (Sparks and Geschwind 1963; Milner, Taylor, and Sperry 1968; Cowey and Dewson 1972). The neural input from the right ear to the left hemisphere should be more influential than that from the left ear and vice versa. According to Kimura's model, the right-ear pathway is more proficient in identifying speech because it has readier access to the left, language-dominant hemisphere. The left-ear advantage of certain nonverbal stimuli, suggesting primarily right hemisphere activity, is consistent with findings that tonal pattern perception is more dependent on processing by the right hemisphere than by the left hemisphere (Milner 1962). Kimura's position is also supported by the results of dichotic studies conducted with split brain patients. Milner, Taylor, and Sperry (1968) and Sparks and Geschwind (1963) have shown that whereas these subjects show no consistent ear performance advantage for speech presented monaurally, their performance on dichotic tasks suggests that they are virtually unable to hear speech in the left ear in the presence of competitive input to the right ear. These data have been interpreted by Studdert-Kennedy (1974) as supporting a conclusion that the left-ear input of a dichotic pair is degraded because of the indirect path it takes to the language-dominant left hemisphere via the right hemisphere and corpus callosum. Recent data gathered in monaural listening studies suggest that active occlusion of ipsilateral input is not always necessary for a laterality effect to appear.

Since subjects generally commit very few errors when listening monaurally, reaction time is used as a more sensitive index of laterality effects. In general, the pattern of results is consistent with that in dichotic experiments: Verbal material yields a right-ear advantage, reflected in shorter right-ear reaction times, whereas nonverbal material produces a left-ear advantage, reflected in shorter left-ear reaction times (Kallman 1977). The difference in reaction times between the two ears generally reported for speech is around 15 msec (Springer 1973a; Morais and Darwin 1974; Catlin, Van Derveer, and Teicher 1976), a differential that apparently represents the amount of additional time necessary to process a signal degraded by transmission over the corpus callosum from the right hemisphere (left ear) to the left hemisphere. Presumably the information is not so degraded that it interferes with absolute identification of the sound; rather, it only prolongs the time necessary to arrive at a suitable response.

Evidence from early investigations of neural lateralization suggested that the left hemisphere was responsible for linguistic activities, whereas the right was specialized for nonlinguistic behaviors (see Bradshaw and Nettleton 1981 for a review). However, several findings suggest that this characterization is inaccurate. First, different classes of phonemes give rise to right-ear advantages (REAs) of varying magnitudes: Stop consonants yield the largest REAs (Shankweiler and Studdert-Kennedy 1967; Studdert-Kennedy and Shankweiler 1970); fricatives (Darwin 1971) and semivowels and liquids (Haggard 1971) elicit somewhat smaller REAs; and vowels (except under special conditions, Darwin 1971; Studdert-Kennedy 1972) generally produce no ear advantage (Shankweiler and Studdert-Kennedy 1967; Studdert-Kennedy and Shankweiler 1970). Second, some aspects of speech sounds produce a left-ear advantage (LEA; right-hemisphere dominance). Paralinguistic features like pitch and intensity, which are introduced simultaneously with phonetic information that is processed predominantly by the left hemisphere, yield strong LEAs, suggesting a preeminent role for the right hemisphere in their analysis (Darwin 1969; Haggard and Parkinson 1971; Carmon and Nachson 1973; Nachson 1973; Blumstein and Cooper 1974). Third, Zaidel (1978) has shown that whereas the right hemisphere may not be as proficient as the left in processing speech, it does nonetheless possess some capabilities for perceiving speech. Finally, certain nonlinguistic acoustic stimuli, particularly those that require resolution of complex spectral and temporal information, are processed primarily by the left hemisphere (Halperin, Nachson, and Carmon 1973; Bever and Chiarello 1974; Cutting 1974; Papcun, Krashen, Terbeek, Remington, and Harshman 1974; Robinson and Solomon 1974; Gordon 1975; Blechner 1976; Natale 1977). This has prompted various authors (e.g., Lackner and Teuber 1973, Nottebohm 1979; Bradshaw and Nettleton 1981) to suggest that the left hemisphere should not be characterized as a site of language dominance but rather as a brain region that possesses property analyzers sensitive to the acoustic and temporal parameters that distinguish various classes of phonemes, as well as the temporal rhythms of running speech (e.g., Schwartz and Tallal 1980). Any complex auditory pattern, linguistic or not, that contains cues best analyzed by the left-hemisphere processor will yield right-ear dominance. The question remains unanswered, however, as to whether this processing strategy represents (1) a specialization, originally evolved to facilitate human speech perception, that can, now that it exists, also be engaged by other acoustic patterns manifesting speechlike properties, or (2) a general property of a system evolved to analyze stimulus properties like those characteristic of, but not unique to, speech. (Studies to determine whether nonhuman primates who employ lateralized networks when listening to their own sounds also use them to analyze human speech could shed some light on this issue.) Since neural lateralization is an important characteristic of human speech perception and since it seems to provide some basis for discriminating among different components of speech sounds, it is useful as a comparative index to study the perception of acoustic communication signals by other primates.

Surprisingly little research has been done on the phylogenesis of lateralization. Several studies involving noncommunicative motor behaviors (e.g., Webster 1972; Butler and Francis 1973; Collins 1977; Warren 1977; Denenberg, Garbanati, Sherman, Yutzey, and Kaplan 1978; Glick, Meibach, Cox, and Maayani 1979) and a small number of studies of visual processing mechanisms (e.g., Gazzaniga 1963; Hamilton 1977) have provided some data for a variety of species that suggest one hemisphere may be used preferentially (see Walker 1980 and Denenberg 1981 for reviews of neural lateralization in animals). The only systematic attempt to investigate the lateralization of neural areas concerned with acoustic communication has been the work of Nottebohm and his colleagues on the control of vocal behavior in various songbirds. Nottebohm (1971, 1972, 1977; Nottebohm and Nottebohm 1976) has found that sectioning a songbird's left hypoglossal nerve (which innervates the bird's sound production apparatus) more severely disrupts singing than the sectioning of the right hypoglossus. Hypoglossal dominance is a reflection of hemispheric dominance in the emission of song (Nottebohm, Stokes, and Leonard 1976). Thus, these findings point to a striking parallel between the lateralization of birdsong and the cerebral asymmetries involved in the production of human speech (e.g., Penfield and Roberts 1959; Ojemann and Mateer 1979).

The work with birdsong lateralization has concentrated, however, on aspects of the neural network responsible for sound production with no attention being directed to the processing mechanisms that underlie the perception of these songs by avian listeners. Indeed, the human literature, which clearly shows that both productive and perceptual aspects of speech are controlled predominantly by the left hemisphere, should alert us to the possibility that the bird's and, by extension, the primate's sound production and perception systems are similarly organized. Other related studies on other aspects of cerebral asymmetries in animals bode well for additional research in this area.

Dewson's (1977) finding of a more severe impairment of monkey auditory memory following ablation of the left superior temporal gyrus than when the corresponding right-lobe tissue was destroyed is reminiscent of the general finding that complex auditory processes, including memory, are disrupted in human patients suffering left temporal lobe damage (Milner 1974; Sperry 1974). This fact, combined with reports that some species of Old World monkeys (Falk 1978) and great apes (LeMay and Geschwind 1975; Yeni-Komshian and Benson 1976) have shown anatomical asymmetries in those regions of the temporal lobe considered homologous to the area (Wernicke's) controlling human speech, offers compelling justification for investigations seeking to determine whether the mechanisms underlying vocal communication are neurally lateralized.

My colleagues and I investigated the question of whether the Japanese macaque (*M. fuscata*) employs neurally lateralized processes during perceptual analysis of its conspecific vocalizations (see Petersen 1978; Petersen, Beecher, Zoloth, Moody, and Stebbins 1978, in preparation a). Green (1975a) determined that *fuscata* has 10 major classes of vocalizations, distinguishable on both functional and acoustic grounds. Following an extensive analysis of the class II tonal, or coo, vocalizations, he determined that 7 subtypes exist within that class. Although all 7 share certain global acoustic properties and occur predominantly in affinitive, contact-seeking situations, the calls fall into clusters composed of vocalizations that are more similar to one another, both acoustically and in terms of the social context in which they are uttered, than they are to calls in other clusters. We obtained from Green several different recorded examples of 2 coo subtypes: smooth early highs (SEHs) and smooth late highs (SLHs). Figure 8.1 contains sonograms of these signals. The primary acoustic distinction between the 2 calls is in the relative temporal position of an ascending-descending frequency modulation, which we call the peak; it occurs before the midpoint of the call in SEHs and after it in SLHs.

Using conventional animal psychophysical techniques, we trained six *fuscata* and, for comparison, five other Old World monkeys (two pigtailed macaques, *M. nemestrina,* two bonnet macaques, *M. radiata,* and one vervet, *Cercopithecus aethiops*), to discriminate between the two call types. All the animals were conditioned to emit a distinct instrumental response to each of the two call types; correct responses produced a food reward. During testing sessions the animals were confined to a restraint chair facing the response manipulanda inside a sound booth, and stimuli were delivered via earphones mounted on the animal's head. To measure laterality effects during performance of the discrimination, we presented vocalizations to one ear only (monaural) with competing wideband noise in the contralateral ear. Over the course of a session, half the stimuli were presented to each ear. But from one stimulus presentation to the next the ear receiving the vocalization was randomly varied. Thus the animals were never certain which ear would receive the discriminative stimulus.

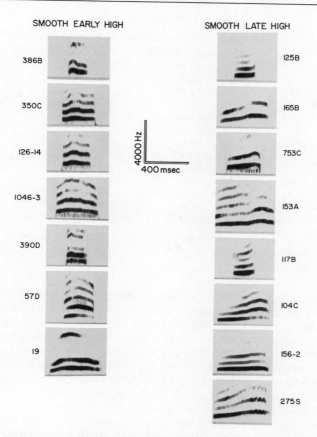

Figure 8.1. Sonagrams of the 15 test stimuli. The dark bands indicate energy present at different frequencies over time. (From Aslin, Alberts, and Petersen, 1981a. Reprinted with permission.)

Initially the animals were trained to discriminate between a single example of each call type. After they mastered that problem, additional stimuli were added progressively to the stimulus set until eventually they were discriminating a total of eight SLHs from seven SEHs. We then computed separate accuracy scores for right-ear and left-ear presentations of the stimuli during the various phases of the discrimination. The right-ear and left-ear scores were then compared with one another to determine superiorities. Table 8.1 presents the results. Five of the six *fuscata* and only one of the five comparison animals showed a right-ear advantage (REA) for the task, suggesting that the peak cue engages neurally lateralized processors located in the left hemisphere. The parallel with human speech perception is obvious.

Green (1975a, 1981) has suggested that, analogous to human speech, the fundamental frequency (pitch) of these vocalizations provides indexi-

Table 8.1. *Laterality index*

Subject	Species	Peak	Pitch
98	*M. fuscata*	.63*	
99	*M. fuscata*	.77*	
100	*M. fuscata*	.62*	
120	*M. fuscata*	.75*	.37*
121	*M. fuscata*	.54	
122	*M. fuscata*	.65*	.51
35	*M. nemestrina*	.53	
58	*C. aethiops*	.63*	
88	*M. radiata*	.54	
93	*M. nemestrina*	.51	.54
133	*M. radiata*	.46	.51

Note: The laterality index (P) = number of REA cases divided by number of REA and LEA cases.
* $p < .05$.

cal, paralinguisticlike information about the age, sex, and affective state of the vocalizer. In a subsequent experiment we determined whether this analogy in production extended to perception. Four animals, two *fuscata* and two comparison monkeys, were trained on a discrimination problem requiring attention to the pitch of the SEH and SLH vocalizations. Of the 15 calls shown in Figure 8.1, 12 were sorted into high-pitch (F_0 (fundamental frequency) > 600 Hz) and low-pitch (F_0 < 600 Hz) categories. Training procedures were identical to those for the peak discrimination: Animals emitted distinct responses to indicate discrimination of the two pitch classes and stimuli were delivered monaurally. Table 8.1 presents the results. As in the peak discrimination, neither comparison animal showed an ear advantge. One *fuscata* had a strong LEA, but the other showed no ear advantage. It is worth noting, in addition, that the strong REA shown by both *fuscata* while processing the peak cue disappeared when they were forced to attend to pitch. Clearly then, at some level the neural processes involved in perceiving the peak cue are somehow distinct from those used to perceive pitch. Moreover, this finding of differential processing lends additional support to Green's (1975a, 1981) suggestion that the peak and pitch dimensions transmit separable types of information.

The species differences in lateralization of the peak dimension are also noteworthy. Earlier, I summarized evidence showing that the verbal–nonverbal dichotomy no longer characterized the functional distinction between the left and right hemisphere. In view of this, one cannot explain the finding of a strong REA in *fuscata* and no ear advantage in the comparison animal by suggesting that the peak cue is of communicative significance to *fuscata* but not to the other species. There is, however, a parallel in comparative studies of certain human languages. Unlike English, pitch assumes important linguistic functions in languages like Thai. Interestingly, native speakers of Thai show a REA for pitch, whereas English speakers show either no ear advantage or a LEA (Van Lancker and Fromkin 1973). Whether pitch is lateralized at birth, and in which direction, is an important, but as yet unanswered, question. However, the implication remains that lateralization of a dimension is dependent not only on its acoustic properties but also on its role in the listener's language as well. By extension, it seems reasonable to conjecture that the species differences we obtained were due to differences in the communicative relevance of the peak cue to the two groups. One might argue that the species differences reflected differences in the cues that the *fuscata* and comparison animals attended to in solving this discrimination problem. That is, perhaps the *fuscata* attended to peak, a temporal cue, which, in humans, seems to require processing by the left hemisphere (e.g., Schwartz and Tallal 1980), whereas the comparison animals listened to some other cue, which required no lateralized processing. The issue thus reduces to a question of whether the *fuscata* and comparison animals were all attending to the peak cue while performing the discrimination task. If they were, the laterality differences are due to differences in the communicative valence of the peak cue to the two groups; if not, their differences might simply reflect differences in the processing requirements of the dimensions the two groups attended to.

To determine whether all animals were attending to the peak cue, we conducted a generalization test with four animals who had completed the peak task, two *fuscata* and two comparison monkeys, using a large number of novel, natural SEH and SLH signals. We reasoned that if they were listening to the peak cue, their responses to the novel stimuli should be controlled by the peak dimension. Figure 8.2 shows that this was indeed the case. The responses of the *fuscata* and comparison animals alike seemed to be controlled by the peak cue: Stimuli with a peak position approximating an SEH elicited an "SEH response" and likewise for the SLH stimuli.

To further test that the animals were attending to the peak cue per se, we synthesized a continuum of SEH and SLH vocalizations with the aid of a computer. Figure 8.3 shows sonagrams of these signals. We were able to vary the peak cue alone while holding other acoustic features constant. Figure 8.4 presents the results of a generalization test with four animals using the synthetic stimuli. Again, the responses of all animals were tightly controlled by the peak cue, implying that they were all listening to it and not to something else.

In sum, our studies demonstrate that *fuscata* use neutrally lateralized processors when perceiving the SEH and SLH vocalizations, whereas comparison animals do not. Moreover, this species difference in lateralization stems at least in part from the communicative relevance of these signals to the *fuscata* and the lack thereof for the comparison species. In addition, we have preliminary evidence that indexical, paralinguisticlike features like pitch and acoustic features like peak, which carry the call's central message, are processed differently. Each finding has a clear paral-

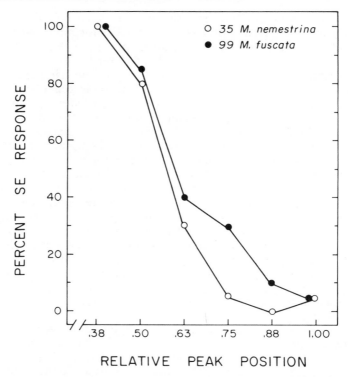

Figure 8.2. Generalization data for the 27 novel, natural signals with relative peak positions ranging from 0.05 to 1.00. (From Aslin, Alberts, and Petersen, 1981a. Reprinted with permission.)

lel in speech perception, suggesting that analogous processes may be involved.

Categorical versus continuous perception of acoustic dimensions

Many distinctive features of certain human phonemes seem to lie at opposite ends of acoustic continua (e.g., Lisker and Abramson 1964) and, to a limited degree, seem to intergrade in their acoustic realization. In addition, paralinguistic features like pitch and intensity seem to vary in a continuous fashion. The oft-cited illustration of this principle of acoustic intergradation is the finding of Lisker and Abramson (1964) that certain classes of phonemes can be distinguished by the modal value of a property referred to as voice onset time (VOT). VOT is a measure of the delay between production of a consonant and the initiation of the laryngeal pulsing known as voicing. The VOT dimension can be envisioned as an acoustic continuum ranging from "prevoiced" consonants, in which voicing precedes consonant onset, to "voiced" consonants, in which voicing is nearly coincident with consonant onset, to "voiceless" consonants, in

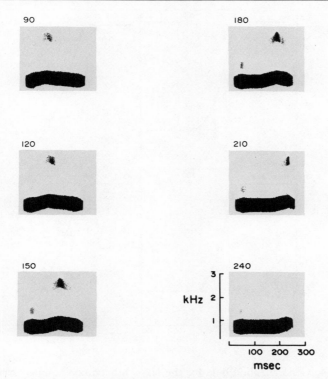

Figure 8.3. Sonagrams of stimuli from the computer-synthesized smooth-early/smooth-late continuum. The number above each sonagram designates, in milliseconds, the position of the peak frequency. (From Aslin, Alberts, and Petersen, 1981a. Reprinted with permission.)

which voicing is either absent or substantially delayed. A plot of the frequency of occurrence of different VOT values reveals distinct peaks in the distribution corresponding to the modal values of the prevoiced, voiced, and voiceless classes, respectively. However, close inspection of such plots shows some overlap between adjacent classes resulting from acoustic variability in the production of sounds.

The acoustic measurements seem at odds, however, with the anecdotal reports of human listeners that speech is composed of discrete, readily identifiable perceptual units, a fact that suggests insensitivity to the variability in the acoustic realization of the phonetic cue. The variability in production is accommodated by perceptual mechanisms that unambiguously sort the sounds into appropriate phonetic categories.

Many researchers have pointed to the high degree of acoustic overlap and continuity between primate calls that seem to have distinct functions (see Marler 1975; Gautier and Gautier 1977; Green and Marler 1979; and Zoloth and Green 1979 for reviews). In fact, Marler (1975) argues that most vocal communication systems can be classified as manifesting either

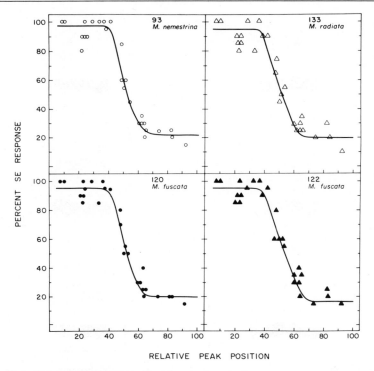

Figure 8.4. Generalization data for the six synthetic coo sounds. Animals were tested with an extinction procedure wherein responses to the generalization signals were not rewarded. (From Aslin, Alberts, and Petersen, 1981a. Reprinted with permission.)

a discrete or graded acoustical organization. The discrete systems are those composed of several acoustically unambiguous categories of vocalizations – because sounds intermediate in acoustic structure between any two categories are rarely produced, each vocalization in such a repertoire can usually be classified without difficulty as belonging to one specific category of sounds. The sounds within a graded communication system are much more difficult to classify because there is considerably more variability in the repertoire. Many vocalizations lie along acoustic continua and individual sounds almost defy assignment to any rigidly defined category. Categories for the graded vocal systems are eventually erected by identifying modal acoustic patterns, and an attempt is made to systematically order the remainder of the repertoire around and between these separate divisions. One can, in general, envision a hierarchy of complexity in the acoustic organization of vocalization systems, ranging from primitive discrete systems, employed by animals whose behavioral ecology and social organization places a premium on specific distinctiveness in their calls, to more complex graded systems, adopted by animals whose social organization and econiche require (or allow) them to use acoustically more variable sounds that have the potential of encoding more information.

The observation that human beings partition continua between speech sounds into perceptual categories raises the question of whether other primate species also make use of a categorical processing strategy. The question is important simply as a test of whether animals and humans make use of similar processing strategies and mechanisms. In addition, the question is relevant to the theory of the evolution of acoustic communication systems which holds that graded communication systems represent an advance over more primitive discrete systems and probably served as precursors to human speech systems (Marler 1975). Since the classification of a species system as graded or discrete is based on a human observer's judgment and not that of the species for whom the signals are biologically significant (Petersen 1978; Snowdon 1979), one is unable to conclude anything about the communicative relevance of subtle variation and intergradation of signals. With the results of comparative perceptual experiments on the categorization of vocal signals from acoustic continua underlying spontaneous vocalizations we will be better able to trace the evolution of communication systems as well as the strategies and the mechanisms responsible for perception of vocal signals.

The empirical test for categorical perception of speech continua consists of two separate experiments conducted with computer-synthesized signals whose acoustic properties can be carefully controlled (e.g., Liberman, Cooper, Shankweiler, and Studdert-Kennedy 1967; Studdert-Kennedy, Liberman, Harris, and Cooper 1970). First, subjects are asked to label individual stimuli from a speech continuum as belonging to one of two or three perceptual categories. The typical labeling function (a plot of the frequency of assignment of each stimulus along the continuum to the different categories) has a quantal form: Subjects report an abrupt, very discrete change from perceiving sounds of one category type to the next with no report of smooth gradations in perception. That is, the continuum is parceled into several distinct categories. Pisoni and Tash (1974) discovered that reaction time is also a reliable indicator of categorical processing in the labeling experiment: Reaction times are longer for stimuli in the vicinity of the category boundary than for stimuli within phonetic categories. In the second experiment subjects are asked to discriminate between individual stimuli drawn from different points on the continuum. The typical finding is that subjects are most accurate when discriminating stimuli from different perceptual categories. Thus, a plot of discrimination performance for pairs of stimuli along the continuum contains performance maxima (peaks) in regions near category boundaries; performance for pairs of stimuli drawn from within a category hovers near chance levels (e.g., Liberman et al. 1967; Studdert-Kennedy et al. 1970). Reaction times measured by Sinnott, Beecher, Moody, and Stebbins (1976) corroborate these findings: They are essentially constant for between-category comparisons but increase dramatically for within-category discriminations.

The categorical mode of perception is in contradistinction to another

188   *Michael R. Petersen*

principal mode of processing known as continuous perception. When a continuum of acoustic stimuli, for example, pure tones, is continuously perceived, one finds it difficult to assign stimuli to fixed or stable categories. Rather, all the stimuli along the continuum seem to blend gradually into one another without abrupt, categorical transitions. Consequently, if one conducts labeling and discrimination tests with signals so perceived, the labeling functions obtained are not quantal, and they show little correlation with discrimination performance.

A review of the categorical perception literature reveals that, as with neural lateralization, different aspects of speech sounds are subject to distinct types of processing. Thus, stop consonants distinguished by place of articulation (Liberman, Harris, Hoffman, and Griffith 1957; Eimas 1963; Mattingly, Liberman, Syrdal, and Halwes 1971; Pisoni 1971) or VOT cues (Lisker and Abramson 1970) are perceived categorically, steady-state vowels presented in isolation are perceived continuously (Fry, Abramson, Eimas, and Liberman 1962; Pisoni 1973, 1975), and the liquids (e.g., /r/ and /l/) show evidence of both categorical and continuous processing (Miyawaki, Strange, Verbrugge, Liberman, Jenkins, and Fujimura 1975). No one has systematically studied the perception of paralinguistic cues with the conventional labeling and discrimination tests. However, because experiments with the intensity and frequency of pure tones demonstrate that they are continuously perceived (Pollack 1952, 1953), it is likely that the corresponding paralinguistic dimensions would evoke similar responses.

Categorical perception was thought to represent a processing mode employed specifically and exclusively for speech sounds (Liberman et al. 1967; Studdert-Kennedy et al. 1970). A revision of this view has been forced by findings that a variety of nonspeech signals are categorically perceived (Cutting and Rosner 1974; Miller, Wier, Pastore, Kelly, and Dooling 1976; Pastore 1976; Pisoni 1977) and that animals, who theoretically should not have access to a speech processing mode, seem to categorically perceive certain phonemes (Kuhl and Miller 1975, 1978; Morse and Snowdon 1975; Waters and Wilson 1976). Speech perception theorists have generally reacted to these findings in one of two ways. Some have suggested that the acoustic information present in the speech signal has been selected, in an evolutionary sense, to produce natural perceptual discontinuities to which the mammalian auditory system is especially sensitive (Stevens 1975; Miller et al. 1976; Kuhl and Miller 1978). On this view, any signal that induces such discontinuities ought to be categorically perceived by most mammals. A corollary of this argument is that the contrasts tested to date are also involved in the natural communication systems of the animals being tested (Petersen 1978; Snowdon 1979). Others have retreated from the position that categorical processing reflects the operation of a speech-specific mode and have instead discovered a whole new collection of perceptual phenomena that they claim are unique to speech (Miller and Liberman 1979; Fitch, Halwes, Erickson, and Liberman 1980; Mann and Repp 1980; see Repp 1982 for a review).

Although categorical perception can no longer be considered a mode of processing unique to speech sounds, it does, nonetheless, represent a characteristic of the manner in which some speech sounds are analyzed. One therefore wonders whether categorical perception is a defining characteristic of the perception of certain types of vocal communication sounds.

Marler (1975) speculates that the evolutionary transition from discrete to graded vocalizations was probably accompanied by a change from purely categorical to continuous perception. That man uses categorical processing mechanisms to perceive his graded system of speech sounds is accounted for by a reverse shift from a continuous to a categorical mode of perception to accomplish the segmentation of speech. The presumed adaptive advantage of graded over discrete systems is that the former permits the transmission of more precise information about the vocalizer's mood and/or probable intentions, and thus allows recipients to plot an appropriate course of behavioral action. To enjoy the selective advantage of the graded repertoire (or any other for that matter), animals must be able to detect and make use of the subtle variations in acoustic structure, hence the need for continuous processing. Alternatively, it is possible that the animals that developed graded repertoires never really abandoned the categorical mode, which was used to perceive discrete signals.

Support for this view comes from the fact that human speech requires both categorical and continuous processing: One mode is responsible for the very brief, highly context-dependent consonants, whereas the other is able to accommodate the steady-state vowels and paralinguistic features (Liberman et al. 1967; Studdert-Kennedy et al. 1970). Thus, man did not forsake the continuous mode in favor of a categorical mode of processing as Marler's (1975) theory seems to imply.

In addition, close examination of published examples of the intergradations between classes of vocalizations in primates reveals that many of the call types also intergrade along multiple dimensions (e.g., Gautier 1974; Green 1975a; Marler 1975; Richman 1976; Struhsaker 1976). For example, Table 8.2 illustrates that the SEH and SLH vocalizations of *M. fuscata* are characterized by continuous variation along the dimensions of peak position and pitch; and the results of our perceptual lateralization experiments with other species suggest that these dimensions undergo different kinds of processing. When this is combined with evidence that those aspects of the speech signal that are categorically perceived are also processed predominantly by the left cerebral hemisphere, and features that are continuously perceived are either processed primarily by the right hemisphere or reveal no hemispheric processing superiority, it leads to an obvious prediction: The *fuscata* and other species with graded repertoires will categorically perceive the primary communicative elements of their vocalizations, like peak, but continuously perceive the paralinguisticlike features, like pitch. Snowdon (chap. 9, this volume) demonstrates an analogous finding with pygmy marmosets. The synthetic version of two trills with different functions are responded to categorically, but animals can still differentiate individual specific versions of calls within a trill class. The "linguistic" variable is responded to categorically, and the "paralinguistic" variable is responded to continuously.

Although we have not conducted studies to explicitly address this ques-

Table 8.2. *Acoustic measurements on SEH and SLH signals*

Stimulus	Relative peak position	Pitch (Hz)
SEH		
386B	0.14	675
350C	0.31	650
1046-3	0.32	700
390D	0.20	550
57D	0.49	675
19	0.18	525
126-14	0.38	750
SLH		
125B	1.00	500
165B	0.61	675
753C	0.81	725
153A	0.77	750
117B	1.00	535
104C	0.80	500
156-2	0.70	500
275S	0.76	700

tion, the results of the generalization tests summarized above are pertinent. The shapes of the generalization gradients for the natural SE and SL signals in Figure 8.2 strongly resemble the quantal form of typical labeling functions obtained for synthetic speech continua. Although we have not yet collected the complementary discrimination data, and therefore have not satisfied all the criteria for concluding that the animals categorically perceive these signals, these generalization data are highly suggestive of such a conclusion.

The results with the synthetic SEH–SLH continuum shown in Figure 8.4 also hint of categorical perception. Note, in particular, the abrupt change in responding between the relative peak positions of 0.50 and 0.63. Interestingly, this corresponds to the boundary between these two call classes reported by Green (1975a) on the basis of his field observations.

Because neither generalization test was constructed or designed to test specifically for categorical versus continuous perception, we simply cannot offer a strong conclusion at this point. I am, however, in the process of conducting a more rigorous study of this question, in collaboration with S. Hopp and J. Sinnott, which is modeled much more closely upon human categorical perception studies. I should have, therefore, a more definitive answer soon.

The only other study to address the nature of the perceptual response to intergradations among primate calls was conducted by Snowdon and Pola (1978). They studied the pygmy marmoset's (*Cebuella pygmaea*) perception of sounds drawn from a synthetic continuum constructed by systematically varying the principal distinction, signal duration, between two marmoset call types: the closed-mouth trill (CMT) and open-mouth trill (OMT). In an earlier study, Pola and Snowdon (1975) observed that occurrence of the CMT reliably elicited antiphonal trill responses from the marmosets. Taking advantage of this fact, Snowdon and Pola (1978) conducted playback experiments with sounds from a synthetic duration continuum and used the antiphonal vocal response as an index of whether a given sound was perceptually equivalent to a CMT. Operationally, the procedure reduced to a labeling task: An antiphonal vocalization labeled the stimulus it followed as a CMT; absence of a vocalization following a stimulus labeled the sound a non-CMT. A plot of the effectiveness of each stimulus in eliciting a vocal response revealed a categorical labeling function: Stimuli from 176 msec (the mean CMT duration) to 249 msec were equally effective in eliciting a response, and stimuli from 257 to 338 msec (the mean OMT duration) were equally ineffective. However, since they were unable to collect discrimination data in their testing situation, they could not show that the pygmy marmosets were using a categorical processing mode per se. According to traditional speech perception literature, one can conclude that a continuum is categorically perceived only when it can be shown that the subject's ability to discriminate stimuli is constrained by his ability to label the stimuli, that is, the discrimination of stimuli from the same category should be more difficult than the discrimination of stimuli of equal physical differences from different categories (Studdert-Kennedy et al. 1970).

One difficulty in concluding anything about categorical perception from the labeling data alone is that the categorical or quantal form of a labeling function is no guarantee that stimuli on either side of the category boundary belong to the same perceptual class. For instance, one might argue that the location of a category boundary simply represents the bisection of the physical or perceptual distance between the two end-point stimuli on the continuum. Other investigators have demonstrated that animals are capable of bisecting intervals from different modalities (Walker 1968; Boakes 1969; Raslear 1975). This explanation argues that stimuli on opposite sides of the boundary are responded to similarly simply because they lie on opposite sides of the bisection point and not because they belong to the same perceptual set. One could easily distinguish between these possibilities by changing the end-point stimuli by some fixed amount, for example, in the Snowdon and Pola (1978) study by shortening one end-point stimulus by 25 msec and then conducting another labeling test (cf. Wright and Cummings 1971). The bisection account predicts that the putative category boundary will shift by some amount in order to bisect the new interval, whereas the categorical perception account predicts no shift because stimuli are responded to according to perceptual similarity, irrespective of the location of the end-point stimuli. In this regard, the special advantage of the conventional discrimination task is that it differentiates between the two explanations while simultaneously satisfying the second criterion, namely, derivation of a discrimination function, for a test of categorical perception.

To summarize, there are parallels between human speech and primate vocalizations with respect to the degree of interclass acoustic variability. Labeling and discrimination tests with human speech suggest that some of this continuous variability is normalized by categorical perceptual responses and some is not. Data from our labortory and from Snowdon and Pola (1978) show that *fuscata* and pygmy marmosets, respectively, also respond categorically to acoustically graded vocalizations. However, neither study provided enough evidence to meet all the criteria used to differentiate continuous from categorical perception. The need for additional studies of animals with discrete and graded vocal repertoires is clear. In addition, we need more data on the perceptual modes used to process the different aspects of primate signals. For instance, it will be important to obtain information about how *fuscata* processes continuous variations in the pitch and/or amplitude of its calls to determine whether there is a link between the pattern of processing seen in our laterality studies for such indexical dimensions and that seen in labeling–discrimination tasks of the sort described earlier.

Perceptual constancy (selective attention)

A fundamental property of the speech perception apparatus is its ability to recognize the phonetic identity of different versions of the same phoneme despite marked variation in acoustic features not relevant to the phonetic identity of the utterance. When the same phonetic segment is embedded in different contexts as when uttered by speakers of different sex, age, or vocal-tract anatomy, its basic acoustic characteristics are altered dramatically. Nonetheless, we perceive a phonetic /p/ as a /p/ in the words pit or tip. The phenomenon of phonetic constancy parallels the ability of the visual system to extract constancies despite minor perturbations in features like orientation, distance, size, luminance, or contrast. For example, we perceive a line as a line whether it is oriented vertically or horizontally, whether it is 1 mm or 1 m long, and whether it is placed against a white or a blue background. Consequently, psycholinguists have adopted the term "perceptual constancy" (Pisoni 1979) from visual system scientists to describe the process of normalizing acoustic variance into a single phonetic percept.

Viewed in its simplest form, perceptual constancy consists of selective attention to the percept(s) produced by some dimension(s) while ignoring irrelevant variations in other aspects of the signal. In this connection, when Kuhl (1976) tested human infants between 1 and 4 months of age, she found that they had little difficulty learning to attend to a phonetic dimension with irrelevant variation of a nonphonetic feature but that they could not selectively attend to a nonphonetic target dimension in the face of variations of a phonetic feature. One interpretation of these results is that the phonetic feature is a much more salient dimension, presumably because of its communicative importance, than the nonphonetic feature. Thus, the salient phonetic feature captures the child's attention in both versions of the constancy test, facilitating performance when the task requires attention to the salient dimension but disrupting performance when the child is required to ignore the salient phonetic feature in favor of a less conspicuous parameter (see Carrell, Smith and Pisoni 1981 for a critical discussion). Fodor, Garrett, and Brill (1975) obtained similar results in a study of perceptual constancy for stop consonants.

The diversity in basic anatomy, body morphology, affective demeanor, sex, and age among members of a primate troop is reflected in considerable variability in the acoustic structure of different versions of functionally identical utterances produced by different animals. Therefore, nonhuman primates are also confronted with the chore of extracting a constant percept from a set of signals that contain variation in acoustic features orthogonal to the primary communicative cues present in the signal. Do they make use of a processing strategy similar to that used by humans to accomplish this task?

The perceptual-constancy paradigm is well suited to the study of whether primates are particularly sensitive to those acoustic elements of conspecific vocalizations that distinguish various classes of functionally different sounds within the repertoire. Using the information provided by field biologists on the acoustic cues that are *likely* to carry the central message and those that function as indexical features, one can conduct experiments, employing call variants from two (putatively) functionally distinct vocal categories, in which conspecifics are required to attend (1) to communicatively prominent acoustic dimensions that distinguish the two categories or (2) to some orthogonally varying indexical dimension. In analogy to the human-constancy experiments, when the target dimension is the central message feature and the orthogonal element is indexical, discrimination performance should be better than when the subjects must attend to the orthogonal dimension and ignore the message feature. This approach would thus permit one to determine whether animals make use of the saliency hierarchies for analyzing communication sounds that have been hypothesized for human speech.

We have conducted just such an experiment with the peak and pitch dimensions of the *M. fuscata* SEH and SLH calls (Petersen 1978; Zoloth, Petersen, Beecher, Green, Marler, Moody, and Stebbins 1979; Petersen, Beecher, Zoloth, Moody, Stebbins, Green, and Marler in preparation b). The laterality data reported earlier for performance on peak and pitch discriminations were obtained serendipitously in the course of perceptual-constancy experiments. Recall that four subjects were trained on two different tasks: One required attention to the peak cue in the face of orthogonal variation in pitch and the second demanded attention to pitch while peak values varied. The 15 stimuli used in the peak task are shown in Figure 8.1. Initially the subjects were trained to discriminate the 2 stimuli in the first row; as they met performance criteria (see Zoloth et al. 1979 and Petersen et al. in preparation b. for details), additional stimuli were added to the stimulus set until the animals eventually reached the final stage of discriminating 8 SLHs from 7 SEHs. For the pitch task, 12 of these same signals were used to form high-pitch and low-pitch categories, each of which contained some SEHs *and* SLHs. Of principal concern was the rate

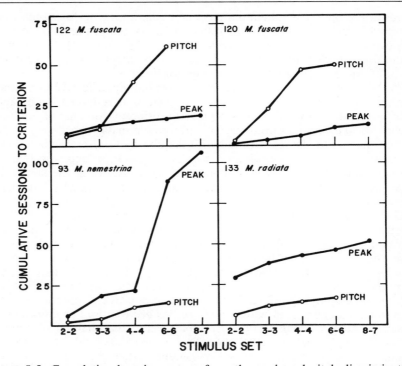

Figure 8.5. Cumulative learning curves from the peak and pitch discrimination tasks for four subjects. Each point represents the cumulative number of sessions necessary to meet the performance criteria and to advance beyond that stage of the problem. The left and right numbers in each dyad along the abscissa indicate the number of SLH–SEH, or low-pitch/high-pitch, stimuli, among which the subjects were required to discriminate at that stage. The subjects on the left were tested on the pitch task first; those on the right experienced the peak task first.

of acquisition of the two discriminations. To test for possible species-specificity in selective attention to the different cues, two of the four subjects were non-*fuscata* comparison animals. The order of testing on the two tasks was counterbalanced across individual subjects and species groups. The results are shown in Figure 8.5 in the form of cumulative learning curves. Both *fuscata* subjects acquired the peak discrimination very quickly, as one would predict. However, they had serious difficulties with the pitch task. The comparison animals showed just the reverse. They encountered little difficulty with the pitch task but acquired the peak discrimination very slowly.

Our interpretation of these findings is that the *fuscata* have a perceptual predisposition to attend to the peak feature of the SEH and SLH calls, whereas the comparison animals do not. Consequently, when performing the peak discrimination *fuscata* had no trouble because they selectively attended to the appropriate cue. However, such a strategy was unproductive for the pitch discrimination: By continuing to attend to peak, they would be expected to show the sort of retarded acquisition that in fact occurred. The comparison animals, on the other hand, came to the experiment with no predispositions for processing the *fuscata* calls (although they undoubtedly were attuned to the critical properties of their own vocalizations) and probably attended to the most salient acoustic feature of the vocalizations. As pitch is a prominent feature of these signals, it is quite conceivable that the comparison animals' attention was arrested by it. Consequently, they performed well on the pitch task but not on the peak discrimination. Incidentally, this account predicts that if the tables were turned so that the comparison animals performed analogous tasks with dimensions of their own calls, they would show the pattern obtained with *fuscata* and vice versa.

These results parallel the findings with human infants (Fodor, Garrett, and Brill 1975; Kuhl 1976). To my knowledge, no one has reported a similar effect with any other primate species. It is likely, however, to be a rather ubiquitous phenomenon with regard to the perception of communication signals. The birdsong literature has many examples of species-specificity in song learning, an important component of which is early perceptual exposure to the species-typical song. For example, Marler and his colleagues (e.g., Marler 1970; Marler and Peters 1977) have discovered that various songbirds are very discriminating about the song they will learn. Depending on the social and ecological pressures (Kroodsma 1978), many refuse to learn anything that deviates substantially from the typical conspecific pattern (e.g., white-crowned sparrows, swamp sparrows), whereas others (e.g., song sparrows) accept as models songs that bear only a superficial resemblance to the species-typical utterance. When Dooling and Searcy (1980) recently extended this work to observations of song and swamp sparrow hatchlings, they found that perceptual selectivity seems to operate very early in life. It will also be important to determine how the attentional selectivity we observed in the *fuscata* develops. Is it largely a genetic trait or is there considerable plasticity in the features to which the young *fuscata* is able to learn to attend? (This prospect is discussed more extensively by Petersen (1981).) Studies of normative patterns of development, the impact of perceptual deprivation, and the effects of cross-species fostering will undoubtedly provide important insights into questions of this sort (Aslin, Alberts, and Petersen 1981a,b).

An important problem facing the field biologist is how to differentiate those aspects of conspecific sounds that represent primary communicative cues from those approximating paralinguistic features (see Petersen 1978 and Snowdon, chap. 9, this volume). Since intergradations between various calls usually involve variations in several acoustic dimensions at once, the discrimination between these two classes of features may not be possible with a strictly field-oriented descriptive approach (however, see Seyfarth and Cheney, chap. 10, this volume). The perceptual constancy task could prove useful as such a differentiator. When contrasting a message cue with an indexical cue (as in the *fuscata* experiments), the message cue should receive perferential attention. Thus the procedure can be used as a tool for identifying those specific acoustic features in a complex, multidimensional vocalization that are of communicative relevance from those that are indexical. Such information would then be useful for evaluating the accuracy of a field researcher's deductions about the gradedness or discreteness of a particular species repertoire by permitting one to concentrate on the "linguistically" important features without being distracted by the "paralinguistic" features.

Interrelations among different processing levels

The information processing account of speech perception holds that speech segments are subjected to processing at several different levels (e.g., Studdert-Kennedy 1976; Pisoni 1978). The major thrust of recent research has been to characterize the auditory and phonetic levels of processing and the interrelations between them (Day and Wood 1972a, b; Studdert-Kennedy, Shankweiler, and Pisoni 1972; Pisoni and Tash 1974; Wood 1974, 1975; Wood and Day 1975). The auditory processing level is responsible for analyzing the speech waveform into a set of neurally coded acoustic parameters that are then transformed, through abstract processes, into features at the phonetic level. Linguistic and nonlinguistic sounds alike are thought to pass through the auditory stage of analysis, whereas the phonetic stage is reserved exclusively for linguistic material. Until recently, justification of the need for the auditory-phonetic distinction was based largely on indirect evidence that phoneme perception was unique and distinct from the perception of other acoustic stimuli. The two sets of findings most frequently cited in support of this distinction were: (1) the tendency for speech sounds to be categorically, rather than continuously, perceived; and (2) results from behavioral studies of cerebral dominance showing that the left hemisphere is specialized for processing speech and not simple acoustic stimuli. We have already reviewed some of the recent work that suggests these phenomena are also characteristic of the perception of certain complex nonspeech sounds. Such findings have been interpreted as undermining the validity of the distinction between auditory and phonetic analysis stages (Pastore 1976; Cutting 1979). The fact remains, however, that certain kinds of complex acoustic signals, notably those that mimic the acoustic complex of cues that correspond to distinctive features (e.g., Cutting and Rosner 1974; Blechner 1976; Miller et al. 1976; Pisoni 1977) seem to enlist a different mode of processing than do either simple or even certain complex acoustic stimuli. Thus, the existing evidence seems to support indirectly a hierarchical organization of processing levels, but perhaps the nominal second stage should be termed "phoneticlike" rather than phonetic.

Recent attempts (Wood 1975; Carrell, Smith, and Pisoni 1981) to more directly explore the nature and interrelation of the auditory and phonetic-like levels of perceptual analysis have used a two-response speeded-classification reaction-time paradigm developed by Garner and his colleagues (Garner and Morton 1969; Garner 1970, 1972, 1974; Garner and Felfoldy 1970). After each stimulus presentation, subjects are required to report which of two values of a target dimension was present in the stimulus. Changes in decision time as a function of various modifications of the test stimulus parameters are interpreted as reflecting a change in the time required to process the stimulus and to reach a response decision. To study the interdependencies among acoustic and phoneticlike levels of processing, listeners are required to perform the binary classification task for stimuli varying along either an acoustic or a phonetic dimension (or both) under several different conditions. The typical finding with consonants is that there is asymmetric interference (measured as an increase in decision time) between the auditory and phonetic dimensions of the same speech stimulus. Imagine a set of four stimuli: /ba/ with a high fundamental frequency (F_0), /ba/–low F_0, /pa/–high F_0, and /pa/–low F_0. When required to classify stimuli from this set, F_0 variation interferes with phonetic classification, but phonetic variation does not interfere with F_0 classifications. Thus phonetic decisions for consonants appear to depend on prior analysis at some acoustic level but not the reverse.

When the same experiment is conducted by contrasting vowels with pitch, one usually finds mutual, symmetric interference effects (Miller 1977), suggesting that vowels and pitch are processed primarily at the acoustic stage. However, a recent study by Carrell, Smith, and Pisoni (1981) found that although pitch does not interfere as much with the processing of vowels as with consonants, it does nonetheless produce some interference. Furthermore, the interference obtained under such conditions is asymmetric: Pitch interferes more with vowel processing than vowels do with the processing of pitch. This suggests then that vowels undergo additional processing at a level separate from the acoustic stage responsible for pitch processing. However, vowels and consonants show different magnitudes of interference, adding yet another item to the list of distinctions between them. When vowels and consonants are contrasted in such studies, they show mutual and symmetric interference with one another (Wood and Day 1975), suggesting that although they are processed somewhat differently, both nonetheless pass through a phoneticlike stage of analysis.

Finally, when simple paralinguistic dimensions like pitch and intensity are pitted against one another in such tasks, they consistently show mutual, symmetric interference effects, suggesting that both are processed primarily at the acoustic stage.

Other manipulations of the acoustic and phoneticlike dimensions permit inferences about whether the processing stages are organized sequentially or in parallel. In one such task subjects are asked to classify signals in which the two dimensions are redundant (e.g., for the stimuli /ba/–high F_0 and /pa/–low F_0, a response decision could be made by attending to either one of the dimensions). The critical question is whether the redundant information yields decision times any different from a condition in which the dimensions are not redundant (e.g., /ba/–high F_0 vs. /pa/–high F_0 or /ba/–high F_0 vs. /ba/–low F_0). A simple serial processing model predicts no difference in decision time between these conditions, since the processing of the acoustic dimension should precede phonetic pro-

cessing under both conditions. A parallel-processing model predicts a
shorter decision time under the redundant condition. This prediction derives from an argument that the mean of the decision-time distribution for
cases in which the components of a binary-valued stimulus are competing
to reach a decision is lower than the means of the distributions for either
individual, noncompeting component alone (Biederman and Checkosky
1970; Lockhead 1972). Wood (1974) obtained a redundancy gain in just
such an experiment with consonants, thus implying that the phonetic and
acoustic features of speech are processed in parallel.

In keeping with trends prevalent in investigations of categorical perception and neural lateralization, two recent demonstrations that certain nonphonetic stimuli produce asymmetric interference and redundancy gain
effects (Blechner, Day, and Cutting 1976, using rise time and intensity as
competing dimensions, and Pastore, Ahroon, Puleo, Crimmins, Golowner, and Berger 1976, using a tone pip and noise buzz) provide additional evidence that the phonetic level does not subserve analysis of phonetic features exclusively. However, since both papers admit that the
signals employed contained acoustic cues resembling those characteristic
of certain classes of phonemes, simply substituting phoneticlike level (as
we have done) for phonetic level should meet this challenge.

In combination with the field biologist's best estimate of which acoustic
features distinguish functionally different classes of vocalizations and the
results of studies of perceptual constancy and categorical perception,
these paradigms should enable one to (1) identify distinct stages in the
processing of conspecific communication sounds (2) determine whether
the stages are organized sequentially or in parallel and (3) specify the nature of the interdependencies and interactions among stages. It is also
noteworthy that asymmetric interference effects and redundancy gains in
human information-processing tasks occur generally for dimensions that
are categorically perceived and that mutual interference with no redundancy gains obtain for continuously perceived dimensions (Wood 1974;
Blechner, Day, and Cutting 1976). One would predict a similar relationship for the continuously and categorically perceived aspects of animal
vocalizations.

In analogy to the human speeded-classification reaction-time task, two
components of primate communication sounds would be of interest: (1)
the acoustic feature (message dimension), which differentiates two functionally distinct vocalization types, and (2) an acoustic feature that is
functionally independent (indexical) of the cue differentiating the two vocalization types. To ensure that the subjects can use information only
from the specific dimensions under study, one should employ synthesized
vocalizations, which are identical in all respects except for the values of
the dimensions of interest. For instance, in the case of the SEH–SLH distinction from the *fuscata* repertoire, one would synthesize a set of four
stimuli varying in peak position and fundamental frequency F_0: a SEH-
high F_0, SEH-low F_0, SLH-high F_0, and SLH-low F_0.

The results of speeded-classification tests of this type also provide information relevant to the distinction between "holistic" and "differen-

tiated'' modes of perception – a topic that is the focus of much research in human experimental psychology (Garner 1974; Smith and Kemler 1978). Holistic perception refers to the observation that the individual dimensions of certain multidimensional stimuli are unanalyzable, that is, all dimensions are seen as forming a unitary whole and cannot be pulled apart. Examples of such unanalyzable, or integral, dimensions that are holistically perceived are saturation and brightness, pitch and loudness, and vowels and consonants within a syllable (Garner and Felfoldy 1970; Wood and Day 1975; Smith and Kemler 1978; see also Garner 1974). Differentiated perception is precisely the opposite of holistic perception and accounts for reports that certain dimensions are analyzable or separable, that is, all dimensions of a stimulus can be easily pulled apart. Examples of separable dimensions that are perceived in a differentiated way are color and geometric form, size and brightness, and size of a circle and orientation of a line within it (Garner and Felfoldy 1970; Garner 1977).

The notion of holistic and differentiated processing is especially relevant to comparative studies of vocal perception because the differences in signal processing observed for different species may simply reflect use of holistic perception by some species and differentiated perception by others. For example, cross-cultural comparisons of speech perception have shown clear differences between the perceptual performances of listeners familiar with a particular phonetic contrast and that of foreign listeners: native listeners have little difficulty labeling and discriminating sounds from their language, whereas foreign listeners unfamiliar with the specific linguistic contrasts have great difficulty (Howie 1972; Zue 1976; Miyawaki, Strange, Verbrugge, Liberman, Jenkins, and Fujimura 1975). Is it possible that these differences in performance simply reflect the differential use of holistic and differentiated processing modes by the native and foreign listeners? It is conceivable, for instance, that foreign listeners apply a differentiated processing mode that results in a set of separated dimensions that confuse the listener who is then unable to select the appropriate linguistic dimension from those available. The native listener might, in contrast, perceive the sounds holistically and is therefore not distracted by a variety of irrelevant, separated dimensions. (Another possibility, of course, is that both listeners apply the same strategy but that the native listener has learned, in the course of language acquisition, which are relevant dimensions.)

Similarly, recall the clear performance differences between *fuscata* and the comparison animals in neural lateralization and perceptual constancy. Is it possible that these species-specific differences reflect differential use of holistic versus differentiated processing of these signals? Further, is it possible that holistic and differentiated perception account for differences in the pattern of perceptual results obtained among species listening to different aspects of their own sounds? Garner (1974) has specified a set of converging operations that allow discrimination between these two modes of processing: holistic processing is characterized by redundancy gains when dimensions provide redundant information and interference effects in speeded classification tests requiring selective attention. In differentiated perception there is no interference between selectively attended dimensions. Thus, through appropriate experimentation, it will be possible to determine which processing mode different species adopt when analyzing complex vocal signals produced by their own species as well as by others.

Short-term memory processes and codes

The role of memorial processes in the perception of speech signals has actually received only a moderate amount of empirical attention. Pisoni (1973) performed a study to assess the role of auditory and phonetic short-term store (STS) in the perception of consonants, which are perceived categorically (e.g., Pisoni 1971), and vowels, which (under some conditions) are perceived continuously (Fry et al. 1962). Subjects were first asked to label stimuli from synthetic vowel and consonant continua to obtain baseline labeling functions. Pisoni (1973) was then able to force a differential use of auditory and phonetic memory processes in a delayed comparison task by selecting pairs of stimuli to be discriminated from either "between" phonetic categories (recruiting primarily phonetic memory) or "within" phonetic categories (activating primariy acoustic memory). The delayed comparison task consisted of presenting a pair of stimuli that were separated by a comparison interval and asking subjects to decide whether the stimuli were the same or different. By varying the length of the comparison interval, it was possible to obtain a memory-decay (forgetting) function from auditory STS and phonetic STS for both consonants and vowels. The experiments clearly showed that auditory STS for vowels differs substantially from that for consonants: Whereas acoustic information from vowels seems to be retained in a STS for about 2 sec, the acoustic information needed to discriminate two physically different but phonetically identical consonants is not retained. In contrast, the phonetic information from both vowels and consonants is retained by, and accessible from, the phonetic STS. These findings led Pisoni (1973) to suggest that the differential role of auditory STS in the discrimination of vowels and consonants is responsible for the continuous perception of vowels and the categorical perception of consonants. Thus, differences in categorical and continuous modes of processing may reduce to differences in the use of auditory versus phonetic memory codes. Whether this explanation accounts for the perception of paralinguistic features is an open but important question. Although no studies directly comparable to Pisoni's have been conducted with paralinguistic cues, in a related experiment Springer (1973b) obtained forgetting functions for the pitch dimension of a phoneme that were as good as those obtained by Pisoni (1973) for short-term acoustic memory of vowels. Her procedures, however, were rather different from Pisoni's, so there is a real need for a study more like his before the issue is closed.

We intend to conduct analogous tests with *M. fuscata* to characterize the role of different kinds of STS in processing the different dimensions of their vocalizations. In addition, memory tests of this sort could determine the source of any perceptual differences that might obtain in cross-species perceptual comparisons. Can the species-specificities in perception obtained in our neural-lateralization and perceptual-constancy studies be traced to the use of different types of memory codes by the two groups? To conduct such experiments I will first need to obtain data on categorical versus continuous processing of the SEH and SLH signals. With those data I can perform the memory experiments by simply introducing different delays between pairs of synthetic stimuli drawn from within and across categories while animals are performing a discrimination task.

SUMMARY

The major points of the preceding sections are summarize in Table 8.3. Shown here are typical findings from the five classes of perceptual phenomena for three aspects – stop consonants, vowels, and paralinguistic features – and the two dimensions of the *fuscata* SEH and SLH calls we have studied thus far – peak position and pitch.

The intent of this survey was to outline a potential framework for the study of vocal-signal perception in nonhuman primates and to illustrate the success obtained by using such an approach in the study of the perception of vocalizations by *fuscata*. Within the context of the five perceptual phenomena shown in Table 8.3, vowels, consonants, and paralinguistic features exhibit distinctly different perceptual profiles. Thus, stop consonants are characterized by (1) left-hemisphere dominance, (2) categorical perception, (3) selective attention, (4) asymmetric interference and redundancy gain effects, and (5) poor memory for acoustic details but good memory for phonetic features. Although vowels are like consonants with respect to (3) and (4), their processing is distinguishable from that of consonants by being generally continuously perceived, by not being reliably lateralized to the left hemisphere, and by the retention of both acoustic and phonetic features for relatively long periods in short-term memory. The processing of paralinguistic features can be contrasted similarly with either vowel or consonant perception. Can this information be used to uncover analogous processes in nonhuman primate vocal communication and, perhaps, to distinguish indirectly among different classes of information in their vocalizations? The success with *fuscata* in this regard has been encouraging and bodes well for additional studies. The acoustic feature coding the central message of a coo vocalization (peak position) evoked selective attention and left-hemispheric lateralization, whereas an indexical feature (pitch) did not attract selective attention and was either lateralized to the right hemisphere or not lateralized at all. Furthermore, there is suggestive, but not definitive, evidence that the peak dimension is categorically perceived. The peak dimension appears to be processed in a manner similar to that of consonants, whereas the pitch dimension is analyzed by processes analogous to those responsible for the perception of paralinguistic dimensions. However, it is important to recognize that although vowels always seem to evoke selective attention, under certain conditions (e.g., when they are rapidly articulated; Pisoni 1971, 1973) they

Table 8.3. *Comparison of five classes of perceptual phenomena for three aspects of human speech and two dimensions of* M. fuscata *SEH and SLH calls*

	Human			M. fuscata	
	Stop consonants	Vowels	Paralinguistics	Peak	Pitch
Neural lateralization Categorical vs. continuous processing	REA Categorical	NEA or REA Continuous or categorical	LEA or NEA Continuous	REA ?(Probably categorical)	LEA or NEA ?
Perceptual constancy (selective attention)	Vary pitch: selective attention to stop	Vary pitch: selective attention to vowel	Vary vowel or consonant: poor attention to pitch	Vary pitch: selective attention to peak	Vary peak: poor attention to pitch
Interference effects and redundancy gains	Pitch vs. stop: asymmetric Stop vs. vowel: mutual Pitch vs. stop: redundancy gain	Pitch vs. vowel: asymmetric Stop vs. vowel: mutual ?	Pitch vs. intensity: mutual ?	?	?
Short-term memory Acoustic Phoneticlike	Poor Good	Good Good	Good ?	? ?	? ?

are lateralized to the left hemisphere and are categorically perceived. Using these three criteria then, vowels and peak position both sometimes appear to undergo consonantlike processing. However, Pisoni (1973) showed that when vowels that are categorically perceived serve as stimuli in memory tests, the pattern of results is more similar to that obtained with continuously perceived vowels than it is to that seen with consonants. Thus, memory tests will probably be necessary before we can unequivocally conclude anything about the processing parallels between consonants, vowels, and peak position.

The conceptual framework described here does appear to be a viable approach to characterizing the perception of primate vocalizations. Although the work is presently in its infancy, I am encouraged by the power of the approach and intend to continue studies with *fuscata* by completing the analysis of the perceptual strategies they employ with the SEH and SLH vocalizations. Given the success to date with the psycholinguistic approach advocated here, prospects for future comparative work are also excellent. By contrasting species differing in such characteristics as eco-niche, social structure and organization, and vocal behavior, I hope to trace a portion of the evolution of those perceptual capacities used in the analysis of vocal communication signals.

REFERENCES

Aslin, R., Alberts, J., and Petersen, M., eds. 1981a. *Development of perception: psychobiological perspectives*, vol. 1: *The auditory, somatosensory and chemosensory systems*. New York: Academic Press.
1981b. *Development of perception: psychobiological perspectives*, vol. 2: *The visual system*. New York: Academic Press.
Beecher, M., Beecher, I., and Nichols, S. 1981. Parent offspring recognition in bank swallows. II. Development and acoustic basis. *Animal Behaviour* 29: 95–101.
Beer, C. 1976. Some complexities in the communication behavior of gulls. *Annals of the New York Academy of Sciences* 280:413–32.
Benton, A. 1965. The problem of cerebral dominance. *Canadian Psychologist* 6a:332–48.
Bever, T., and Chiarello, R. 1974. Cerebral dominance in musicians and nonmusicians. *Science* 185:537–9.
Biederman, I., and Checkosky, S. 1970. Processing redundant information. *Journal of Experimental Psychology* 83:486–90.
Blechner, M. 1976. Right-ear advantage for musical stimuli differing in rise time. *Haskins Laboratories Status Report on Speech Research* 47:63–70.
Blechner, M., Day, R., and Cutting, J. 1976. Processing two dimensions of non-speech stimuli: the auditory-phonetic distinction reconsidered. *Journal of Experimental Psychology: Human Perception and Performance* 2:257–66.
Blough, D. 1977. Visual search in the pigeon: hunt and peck method. *Science* 196:1013–14.
1979. Effects of number and form of stimuli on visual search in the pigeon. *Journal of Experimental Psychology: Animal Behavior Processes* 5:211–23.

Blumstein, S., and Cooper, W. 1974. Hemispheric processing of intonation contours. *Cortex* 10:146–58.

Boakes, R. 1969. The bisection of a brightness interval by pigeons. *Journal of the Experimental Analysis of Behavior* 12:201–9.

Bradshaw, J., and Nettleton, N. 1981. Pre-nature of hemispheric specialization in man. *Behavioral and Brain Sciences* 4:51–91.

Butler, C., and Francis, A. 1973. Specialization of the left hemisphere in baboon: evidence from directional preferences. *Neuropsychologia* 11:351–4.

Carmon, A., and Nachson, I. 1973. Ear asymmetry in perception of emotional nonverbal stimuli. *Acta Psychologica* 37:351–7.

Carrell, T., Smith, L., and Pisoni, D. 1981. Some perceptual dependencies in speeded classification of vowel color and pitch. *Perception & Psychophysics* 29:1–10.

Catlin, S., Van Derveer, N., and Teicher, R. 1976. Monaural right-ear advantage in a target identification task. *Brain & Language* 3:470–81.

Cohn, R. 1971. Differential cerebral processing of noise and verbal stimuli. *Science* 172:599–601.

Collins, R. 1977. Toward an acceptable model for the inheritance of the degree and direction of asymmetry. In *Lateralization of the nervous system,* ed. S. Harnad, R. Doty, L. Goldstein, J. Jaynes, G. Krauthamer, pp. 137–50. New York: Academic Press.

Cowey, A., and Dewson, J. 1972. Effects of unilateral ablation of superior temporal cortex on auditory sequence discrimination in *Macaca mulatta. Neuropsychologia* 10:279–89.

Cutting, J. 1974. Two left-hemisphere mechanisms in speech perception. *Perception & Psychophysics* 16:610–12.

1979. There may be nothing peculiar to perceiving in a speech mode. In *Attention and Performance VII,* ed. J. Requin, pp. 229–43. Hillsdale, NJ: Erlbaum.

Cutting, J., and Rosner, B. 1974. Categories and boundaries in speech and music. *Perception & Psychophysics* 16:564–70.

Darwin, C. 1969. Auditory perception and cerebral dominance. Doctoral dissertation, University of Cambridge.

1971. Early differences in recall of fricatives and vowels. *Quarterly Journal of Experimental Psychology* 23:46–62.

Day, R., and Wood, C. 1972a. Interactions between linguistic and nonlinguistic processing. *Journal of the Acoustical Society of America* 51(A):79.

1972b. Mutual interference between two linguistic dimensions of the same stimuli. *Journal of the Acoustical Society of America* 52(A):175.

Denenberg, V. 1981. Hemispheric laterality in animals and the effects of early experience. *Behavioral and Brain Sciences* 4:1–49.

Denenberg, V., Garbanati, J., Sherman, G., Yutzey, D., and Kaplan, R. 1978. Infantile stimulation induces brain lateralization in rats. *Science* 201:1150–1.

Desmedt, J., ed. 1977. *Language and hemispheric specialization in man: cerebral ERPs.* Basel: Karger.

Dewson, J. 1977. Preliminary evidence of hemispheric asymmetry of auditory function in monkeys. In *Lateralization in the nervous system,* ed. S. Harnad, R. Doty, L. Goldstein, J. Jaynes, and G. Krauthamer, pp. 63–74. New York: Academic Press.

Dooling, R., and Searcy, M. A. 1980. Early perceptual selectivity in the swamp sparrow. *Developmental Psychobiology* 13:499–506.

Eimas, P. 1963. The relation between identification and discrimination along speech and nonspeech continua. *Language and Speech* 6:206–17.

Falk, D. 1978. Cerebral asymmetry in Old World monkeys. *Acta Anatomica* 101:334–9.

Fitch, H., Halwes, T., Erickson, D., and Liberman, A. 1980. Perceptual equivalence of two acoustic cues for stop-consonant manner. *Perception & Psychophysics* 27:343–50.

Fodor, J., Garrett, M., and Brill, S. 1975. The perception of speech sounds by prelinguistic infants. *Perception & Psychophysics* 18:74–8.

Fry, D., Abramson, A., Eimas, P., and Liberman, A. 1962. The identification and discrimination of synthetic vowels. *Language and Speech* 5:171–89.

Galaburda, A., LeMay, M., Kemper, T., and Geschwind, N. 1978. Right-left asymmetries in the brain. *Science* 199:852–6.

Garner, W. 1970. The stimulus in information processing. *American Psychologist* 25:350–8.

1972. Information integration and form of encoding. In *Coding processes in human memory*, ed. A. Melton and E. Martin, pp. 261–81. Washington, D.C.: Winston.

1974. *The processing of information and structure.* Hillsdale, NJ: Erlbaum.

1977. The effect of absolute size on the separability of dimensions of size and brightness. *Bulletin of the Psychonomic Society* 9:380–382.

Garner, W., and Felfoldy, G. 1970. Integrality of stimulus dimensions in various types of information processing. *Cognitive Psychology* 1:225–41.

Garner, W., and Morton, J. 1969. Perceptual independence: definitions, models, and experimental paradigms. *Psychological Bulletin* 72:233–59.

Gautier, J.-P. 1974. Field and laboratory studies of talapoin monkeys (*Miopithecus talapoin*). *Behaviour* 51:209–73.

Gautier, J.-P., and Gautier, A. 1977. Communication in Old World monkeys. In *How animals communicate*, ed. T. Sebeok, pp. 890–964. Bloomington: Indiana University Press.

Gazzaniga, M. 1963. Effects of commisurotomy on a preoperatively learned visual discrimination. *Experimental Neurology* 8:14–19.

Geschwind, N. 1970. The organization of language and the brain. *Science* 170:940–4.

Geschwind, N., and Levitsky, W. 1968. Human brain: left-right asymmetries in temporal speech region. *Science* 161:186–7.

Glick, S. Meibach, R., Cox, R., and Maayani, S. 1979. Multiple and interrelated functional asymmetries in rat brain. *Life Sciences* 25:395–400.

Gordon, H. 1975. Hemispheric asymmetry and musical performance. *Science* 189:68–9.

Green, S. 1975a. Variation of vocal pattern with social situation in the Japanese monkey (*Macaca fuscata*): a field study. In *Primate behavior: Vol 4, Developments in field and laboratory research*, ed. L. Rosenblum, pp. 1–102. New York: Academic Press.

1975b. Dialects in Japanese monkeys: vocal learning and cultural transmission of locale-specific behavior? *Zeitschrift für Tierpsychologie* 38:304–14.

1981. Sex differences and age gradations in vocalizations of Japanese and lion-tailed monkeys. *American Zoologist* 21:165–83.

Green, S., and Marler, P. 1979. The analysis of animal communication. In *Handbook of behavioral neurobiology*, Vol. 3: *Social behavior and communication*, ed. P. Marler and J. G. Vandenbergh, pp. 73–158. New York: Plenum.

Haber, R. 1969. *Information-processing approaches to visual perception.* New York: Holt, Rinehart and Winston.

Haggard, M. 1971. Encoding and the REA for speech signals. *Quarterly Journal of Experimental Psychology* 23:34–45.

Haggard, M., and Parkinson, A. 1971. Stimulus and task factors as determinants of ear advantages. *Quarterly Journal of Experimental Psychology* 234: 168–77.

Hall, J., and Goldstein, M. 1968. Representation of binaural stimuli by single units in primary auditory cortex of unanesthetized cats. *Journal of the Acoustical Society of America* 43:456–61.

Halperin, Y., Nachson, I., and Carmon, A. 1973. Shift of ear superiority in dichotic listening to temporally patterned nonverbal stimuli. *Journal of the Acoustical Society of America* 53:46–50.

Hamilton, C. 1977. Investigations of perceptual and mnemonic lateralization in monkeys. In *Lateralization in the nervous system*, ed. S. Harnad, R. Doty, L. Goldstein, J. Jaynes, and G. Krauthamer, pp. 45–62. New York: Academic Press.

Hockett, C. 1960. Logical considerations in the study of animal communication. In *Animal sounds and communication*, ed. W. Lanyon and W. Tavolga, pp. 392–430. Washington, D.C.: American Institute of Biological Sciences.

Howie, J. 1972. Some experiments on the perception of Mandarin tones. In *Proceedings of the seventh international congress of phonetic sciences*, ed. A. Riganlt and R. Charbonneau, pp. 900–4. The Hague: Mouton.

Hulse, S., Fowler, H., and Honig, W., eds. 1978. *Cognitive processes in animal behavior*. Hillsdale, N.J.: Erlbaum.

Kallman, H. 1977. Ear asymmetries with monaurally-presented sounds. *Neuropsychologia* 15:833–5.

Kimura, D. 1961a. Some effects of temporal lobe damage on auditory perception. *Canadian Journal of Psychology* 15:156–65.

1961b. Cerebral dominance and the perception of verbal stimuli. *Canadian Journal of Psychology* 15:166–71.

1964. Left-right differences in the perception of melodies. *Quarterly Journal of Experimental Psychology* 16:355–8.

1980. Neuromotor mechanisms in the evolution of human communication. In *Neurobiology of social communication in primates*, ed. H. D. Steklis and M. J. Raleigh, pp. 197–219. New York: Academic Press.

Kroodsma, D. 1978. Aspects of learning in the ontogeny of bird song: where, from whom, when, how many, which and how accurately? In *The development of behavior: comparative and evolutionary aspects*, ed. G. Burghardt and M. Bekoff, pp. 215–30. New York: Garland Press.

Kuhl, P. 1976. Speech perception in early infancy: the acquisition of speech categories. In *Hearing and Davis: essays honoring Hallowell Davis*. ed. S. Hirsh, D. Eldredge, I. Hirsh, and S. Silverman, pp. 265–80. St. Louis: Washington University Press.

Kuhl, P., and Miller, J. 1975. Speech perception by the chinchilla: voiced-voiceless distinction in alveolar plosive consonants. *Science* 190:69–72.

1978. Speech perception by the chinchilla: identification functions for synthetic VOT stimuli *Journal of the Acoustical Society of America* 63:905–17.

Lackner, J., and Teuber, H.-L. 1973. Alterations in auditory fusion thresholds after cerebral injury in man. *Neuropsychologia* 11:409–15.

LeMay, M., and Geschwind, N. 1975. Hemispheric differences in the brains of great apes. *Brain, Behavior and Evolution* 11:48–52.

Liberman, A., Cooper, F., Shankweiler, D., and Studdert-Kennedy, M. 1967. Perception of the speech code. *Psychological Review* 74:431–61.

Liberman, A., Harris, K., Hoffman, H., and Griffith, B. 1957. The discrimination of speech sounds within and across phoneme boundaries. *Journal of Experimental Psychology* 54:358–68.

Lisker, L., and Abramson, A. 1964. A cross-language study of voicing in initial stops: acoustical measurements. *Word* 20:384–422.

1970. The voicing dimension: some experiments in comparative phonetics. In *Proceedings of the Sixth International Congress of Phonetic Sciences* (Prague, 1967), pp. 563–7.

Lockhead, G. 1972. Processing dimensional stimuli: a note. *Psychological Review* 79:410–19.

McAdam, D., and Whitaker, H. 1971. Language production: electroencephalographic localization in the normal human brain. *Science* 172:499–502.

Mann, V., and Repp, B. 1980. Influence of vocalic context on perception of the [ʃ] – [s] distraction. *Perception & Psychophysics* 28:213–28.

Marler, P. 1970. A comparative approach to vocal development: song learning in the white-crowned sparrow. *Journal of Comparative and Physiological Psychology* 71, No. 2, p. 2:1–25.

1975. On the origins of speech from animals sounds. In *The role of speech in language,* ed. J. Kavanagh and J. Cutting, pp. 11–40. Cambridge, Mass.: M.I.T. Press.

Marler, P., and Peters, S. 1977. Selective vocal learning in a sparrow. *Science* 98:519–21.

Massaro, D. 1972. Preperceptual images, processing time, and perceptual units in auditory perception. *Psychological Review* 79:124–45.

Matsumiya, Y., Tagliasco, V., Lombroso, C., and Goodglass, H. 1972. Auditory evoked response: meaningfulness of stimuli and interhemispheric asymmetry. *Science* 175:790–2.

Mattingly, I., Liberman, A., Sydral, A., and Halwes, T. 1971. Discrimination in speech and nonspeech modes. *Cognitive Psychology* 2:131–57.

Miller, J. D., Wier, C., Pastore, R., Kelly, W., and Dooling, R. 1976. Discrimination and labeling of noise-buzz sequences with varying noise-lead times: an example of categorical perception. *Journal of the Acoustical Society of America* 60:410–17.

Miller, J. L. 1977. Interactions in processing segmental and suprasegmental features of speech. *Perception & Psychophysics* 24:175–80.

Miller, J. M., and Liberman, A. 1979. Some effects of later-occurring information on the perception of stop consonant and semivowel. *Perception & Psychophysics* 25:457–65.

Milner, B. 1962. Laterality effects in audition. In *Interhemispheric relations and cerebral dominance,* ed. V. Mountcastle, pp. 177–95. Baltimore: Johns Hopkins University Press.

1974. Hemispheric specialization: scope and limits. In *The neurosciences: third study program,* ed. F. Schmitt and F. Worden, pp. 75–89. Cambridge, Mass.: M.I.T. Press.

Milner, B., Taylor, L., and Sperry, R. 1968. Lateralized suppression of dichotically presented digits after commissural section in man. *Science* 161: 184–6.

Miyawaki, K., Strange, W., Verbrugge, R., Liberman, A., Jenkins, J., and Fujimura, O. 1975. An effect of linguistic experience: the discrimination of /r/ and /l/ by native speakers of Japanese and English. *Perception & Psychophysics* 18:331–40.

Molfese, D. 1978. Left and right hemisphere involvement in speech perception: electrophysiological correlates. *Perception & Psychophysics* 23:237–43.

1979. Cortical involvement in the semantic processing of coarticulated speech cues. *Brain and Language* 7:86–100.

Molfese, D., and Molfese, L. 1979. Hemisphere and stimulus differences as reflected in the cortical responses of newborn infants to speech stimuli. *Developmental Psychology* 15:505–11.

Molfese, D., Freeman, R., and Palermo, D. 1975. The ontogeny of brain lateralization for speech and non-speech stimuli. *Brain & Language* 2:356–68.

Morais, J., and Darwin, C. 1974. Ear differences for same–different reaction times to monaurally presented speech. *Brain and Language* 1:383–90.

Morrell, F., and Salamy, J. 1971. Hemispheric asymmetry of electrocortical responses to speech stimuli. *Science* 174:164–6.

Morse, P., and Snowdon, C. 1975. An investigation of categorical discrimination by rhesus monkeys. *Perception & Psychophysics* 17:9–16.

Nachson, I. 1973. Effects of cerebral dominance and attention on dichotic listening. *Journal of Life Sciences* 3:107–14.

Natale, M. 1977. Perception of nonlinguistic auditory rhythms by the speech hemisphere. *Brain and Language* 4:32–44.

Neisser, U. 1967. *Cognitive psychology.* New York: Appleton-Century-Crofts.

Nottebohm, F. 1971. Neural lateralization of vocal control in a passerine bird. I. Song. *Journal of Experimental Zoology* 177:229–62.

1972. Neural lateralization of vocal control in a passerine bird. II. Subsong, calls, and a theory of vocal learning. *Journal of Experimental Zoology* 179:35–50.

1977. Asymmetries in neural control of vocalization in the canary. In *Lateralization in the nervous system*, S. Harnad, R. Doty, L. Goldstein, J. Jaynes, and G. Krauthamer, pp. 23–44. New York: Academic.

1979. Origins and mechanisms in the establishment of cerebral dominance. In *Handbook of behavioral neurobiology*, Vol. 2., ed. M. Gazzaniga, pp. 295–344. New York: Plenum.

Nottebohm, F., and Nottebohm, M. 1976. Left hypoglossal dominance in the control of canary and white-throated sparrow song. *Journal of Comparative Physiology: Sensory, Neural, and Behavioral Physiology* 108:171–92.

Nottebohm, F., Stokes, T., and Leonard, C. 1976. Central control of song in the canary, *Serinus canarius. Journal of Comparative Neurology* 165:457–86.

Ojemann, G., and Mateer, C. 1979. Human language cortex: localization of memory, syntax and sequential motor-phoneme identification systems. *Science* 205:1401–3.

Papcun, G., Krashen, S., Terbeek, D., Remington, R., and Harshman, R. 1974. Is the left hemisphere specialized for speech, language and/or something else? *Journal of the Acoustical Society of America* 55:319–27.

Pastore, R. 1976. Categorical perception: a critical re-evaluation. In *Hearing and Davis: Essays honoring Hallowell Davis*, ed. S. Hirsh, D. Eldredge, I. Hirsh and S. Silverman, pp. 253–64. St. Louis: Washington University Press.

Pastore, R. Ahroon, W., Puleo, J., Crimmins, D., Golowner, L., and Berger, R. 1976. Processing interaction between two dimensions of nonphonetic auditory signals. *Journal of Experimental Psychology: Human Perception and Performance* 2:267–76.

Penfield, W., and Roberts, L. 1959. *Speech and brain mechanisms.* Princeton: Princeton University Press.

Petersen, M. 1978. The perception of species-specific vocalizations by Old World monkeys. Doctoral dissertation, University of Michigan.

1981. The perception of species-specific vocalizations by animals: developmental perspectives and implications. In *Development of perception: psychobiological perspectives*, vol. 1: *Auditory, chemosensory and somatosensory systems*, ed. R. Aslin, J. Alberts, and M. Petersen, pp. 67–109. New York Academic Press.

Petersen, M., Beecher, M., Zoloth, S., Moody, D., and Stebbins, W. 1978. Neural lateralization of species-specific vocalizations by Japanese macaques (*Macaca fuscata*). *Science* 202:324–7.

Petersen, M., Beecher, M., Zoloth, S., Moody, D., Stebbins, W., Green, S., and Marler, P. In preparation a. Monkeys employ different neural mechanisms when processing their own versus another species' vocalizations.

In preparation b. Comparative studies of the perception of vocalizations by primates: neural lateralization, perceptual constancy and categorization.

Pisoni, D. 1971. On the nature of categorical perception of speech sounds. Doctoral dissertation, University of Michigan.

1973. Auditory and phonetic codes in the discrimination of consonants and vowels. *Perception & Psychophysics* 13:253–60.

1975. Auditory short-term memory and vowel perception. *Memory and Cognition* 3:7–18.

1977. Identification and discrimination of the relative onset time of two-component tones: implications for voicing perception in stops. *Journal of the Acoustical Society of America* 61:1352–61.

1978. Speech perception. In *Handbook of learning and cognitive processes*, ed. W. K. Estes, pp. 167–233. Hillsdale, N.J.: Erlbaum.

1979. On the perception of speech sounds as biologically significant signals. *Brain, Behavior and Evolution* 16:330–50.

Pisoni, D., and Tash, J. 1974. Reaction times to comparisons within and across phonetic categories. *Perception & Psychophysics* 15:285–90.

Pola, Y., and Snowdon, C. 1975. The vocalizations of pygmy marmosets. *Animal Behaviour* 23:826–42.

Pollack, I. 1952. The information in elementary auditory displays. *Journal of the Acoustical Society of America* 24:745–9.

1953. The information in elementary auditory displays, II. *Journal of the Acoustical Society of America* 25:765–9.

Posner, M. 1978. *Chronometric explorations of the mind.* Hillsdale, N.J.: Erlbaum.

Raslear, T. 1975. The effects of varying the distribution of generalization stimuli within a constant range upon the bisection of a sound-intensity interval by rats. *Journal of the Experimental Analysis of Behavior* 23:369–75.

Repp, B. 1982. Phonetic trading relations and context effects: new experimental evidence for a speech mode of perception. *Psychological Bulletin* in press.

Richman, B. 1976. Some vocal distinctive features used by gelada baboons. *Journal of the Acoustical Society of America* 60:718–24.

Riley, D., and Leith, C. 1976. Multidimensional psychophysics and selective attention in animals. *Psychological Bulletin* 83:135–60.

Robinson, G., and Solomon, D. 1974. Rhythm is processed by the speech hemisphere. *Journal of Experimental Psychology* 102:508–11.

Robinson, J. G. 1979. An analysis of the organization of vocal communication in the titi monkey, *Callicebus moloch. Zeitschrift für Tierpsychologie* 49:381–405.

Rosenzweig, M. 1951. Representations of the two ears at the auditory cortex. *American Journal of Physiology* 167:147–58.

Schwartz, J., and Tallal, P. 1980. Rate of acoustic change may underlie hemispheric specialization for speech perception. *Science* 207:1380–1.

Seyfarth, R., Cheney, D., and Marler, P. 1980a. Vervet monkey alarm calls: semantic communication in a free-ranging primate. *Animal Behaviour* 28:1070–94.

1980b. Monkey responses to three different alarm calls: evidence of predator classification and semantic communication. *Science* 210:801–3.

Shankweiler, D., and Studdert-Kennedy, M. 1967. Identification of consonants and vowels presented to left and right ears. *Quarterly Journal of Experimental Psychology* 19:59–63.

Sinnott, J., Beecher, M., Moody, D., and Stebbins, W. 1976. A comparison of speech sound discrimination in man and monkey. *Journal of the Acoustical Society of America* 60:687–95.

Smith, L., and Kemler, D. 1978. Levels of experienced dimensionality in children and adults. *Cognitive Psychology* 10:502–32.

Smith, W. 1968. Message-meaning analysis. In *Animal communication*, ed. T. A. Sebeok, pp. 44–60. Bloomington: Indiana University Press.

1977. Communication in birds. In *How animals communicate*, ed. T. Sebeok, pp. 545–74. Bloomington: Indiana University Press.

Snowdon, C. T. 1979. Response of animals to speech and to species-specific sounds. *Brain, Behavior and Evolution* 16:409–29.

Snowdon, C. T., and Pola, Y. 1978. Interspecific and intraspecific responses to synthesized pygmy marmoset vocalizations. *Animal Behaviour* 26:192–206.

Sparks, R., and Geschwind, N. 1963. Dichotic listening in man after section of neocortical commissures. *Cortex* 4:3–16.

Sperry, R. 1974. Lateral specialization in the surgically separated hemispheres. In *The neurosciences: third study program*, ed. F. Schmitt and F. Worden, pp. 5–19. Cambridge, Mass.: M.I.T. Press.

Springer, S. 1971. Ear asymmetry in a dichotic detection task. *Perception & Psychophysics* 10:239–41.

1973a. Hemispheric specialization for speech opposed by contralateral noise. *Perception & Psychophysics* 13:391–3.

1973b. Memory for linguistic and nonlinguistic dimensions of the same acoustic stimulus. *Journal of Experimental Psychology* 101:159–63.

Stevens, K. 1975. The potential role of property detectors in the perception of consonants. In *Auditory analysis and perception of speech*, ed. G. Fant and M. A. A. Tatham, pp. 303–36. London: Academic Press.

Struhsaker, T. 1976. *Behavior and ecology of red colobus monkeys*. Chicago: University of Chicago Press.

Studdert-Kennedy, M. 1972. A right-ear advantage in choice reaction time to monaurally presented vowels: a pilot study. *Haskins Laboratories Status Report on Speech Research* 31/32:75–82.

1974. The perception of speech. In *Current trends in linguistics*, ed. T. Sebeok, 12:2349–85. The Hague: Mouton.

1976. Speech perception. In *Contemporary issues in experimental phonetics*, ed. N. Lass, pp. 243–93. New York: Academic Press.

Studdert-Kennedy, M., and Shankweiler, D. 1970. Hemispheric specialization for speech perception. *Journal of the Acoustical Society of America* 48:579–94.

Studdert-Kennedy, M., Shankweiler, D., and Pisoni, D. 1972. Auditory and phonetic processes in speech perception: evidence from a dichotic study. *Cognitive Psychology* 3:455–66.

Studdert-Kennedy, M., Liberman, A., Harris, K., and Cooper, F. 1970. The motor theory of speech perception: a reply to Lane's critical review. *Psychological Review* 77:234–9.

Teszner, D., Tzavaras, A., Gruner, J., and Hécaen, H. 1972. L'asymmetric droit–gauche du planum temporale: a propos de l'etude anatomique de 100 cerveaux. *Revue Neurologie* 126:444–9.

Van Lancker, K., and Fromkin, V. 1973. Hemispheric specialization for pitch and "tone": evidence from Thai. *Journal of Phonetics* 1:101–9.

Wada, J., Clarke, R., and Hamm, A. 1975. Cerebral hemispheric asymmetry in humans. *Archives of Neurology* 32:239–46.

Walker, J. 1968. The bisection of a spatial interval by the pigeon. *Journal of the Experimental Analysis of Behavior* 11:99–105.

Walker, S. 1980. Lateralization of functions in the vertebrate brain: a review. *British Journal of Psychology* 71:329–67.

Warren, J. 1977. Handedness and cerebral dominance in monkeys. In *Lateralization of the nervous system,* ed. S. Harnad, R. Doty, L. Goldstein, J. Jaynes, and G. Krauthamer, pp. 151–72. New York: Academic Press.

Waters, R., and Wilson, W. 1976. Speech perception by rhesus monkey: the voicing distinction in synthesized labial and velar stop consonants. *Perception & Psychophysics* 19:285–9.

Webster, W. 1972. Functional asymmetry between the cerebral hemispheres of the cat. *Neuropsychologia* 10:75–87.

Witelson, S. 1977. Anatomic asymmetry in the temporal lobes: its documentation, phylogenesis and relationship to functional asymmetry. In *Evolution and laterization of the brain,* ed. S. Diamond and D. Blizard. *Annals of the New York Academy of Sciences* 299:328–54.

Witelson, S., and Pallie, W. 1973. Left hemisphere specialization for language in the newborn: neuroanatomical evidence of asymmetry. *Brain* 96:641–7.

Wood, C. 1974. Parallel processing of auditory and phonetic information in speech perception. *Perception & Psychophysics* 15:501–8.

  1975. Auditory and phonetic levels of processing in speech perception: neurophysiological and information-processing analyses. *Journal of Experimental Psychology: Human Perception and Performance* 104:3–20.

Wood, C., and Day, R. 1975. Failure of selective attention to phonetic segments in consonant-vowel syllables. *Perception & Psychophysics* 17:346–50.

Wood, C., Goff, W., and Day, R. 1971. Auditory evoked potentials during speech perception. *Science* 173:1248–51.

Wright, A., and Cummings, W. 1971. Color-naming functions for the pigeon. *Journal of the Experimental Analysis of Behavior* 15:7–17.

Yeni-Komshian, G., and Benson, D. 1976. Anatomical study of cerebral asymmetry in the temporal lobe of humans, chimpanzees and rhesus monkeys. *Science* 192:387–9.

Zaidel, E. 1978. Lexical organization in the right hemisphere. In *Cerebral correlates of conscious experience,* ed. P. Buser and A. Rougeul-Buser, pp. 177–97. Amsterdam: Elsevier/North-Holland.

Zoloth, S., and Green, S. 1979. Monkey vocalizations and human speech: parallels in perception? *Brain, Behavior and Evolution* 16:430–42.

Zoloth, S. R., Petersen, M. R., Beecher, M. D., Green, S., Marler, P., Moody, D. B., and Stebbins, W. 1979. Species-specific perceptual processing of vocal sounds by Old World monkeys. *Science* 204:870–3.

Zue, V. 1976. Some perceptual experiments on the Mandarin tones. *Journal of the Acoustical Society America* 60(A):S45

9 · Linguistic and psycholinguistic approaches to primate communication

CHARLES T. SNOWDON

Traditional ethology conceived of animal communication as a genetically fixed, developmentally immutable, stereotyped activity. Within the communicative repertoire of a species there were said to be only a relatively small number of invariant signals (Moynihan 1970a), which were used in an equally small number of motivational, or contextual, situations (see Smith 1969, 1977). Although the critical importance of context in the interpretation of signals has been recognized for many years (Smith 1965, 1977), the prevailing view of communication in nonhuman animals has been one of restricted signal repertoire and restricted communicative referents.

According to the traditional ethological view, which poses a dichotomy between human and animal communication, human communication is not stereotyped, is subject to considerable developmental modification, has a signal repertoire of enormous size compared with those of nonhuman animals, and has signal invariants that are easily perceived by human recipients even though they are often difficult to discern in the physical structure of signals.

The major argument of this chapter and indeed of this entire volume is that the so-called dichotomy between simple, stereotyped, fixed communication systems in animals and complex, variant, and open communication systems in human beings just does not exist. We have had evidence for two decades that the song of some bird species is capable of some developmental modification (Thorpe 1958; Marler 1970a) and more recent data on the organizational complexity of vocal communication in certain birds is available (Smith, Pawlukiewicz, and Smith 1978; Krebs and Kroodsma 1980; J. Baylis, in preparation). Moreover, nonhuman primates offer even clearer evidence against the dichotomization of human and animal communication.

In the first study of a primate vocal repertoire to use spectrographic

The preparation of this chapter and the work described herein were supported by Research Scientist Development Award MH 00177 and Research Grant MH 29,775 from the U.S. Public Health Service, National Institute of Mental Health. I am grateful for the assistance, advice, and colleagueship of Jayne Cleveland, Jeffrey French, Alexandra Hodun, Robyn Lillehei, Philip Morse, and Yvonne Pola.

analysis, Rowell and Hinde (1962) noted the complex call structure of rhesus macaques, which is highly variable both between animals and within the repertoire of a single animal. Many calls could not be easily categorized into discrete classes; rather, call structures seemed to intergrade with one another. In many cases these intergradations corresponded to posited underlying motivational continua and thus the intergrading call structure was said to map a continuous motivational system.

Subsequent studies have confirmed Rowell and Hinde's findings about the complexity of primate vocalizations. In the face of the complexity of primate call structure and function, the traditional ethological techniques for the study of communication have proved inadequate, and primate behaviorists have had to seek models from other disciplines.

SEEKING ANSWERS THROUGH LINGUISTIC AND PSYCHOLINGUISTIC APPROACHES

At present, the major source for alternatives to the simple ethological model of communication is the field of linguistics. Linguists and ethologists interested in animal communication are similar in their approach. Both are essentially natural historians who seek out and describe as accurately as possible the signal structure of communication and the sequences in which signals are emitted. Once the signal structure is described, they try to develop a formalized rule system, or grammar, that is predictive of the structure and the sequence of future utterances. Finally, they are each concerned with the context in which a given signal is emitted by the communicator and with the interpretation given to a signal by the recipients.

Given that primate communication is more complex, variable, and open than previously assumed by traditional ethologists, it is reasonable to adopt the techniques devised by linguists and psycholinguists for human language study to the study of primate communication. (See Petersen, chap. 8, and Seyfarth and Cheney, chap. 10, this volume, for other discussions of this approach to primate communication studies.) The adoption of these techniques does not assume that primate communication equals, or even resembles, human speech and language; it simply provides more powerful tools.

The techniques developed by linguists and psycholinguists are capable of precisely analyzing the acoustic structure of the complex and variable sounds in speech (see Lieberman 1975). They are able to distinguish the acoustic components present in the same sounds that indicate gender, dialect, and individual identity. They have generated synthetic speech sounds and, through identification and discrimination tests, determined the perceptual invariants of phonemes even when the structural invariants were not readily apparent (Liberman, Cooper, Shankweiler, and Studdert-Kennedy 1967; Liberman 1970). Linguists have also successfully analyzed sequences of utterances to determine the rules of sequence formation (Chomsky 1957). Through tests involving presentations of normal and distorted utterances, psycholinguists have learned much about the semantic and syntactic functions of language and our understanding of grammar (see Fodor, Bever, and Garrett 1974). Some examples of these distortions are the famous "colorless green ideas sleep furiously" and Lewis Carroll's "Jabberwocky," which are semantically meaningless but syntactically valid, and "they are pushing policemen," which is ambiguous both semantically and syntactically. Finally, psycholinguists have generated a large body of evidence on the ontogeny of language. In particular they have collected data on phonetic perception by very young infants (Eimas, Siqueland, Jusczyk, and Vigorito 1971; Miller and Morse 1976), on the production of adultlike phonemes in infant babbling (see Miller 1951), and on the acquisition of the syntactic and semantic rule systems of language and their relationship with the stages of intellectual development proposed by Piaget.

The remainder of this chapter will discuss the application of some linguistic and psycholinguistic methods to the study of primate vocalizations. The results obtained using such methods with three species of the family Callitrichidae (marmosets and tamarins) will be described with particular reference to the analysis of the source and function of vocalization variability, nonhuman versus human perception of calls, the syntactic structure of call sequences within an individual animal's vocalizations and the syntax of dialogues (vocal exchanges) between animals, and, finally, the ontogeny of vocal communication in monkeys.

ANALYTICAL TECHNIQUES

Before presenting the results of our studies, I want to mention a few points about the techniques involved in the analysis of the structure of primate calls, the determination of the function of these calls, and the perception of these calls.

In general, studies of animal vocalizations have been dependent upon sound spectrographs, which provide a visual analog of vocalizations. An investigator chooses the set of characteristics most salient to him or her and proceeds to sort the spectrograms into categories based on those particular structural differences. Thus, two investigators might obtain different numbers of vocal signals according to the criteria each has set for sorting. One investigator may sort only according to obvious differences in call type; another may sort according to minor, subtle differences in call structure. The repercussions of this technique can be striking. As will be shown, much of the evidence for the complexity of primate vocal communication has come from recent studies that have found variant, functionally meaningful forms within a category formerly classified as unitary (Green 1975a).

There are two points to be made about sorting techniques. First, most studies of primate vocalizations have used only qualitative sorting techniques. Now quantitative techniques (see Smith, Newman, and Symmes, chap. 2, this volume) are being used to supplement qualitative ones. Quantification does have important benefits. When looking at subcategories or variants within a larger category, quantitative statistical techniques can be useful. Quantified techniques can identify the acoustic parameters that differentiate call variants and demonstrate that the differences the investigator has perceived in sorting are indeed valid. It is easy to overclassify call types when no quantitative basis is available to support a finer division of types. It is equally easy to overlook true call variants that may be from different populations. Second, the use of quantitative techniques allows one to separate the parameters that determine differences in the content of the call from the parameters that identify the individuals or populations making the call (see Lillehei and Snowdon 1978; Symmes, Newman, Talmage-Riggs, and Lieblich 1979).

To obtain our parameters, we use a calibrated graticule to measure the temporal and frequency parameters of each spectrogram. The resulting parameters can be subjected to single analysis of variance, multiple analysis of variance, or discriminant analysis (see Smith, Newman, and Symmes, chap. 2, this volume), using the situations in which calls are given as one variable and the individual animal as another independent variable. This allows the determination of which parameters, if any, are varied in different behavioral situations and which, if any, are varied between individual animals.

However, suppose statistical techniques reveal a degree of complexity in vocal structure that qualitative sorting ignored. This does not automatically mean that the greater acoustical complexity is functional in the communication system of the animals. That determination must be made empirically. Three steps are involved: First, structure must be correlated with the behavioral and social context in which a call is given; second, hypotheses must be formed about call function; and third, these hypotheses must be tested using playbacks of calls in appropriate and inappropriate contexts. In the ideal situation, the identification of five categories or subcategories of vocal structure will be accompanied by the identification of five types of social situations in which calls are given with a perfect correlation between one given call structure and one social–behavioral situation. But the normal state of affairs is less than ideal; one call type may appear in more than one situation, and one situation may be associated with more than one call type. (See Green and Marler, 1979, for an extended discussion of this point.)

When it is difficult to distinguish between two alternative call functions or when one wants to establish experimentally the relationship between a call structure and a given behavioral setting, the playback technique is useful. One hypothesizes which behavior should occur following a call in situation A and which behavior should occur following a call in situation B. One can establish that call X is given functionally in situation A by playing back call X in situations A and B. Animals should give their normal responses to call X only in situation A and not in situation B if call X is most closely associated with situation A. Playback of call X in situation A should also be more effective at eliciting appropriate behavior than playback of call Y.

This playback technique has several advantages. It provides an experimental resolution to questions that could otherwise be answered only by the use of correlational techniques. It allows the testing of animals in more or less normal social settings, since it is not necessary to condition a special or novel response or to maintain strict environmental control. On the other hand, there are serious problems with the use of playback techniques. The audio speaker unit must be well hidden from the animals and the sound must appear to come from locations where animals normally are active. Thus if the animals live high above the ground, playbacks from ground level are rarely effective. In addition, excessive playbacks provoke habituation, and animals no longer respond. We have rarely played more than five playback stimuli in any one-hour session per day and currently present no more than three playbacks per day. With these cautions, we have found playbacks of vocalizations to be valuable. However, when playbacks do not elicit normal responses, one must resort to conditioning techniques (Petersen, chap. 8, this volume) or the use of evoked potentials (Molfese 1978). A more complete discussion of playback and alternative techniques for studying the function and perception of sounds in monkeys can be found in Snowdon (1979).

ANALYSIS OF VARIABILITY IN PRIMATE VOCALIZATIONS

I have argued so far that primate vocalizations are complex and variable, and I have described some techniques with which this variability might be analyzed. Let us now consider four types of variability within the structure of a single call type: "Phoneticlike" variability, individual variability, population variability, and localization variability.

"Phoneticlike" variability

For lack of a better term, I have borrowed the word "phonetic" from linguistics to describe calls that seem similar but are acoustically different. An alien looking at spectrograms of spoken English would would at first think that the sounds /ba/ and /pa/ were members of the same class with more or less continuous variation among different tokens. With the help of an informed speaker of English, the alien could learn to appreciate the fine distinctions in acoustic structure (the relative onset of voicing) that differentiate between two sounds, sounds recognized and labeled as quite different and used in quite different contexts. Similarly, we as aliens looking at primate spectrograms have seen only a highly variable version of a single call type, but after increasing our knowledge of the animals' repertoire and of the situations in which calls are used, we have been able to discover similar phoneticlike variability in primate vocalizations.

The most impressive demonstration of phoneticlike variability in nonhuman primates was the analysis of the "coo" vocalization of Japanese macaques (*Macaca fuscata*) made by Green (1975a). Green identified seven coo variants given in ten behavioral and social settings. When

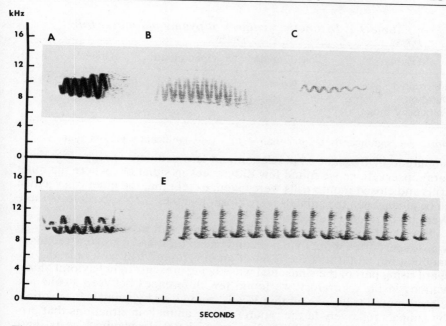

Figure 9.1. Trill variants of pygmy marmosets: (A) closed-mouth trill, (B) open-mouth trill, (C) quiet trill, (D) juvenile trill, (E) J-call. (Reprinted from Snowdon 1979, with permission.)

Green mapped call structure of variants against the situations he found a close, but not perfect, correlation. That is, one variant was rarely used in more than two different situations, and in any behavioral setting more than two variants were rarely given. Similar findings have been reported by Gautier (1974) and by Eisenberg (1976).

In our laboratory we found similar phoneticlike variability in the calls of two species of callitrichids. The first case involves a complex of trill vocalizations produced by the pygmy marmoset (*Cebuella pygmaea*) (Figure 9.1). We have established through quantitative analysis that adult animals use four distinct trill variants (Pola and Snowdon 1975; Snowdon and Pola 1978). See Table 9.1 for a distinctive-feature analysis of the trills. The most commonly given trill in captivity is the closed-mouth trill, which will serve as the basis of comparison for the other variants. The open-mouth trill differed from the closed-mouth trill only by being significantly longer in duration (a mean of 176 msec for the closed-mouth trill and 334 msec for the open-mouth trill). From our observations, the open-mouth trill was followed by a significantly greater amount of agonistic behavior (threats, approach–avoidance, piloerection) than the closed-mouth trills. Playbacks of the two calls elicited antiphonal calling to the closed-mouth trills and no vocal response to the open-mouth trill.

The quiet trill differed from the closed-mouth trill by having a smaller

Table 9.1. *Distinctive features of pygmy marmoset trills*

Variable	Closed mouth	Open mouth	Quiet	J-call
Interrupted	No	No	No	Yes
Duration	−	+	−	++
Frequency range	+	+	−	++

Note: ++ significantly greater than + or −; + significantly greater than −.

amplitude of frequency modulation and less acoustic energy. In behavioral observations we found few differences in situations where the quiet trills and closed-mouth trills were given, except that the quiet trill seemed to be given when the animals were closer together. In playback tests there were no differences in responses to the calls.

The J-call differed from the closed-mouth trill in several respects. It was longer, often more than a second in length; it had a greater frequency modulation; and it had the form of an interrupted trill with only the upward rising part of the sinusoidal wave being present. In behavioral observations in the laboratory we found few differences between J-calls and closed-mouth trills, except that J-calls were given by animals in isolation, by animals relatively farther apart, and by animals in situations that provoked arousal (e.g., while feeding on crickets). In playback tests there was no differential response to J-calls and closed-mouth trills.

In short, of the four statistically definable variants of the trill in the pygmy marmoset, only one, the open-mouth trill, seemed to be functionally distinct from the others. Whereas it was given in situations where agonistic behavior was about to occur, the other trills were given in situations with low or moderate arousal and no agonistic activity. Thus, it seemed that the four different trills served only two different functions.

The second example of phoneticlike variability comes from our studies of the cotton-top tamarin (*Saguinus oedipus oedipus*) (Cleveland and Snowdon 1982). The major part of the repertoire of these animals is based on the variation of two basic elements: "chirps" (short, frequency-modulated calls) and "whistles" (longer, nonmodulated calls). We have found 35 distinct call structures or sequences formed either by variation of the elements given singly or by different combinations of the elements into sequences.

Let us look at single chirps (Figure 9.2). We were able to distinguish eight variations of chirps based on significant differences in four parameters (presence or absence of upswing, peak frequency, duration of downsweep, difference between peak and end frequency). Figure 9.3 presents a distinctive-feature analysis of these eight chirps. When we analyzed each call type with respect to the behavioral conditions in which they were given, we found no overlap in behavioral categories. Each call was given in one clearly definable behavioral setting. Type A chirps were used in mobbing; type B chirps were given during investigation; type C chirps were given while approaching food; type D chirps were given after taking

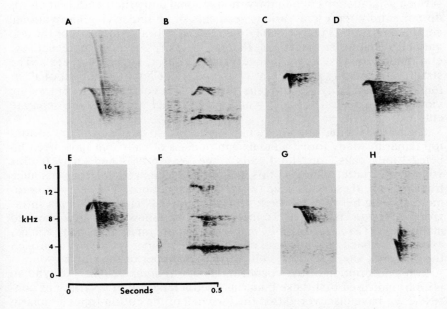

Figure 9.2. Chirp variants of cotton-top tamarins: Types A–H. (From Snowdon, 1979. Reprinted with permission.)

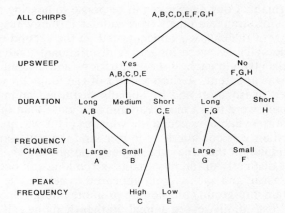

Figure 9.3. Distinctive-features analysis of eight chirps of cotton-top tamarins. All differences significant at $p < .05$.

food; type E chirps were given as general alarm calls; type F chirps were used when calls from other groups were audible; type G chirps were given while animals were in a state of mild attentiveness to their environment; type H chirps were used as mild alarms to novel visual stimuli. Furthermore, the situations in which these chirps occurred cannot easily be de-

scribed as gradations of a unitary motivational construct. Different chirps
do not simply represent various segments of an underlying motivational
continuum, such as that which Green (1975a) was able to posit for the coo
variants of Japanese macaques. Rather, the chirps seem to represent several motivational systems (hunger, fear, arousal, aggression, etc.). The
structural similarity of different chirps does not imply a motivational, or
functional, similarity (see Jürgens, chap. 3, this volume, for a contrasting
view); the structural variability and motivational variability are separate
entities.

The third example of phoneticlike variation also comes from the cotton-
top tamarin. Many forest animals emit a class of calls that have been labeled "long calls" or "loud calls" (see Waser 1977 and chap. 6, this
volume; Deputte, chap. 4, this volume). Specialized for long-distance
transmission, these calls have two different functions: They serve to promote spacing between adjacent groups of animals and to maintain intragroup cohesion. Waser (1977), for example, has shown that gray-cheeked
mangabeys (*Cercocebus albigena*), other than "rapid approach" males,
avoid a playback of a long call coming from another group but approach
the playback site if the long call is from a member of their own group.

In his description of the vocalizations of *Saguinus oedipus*, Moynihan
(1970b) indicated that these tamarins used a long call in a variety of contexts. We have discovered that the long call of the cotton-topped tamarin
has at least three variant forms. The basic structure is that of a series of
whistles. However, in calls given rather softly within a group, when other
groups of animals could not be heard, acoustic energy was concentrated
in the first formant. Long calls given in response to hearing the vocalization of groups of strange animals were similar in many respects except
that the energy of the call was concentrated on the second to fourth formants and the syllables were longer. A discriminant analysis was performed on calls given in each of these situations and a significant separation of calls was apparent. A third variant of the long call was given by
animals placed in isolation and by socially disturbed animals. In these
long calls, which were much more variable as a class than the other two,
at least one chirp syllable preceded the whistles (Thürwächter 1980). The
total call duration was considerably longer, and the call contained more
frequency modulation than the other types of long calls.

Thus, three distinctly different variants of the long call occurred in the
cotton-top tamarin, and each variant appeared to be specific to one function: to maintain intragroup cohesion, to promote intergroup spacing, and
to identify a "lost" or distressed animal. Rather than one call type serving
three functions, there are really three structural variants, one for each
function.

To further examine these long calls to determine whether they were
being distinguished by the animals themselves, we carried out a playback
study (Snowdon, Cleveland, and French, in press). We recorded examples of both the intragroup-cohesion variant and the intergroup-spacing
variant from each of our adult cotton-top tamarins. We played back these
calls to each group with the following design: The animals heard their own

calls, their mates' calls, a call from another group of animals from the same colony that they had heard previously, and a call from a group of animals that they had never heard before. Each type of stimulus was presented six times to each animal. To avoid habituation, calls were presented once every 15 min with no more than three trials per day per group. Behavior was observed for the 5 min prior to the playback stimulus and the 5 min following the stimulus. During the study several "null" trials were interposed in which no vocal stimulus was presented.

The results showed that animals responded differently to the playbacks of the intergroup-spacing and the intragroup-cohesion calls. There was more scent marking and piloerection (behavior shown by animals in response to unfamiliar intruders; French and Snowdon 1981) with intergroup-spacing playbacks than with the intragroup-cohesion calls. All the significant increases in responses (alarms, whistles, scanning, piloerection, and scent marking) were given to the calls of nongroup members. Thus, the animals differentiated not only between the two types of calls but between calls given by members of their own group and calls given by animals from other groups.

To summarize, there are now several well-established examples of calls similar in structure with identifiable variants that seem to be closely correlated with different functions. Thus, phoneticlike variability seems to be an important component of primate communication.

Individual variability

The most obvious source of variability in primate vocalizations is variability that indicates the identity of the individual caller. In any social group of animals, the members of the group must be able to readily identify one another since responses must often be quite different to different members of the group. Therefore, one is not surprised by the finding that the vocalizations of monkeys encode specific individual variations.

Marler and Hobbett (1975) demonstrated individual differences in the pant-hoot vocalization of chimpanzees. Symmes, Newman, Talmage-Riggs, and Lieblich (1979) and Smith, Newman, and Symmes (chap. 2, this volume) have demonstrated individual differences in several calls of the squirrel monkey. Waser (1977) showed individual differences in the structure of the whoopgobble display of the gray-cheeked mangabey and showed individual recognition of these calls through playback experiments.

We also have been involved in the study of individual variability in call structure and its recognition by monkeys. Lillehei and Snowdon (1978) measured 18 parameters of the coo vocalizations of infant stump-tailed macaques (*Macaca arctoides*) and analyzed these according to contextual situation and individual callers. We found two different social contexts that were encoded by consistent differences in the structure of coos: young-alone and young-calling-to-mother. Young-alone coos showed a rapid rise to the peak frequency of the syllable, whereas young-to-mother coos showed a slow rise and late peak frequency. These correspond exactly to Green's (1975a) findings with the related species of Japanese macaques. Whereas 1 acoustic parameter distinguished between the two contexts, 13 other parameters served to distinguish individuals from one another. We could identify a unique set of parameters to differentiate 13 of the 15 possible pairs of animals from each other. Not only did individual differences constitute one component of the variability of call structure of monkeys, but the parameters indicating individual identification were different from those indicating situational differences.

This study with stump-tailed macaques also indicated a way in which individual differences might indicate situational differences. The probability with which animals might call in different situations greatly varied. Whereas 60% of animal A's calls might occur when it was alone, only 10% of animal B's calls would occur in that situation. By keeping track of the probability of an animal calling in a given situation and by being able to identify that individual, a recipient could predict a calling animal's situation simply by identifying that individual's call characteristics.

We have also studied individual differences and individual recognition of the closed-mouth trills and J-calls of pygmy marmosets (Snowdon and Cleveland 1980). We measured 30 examples of each type of call from each individual and performed analyses of variance on each of the parameters we measured. From these data we constructed a distinctive-features table for each individual. There was a unique pattern of features that discriminated every animal from every other animal. We then observed the animals while calls were given spontaneously and focused on the responses of each of the other animals in the colony. For example, we would wait until animal A gave a trill and then observe the responses of animal B, another time the responses of animal C, and so forth until 30 observations were made of each call type with each caller–recipient pair. Analyses showed that there were consistent individual patterns of responses. With closed-mouth trills, all animals responded most to the dominant male of the colony regardless of whether or not they were in his family group. With the J-calls, animals responded most to their mates and to the same-sexed members of the other family groups.

Next we played back the J-calls of male pygmy marmosets through hidden speakers. Speakers were located either in the normal compartment where the stimulus animal lived or in another animal's compartment. Animals responded significantly only to calls played back through the speakers in the stimulus animal's own compartment. For instance if animal A lived in compartment 1, playback of its call would elicit responses from other animals only if the playback was transmitted through the speaker in compartment 1. When the playback was transmitted from compartments 2 or 3, there was no response. To our surprise, individual animals responded most to the playback of their own vocalizations. Three possible reasons for this response are: First, one's own sounds might be perceived as different and novel, since the normal bone conduction that is present when emitting a call is absent from the playback call. The sound would be perceived as coming from a strange animal and this might increase the stimulus animal's responsiveness. Second, habituation might occur to familiar vocal stimuli (Verner and Milligan 1971; Petrinovich and Peeke 1973). The sudden appearance of one's own sound would be familiar, yet different, producing dishabituation and thus increase response. Third, it is possible that these responses constitute a form of vocal self-recognition analogous to the visual self-recognition demonstrated by Gallup (1977) for chimpanzees. Studies are currently underway to determine whether pygmy marmosets recognize their own vocalizations.

A recent study (Snowdon and Hodun, in preparation) has shown that wild white-lipped tamarins (*Saguinus mystax*) respond to the calls of an isolated animal only if the animal is from their own group. There are individual- but not group-specific characteristics in the long-call structure. Individual variability seems to be quite well established in primate vocalizations, and it does constitute a major component of the variability of the calls. There have been some interesting uses of this variability to identify individual callers during vocal exchanges between several animals (see Gautier and Gautier-Hion, chap. 1, this volume; Smith, Newman, and Symmes, chap. 2, this volume). A clear understanding of individual variability will be of great use in the future as we try to understand the relationships between communication and social behavior in monkeys.

Population variability

Vocal variability has been correlated with different populations in species of birds and primates. Green (1975b) provided suggestive evidence of "dialects" among different populations of the Japanese macaque. Marshall and Marshall (1976) have similarly shown vocal variability among different populations of the gibbon. Perhaps the best studied primate example of vocal variability are the calls of the squirrel monkey, which differ between the "Roman arch" and "Gothic arch" varieties (Winter 1969). These two varieties clearly differ in the structure of isolation peeps and other vocalizations, and there is evidence that animals of one population respond most to the calls of their own kind, indicating a populational discrimination based on vocal cues (Newman and Symmes, chap. 11, this volume; Snowdon, Coe, and Hodun, in preparation). Unfortunately, little is known about how these two populations are distributed in the wild, though there is some evidence that animals of similar morphs tend to group together.

In the course of a field study in the Amazon we were located in an area that provided the habitat for three subspecies of the saddleback tamarin, *Saguinus fuscicollis* ssp. (Hodun, Snowdon, and Soini, 1981). The Rio Tapiche in the headwaters of the Amazon served as the dividing line, with *S.f. illigeri* on the west bank, *S.f. nigrifrons* on the lower east bank, and *S.f. fuscicollis* on the upper east bank. We were able to record the long-call vocalizations of recently trapped animals collected for a joint project of the Republic of Peru and the Pan American Health Organization. We placed individually caged animals out of eyesight but within hearing distance of their family group and recorded their vocalizations over a 20 min period. Analysis of the long-call structure showed differences in the syllable structure that were consistent within each subspecies. Thus, it was possible to identify a series of variables that provided subspecific markers in the long calls. We do not yet know the function (if any) of these subspecies differences, since it is not clear whether adjacent groups of different subspecies maintain strict segregation in their home ranges and in breeding. The various subspecies do interbreed readily in captivity (G. Epple, personal communication).

Of interest is an observation that we made during our field study. One animal, *S.f. nigrifrons* morphologically, was trapped directly across the river from a group of *S.f. illigeri*. The long call of this *nigrifrons* was similar to *nigrifrons* on two variables (peak–start frequency and start–end frequency of the last part) but similar to *illigeri* on two other variables (duration and start–end frequency of the intial part of the syllable). This could represent a hybrid vocalization. However, when we recorded the long calls of known hybrids of *S.f. nigrifrons* and *S.f. illigeri* in the laboratory of Dr. Gisela Epple, the hybrid calls resembled *nigrifrons* on most variables; they did not resemble *illigeri* calls. In recording the calls of other subspecies in Dr. Epple's laboratory, we observed an incorporation of features of long calls from those other subspecies with which an animal is in vocal contact. This suggests a possible role for vocal learning of long-call structure in this species. It may eventually be possible to perform a study with *S. fuscicollis* ssp. analogous to Marler's (1970a) study on song learning in white-crowned sparrows. Because the evidence on vocal learning in primates is scanty and because the phenomenon is difficult to demonstrate (Newman and Symmes, chap. 11, this volume), the subspecies *S. fuscicollis* could be valuable to an understanding of vocal ontogeny.

Localization variability

The final form of variability within call structure reported here concerns changes in call structure that are correlated with the distance of the caller from other animals of his group. Several recent papers have addressed the issue of the design features of vocal communication, that is, those characteristics of sounds that maximize or minimize detectability in a given environment, those acoustic features that are suitable for long-distance transmission, and those features that are suitable for shorter distances (Waser and Waser 1977; Wiley and Richards 1978). Recent work on the mechanisms of sound localization by monkeys also indicates that frequency modulation is very important for sound localization (Brown, Beecher, Moody, and Stebbins 1979; Brown, chap. 7, this volume).

We were somewhat puzzled that three of the trill variations of the pygmy marmoset (closed-mouth trill, quiet trill, and J-call) did not seem to differ greatly in the context in which they were used nor in the responses that animals gave to playbacks of the calls. However, the structure of these calls varied greatly in the degree to which they should be locatable according to psychophysical studies on the cues involved in auditory localization. With minimal frequency modulation and a continuous amplitude structure, the quiet trill should be the least localizable. The J-call, on the other hand, should be the most localizable, since it has both a large frequency modulation and is interrupted, which provides repeated time-of-arrival cues for localization. That these calls had similar functions and differed only in their relative localizability suggested that the quiet trill should be given at short distances between caller and recipient when cues from other sensory modalities would be available for localization. In contrast, the J-call should be given at far distances when cues from other sensory modalities would not be available and when an error in localization angle would produce a far greater error in absolute localization than the same error in localization angle would make when the animals were close.

During our field trip to the Amazon we studied a family group of pygmy marmosets (Snowdon and Hodun 1981). Initially we observed the movement patterns of the animals and constructed a map of their use of the home range. We then recorded incidents of trills and simultaneously identified the location of both caller and recipient. The calls were sonographed, analyzed, and sorted into the three trill classes. From these data we produced a distribution of call type by the distance between caller and recipient. Each call type had a distance distribution that was different from that of every other call type, and the distributions were in the order that we had predicted. Quiet trills were never used between animals more than 5 m apart, whereas J-calls were the only calls given when animals were more than 15 m apart. Thus, the pygmy marmoset used the most localizable trill at greater distances, and the least localizable trill at shorter distances, from other animals.

A similar variation in call structure with distance between animals has been observed in the calls of our captive cotton-top tamarins (Cleveland and Snowdon 1982). The three types of long calls discussed previously differ in their acoustic cues for sound localization. Most of the energy of the intragroup-cohesion variant is in the first formant and there is little frequency modulation. The energy of the intergroup-spacing variant is distributed across several formants (many bands of energy are more localizable than a few; Simmons 1973). Finally, the disturbed-animal variant is preceded by several short, frequency-modulated chirps, and the whistles are repeated with greater frequency than with the other call types (providing more time-of-arrival cues). In another cotton-top tamarin example a series of contact-call variants displayed highly frequency-modulated forms when animals were separated from each other. As animals approached each other, calls were emitted with progressively less frequency modulation.

The results from the pygmy marmoset field study and from the study of captive cotton-top tamarins indicate that variations in call structure occur when animals are at different distances from each other. These variations are correlated with the cues that we know aid in sound localization. The function of each variant is the same; only the relative localizability of the call alters.

So far four types of variability in signal structue have been described and documented: phoneticlike variability, individual variability, populational variability, and locational variability. These probably do not represent all the possible sources of variability within monkey calls, but the analysis presented here suggests how sources of variance in primate calls can be identified and separated from one another.

PERCEPTION OF CALLS

How animals perceive their own calls is an important question. First, we should investigate whether the way an animal perceives its own calls is at all isomorphic with the way in which human observers perceive those calls. Studies on pheromones that we cannot smell, discoveries of ultrasounds that we cannot hear, and evidence of electrical signals that we cannot feel have cautioned us to be aware of the existence of animal signals that we cannot perceive. We also need to be cautious of nuances in animal signals that do fall within our perceptual capacities; we could be led astray by attending different features of the signal or organizing signals differently.

Second, we should investigate the possible existence in human beings and other species of parallel specializations for perceiving species-specific vocalizations. Although the form of signal structures and the complexity of signal function may be quite different in human beings and animals, the basic mechanisms of perceiving signals may be similar.

Let us consider some human perceptual phenomena that may pertain to this important question of how animals perceive their own vocalizations. In human speech perception an important finding has been that of "categorical perception." Looking at the acoustical structure of different speech sounds, we discover that some can be placed on a physical continuum (for example, /b/ and /p/ represent points on a continuum of voice onset time, that is, whether voicing of the larynx occurs simultaneously with plosion of air from the lips, /b/, or whether the voicing of the larynx is delayed, /p/). If we then create synthetic speech stimuli representing equal steps on this voice-onset-time continuum, ranging from simultaneous voicing to delayed voicing, and play these stimuli to English-speaking human subjects, an interesting result occurs. The subjects report the experience of hearing either /ba/ or /pa/, that is, they perceive several different stimuli as /ba/ and certain other stimuli all as /pa/. There is no apparent ambiguity between /ba/ and /pa/. A given stimulus from any point on the continuum is labeled as one or the other phoneme. The two phonemes are thus strictly categorized. If we select pairs of stimuli and ask subjects to say whether they are the same or different, the classical results are that stimuli are discriminated as different only if they are drawn from categories that are labeled as different (Liberman et al. 1967). This finding of categorization was thought to be unique to speech. For instance, when the sounds can no longer be recognized and labeled as speech (by removing the vowel component from the synthesized syllable), each separate stimulus can be discriminated from every other stimulus. Only when sounds are perceived as speech do subjects successfully discriminate variants along a physical continuum at certain category boundaries. This categorical discrimination of speech is a rather unusual finding, since it appears to be at odds with perception studies using other modalities. For example, we label a larger number of stimuli as red. However, for most of those stimuli labeled red we are able to discriminate different types of red. In nonspeech sensory modalities, then, the labeling function is quite often categorical, whereas the discrimination function is continuous.

The restriction of this categorization phenomena to speech sounds has led to arguments that categorical perception is a unique function of those organisms that use speech (i.e., human beings). An obvious step is to determine whether nonhuman animals perceive speech categorically. There have been two monkey studies to test their perception of the-place-of-articulation continuum (/ba/, /da/, /ga/) (Morse and Snowdon 1975; Sinnott, Beecher, Moody, and Stebbins 1976), and one chinchilla study and one monkey study to test their perceptions of the voice-onset-time continuum (/ba/, /pa/; /da/, /ta/) (Kuhl and Miller 1975; Waters and Wilson 1976). In each study (except for Sinnott et al. 1976) at least some data indicated that nonhuman animals could perceive speech sounds in a categorical fashion and that the category boundaries were similar to those found with human subjects. However, it has been argued (Snowdon 1979) that these studies might have been coincidentally tapping important discrimination boundaries in each species' own communication system. For example, the Morse and Snowdon (1975) study used three formant stimuli where the third formant change between two human phonetic categories also coresponded to an important featural difference in rhesus monkey communication. The frequency change of the third formant of human speech is not important in our understanding of speech sounds, but it is an important frequency change in rhesus monkey vocalizations. Similarly, the voice-onset-time results reported by Waters and Wilson (1976) may represent an acoustic distinction in the natural calls of macaques. Related species of Japanese macaques (Green 1975a) and stump-tailed macaques (Lillehei and Snowdon 1978) show a voicing distinction in their calls. Until we know more about the natural communication signals of macaques we cannot rule out the possibility that their seemingly humanlike categorization might be incidental to their own signal perception mechanisms.

Several recent studies have also challenged the notion that categorical perception is something special and speech-specific. For example, Pisoni (1977) has shown that voice-onset-time categorization occurs with pure tones; thus phonemes of speech are not necessary for categorical perception. It has also been shown that the categorical discrimination can be made continuous if subjects are given the appropriate set (e.g., by sequencing the presentation order of stimuli as they would occur on the relevant acoustic continuum). Moreover, continuous perception appears by minimizing the memory load during the discrimination task – when subjects are simply asked to detect a change in stimuli rather than whether the last sound is more similar to the first or the second (Pisoni and Lazarus 1974).

In short, these latter findings make speech perception similar to perceptual categorization in other modalities. That is, there is a labeling function that is categorical and a discrimination function that may be categorical or continuous according to the method of stimulus presentation and to the demands of the experimental task.

The next step is to determine whether nonhuman animals perceive their own vocalizations categorically. There have been at least two efforts to study the categorization and perception of the sounds of an animal's own repertoire. Petersen (chap. 8, this volume) has described his work on the perception of sounds by Japanese macaques. (See also Beecher, Petersen, Zoloth, Moody, and Stebbins 1979). In our study on the responses of pygmy marmosets to synthesized versions of their own calls (Snowdon and Pola 1978), we first constructed an electronic synthesizer to reproduce each of the various trills. This synthesizer permitted us to manipulate the parameters of duration, center frequency, rate of frequency modulation, and amplitude of frequency modulation independently from one another. We played back more than 1,800 synthesized calls to our animals, using as our major response measure whether or not there was an antiphonal response to the synthesized calls. Between 60 and 70% of normally emitted closed-mouth trills were followed by antiphonal calls. We found we could vary the structure of the closed-mouth trill over a broad range of duration, center frequency, and amplitude of frequency modulation and still elicit a normal rate of antiphonal response. Thus, the animals responded to many variants of the prototype closed-mouth trill as though it were a normal call. However, when we increased the duration of the call beyond the normal range of the closed-mouth trill into the range of the open-mouth trill, the animals' response immediately ceased (Figure 9.4). The dividing line between these two responses was quite precise. A call duration of 248 msec elicited a normal antiphonal response, whereas a call that was 257 msec in duration elicited no response. Coincidentally, we found in our sample that a spontaneous closed-mouth trill never exceeded 250 msec in duration, whereas a spontaneous open-mouth trill was never less than 250 msec. Thus there was a sharp and precise categorical boundary in responding to calls that was coincident with the boundary in the production of these calls.

An obvious question is whether human beings would respond to these pygmy marmoset calls categorically. Using three different discrimination paradigms we tested several human subjects for their ability to differentiate between pygmy marmoset trills. With each paradigm and with each stimulus continuum, we found that finer, more precise discriminations between monkey sounds were made by human subjects than by monkeys. Why should monkeys be less accurate than human beings in discriminating the variations of their own calls? This raises the question of the reliability of species-specific categorization. Remember that human beings failed to categorize components of speech sounds when they could not be labeled (see earlier discussion). For a human observer, the various pygmy marmoset trills do not signify; there is no basis for making decisions about categorization. So the human inability to categorize unfamiliar sounds is really not surprising. But why do pygmy marmosets fail to distinguish be-

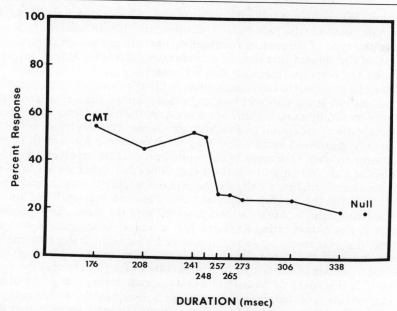

Figure 9.4. Categorical labeling function of pygmy marmosets measured by antiphonal responses to synthesized trills varying in duration. CMT: mean duration of natural closed-mouth trills. Null: response level when no stimulus is present.

tween minor variations in their call structures, variations that human beings seem to be able to discriminate easily? To answer this we must consider the task demands upon both humans and monkeys and the function of variability in the call structure for the marmoset. The task we gave the pygmy marmoset with our playback technique was basically a labeling task. The monkeys were not forced to discriminate between subtle variations but simply to indicate through antiphonal calling whether the playback sounds were of one class or another. The human subjects, in contrast, were asked to make pairwise discriminations between the various stimuli; they had no basis for labeling the calls at all. Therefore, the human beings and the monkeys were performing different tasks, and their response functions might be expected to differ.

Subsequently we have produced a functional analog to a discrimination task with the pygmy marmoset. One source of variabiity within a variant call type is individual variability. The values of the acoustic parameters that differentiated between the calls of individual animals fell within the range of sounds responded to equivalently as normal closed-mouth trills in the playback study. Since pygmy marmosets have demonstrated individually different response patterns to one another's closed-mouth trills (Snowdon and Cleveland 1980), it must be true that these animals can discriminate within the class of closed-mouth trills. When we play synthesized calls and measure a classifying response, we obtain a categorization function. When we use individually different calls and ask animals to discriminate individuals, we find a continuous discrimination function. Again, the type of perceptual function we obtain depends on the question we ask of the subject, the type of stimulus we present (synthetic vs. natural), and the response measure that we select.

If we can manipulate conditions so as to favor continuous or categorical discrimination in animals and human beings, can we predict where either might be naturally most useful? We have hypothesized a two-stage model of vocalization perception to deal with this issue. Our model accounts for the pygmy marmoset results and may be appropriate for human speech perception as well. One stage of perceptual processing involves identification of the call. When an important functional difference exists between two similarly structured calls (as between the closed-mouth and open-mouth trill for the pygmy marmoset and between /ba/ and /pa/ for the human being), the incoming sound is sorted into the most likely phonetic category. Once this sorting has been performed or simultaneous with this sorting, the perceptual system can proceed to a second stage to analyze other aspects of the signal structure – Is this Monkey A or Monkey B? Is this call from one of my family or from another group? Where is the caller located? Is this Mary or Susan? The first stage of perceptual processing is the identification of the phonetic category. Subsequently, within-category discriminations that may be of functional importance to the perceiver may be performed. In this sense the processing of speech or of an animal's own vocalizations is no more special or different than the processing of any other sensory stimuli.

In summary, there are several similarities between the perception of animal calls by animals and of speech sounds by human beings; in addition, the study of animal perceptual abilities may be of importance to an understanding of the human speech perceptual system. The study of how animals perceive their own calls is of importance for yet another reason. A major problem in studying animal vocalizations is how to categorize animal sounds. Are we observers sorting animal sounds into the same categories that are functional to the animals? Only with perceptual studies using both synthesized and natural sounds can we determine the true functional categories of animal vocalizations. Only then can we be reasonably confident that our division of call types is isomorphic with the species' own division of call types.

SYNTAX IN COMMUNICATION

Two very different uses of the word syntax need to be discussed before we can consider its application to primate communication. Linguists (e.g., Chomsky 1957; Umiker-Sebeok and Sebeok 1981) have defined syntax as a rule system that accounts for the ability to produce an infinite variety of sequences from a fixed number of phonemes. The existence of this type of generative syntax has been proved in theory for human beings, although it cannot be demonstrated empirically for either human speakers or animal communicators. This restricted definition of syntax is not relevant for our purposes.

A more general definition of syntax is simply: any system of rules that allows us to predict sequences of signals. If all subsequent signals are equally probable following the emission of one signal, there is no syntax. In contrast, several studies have shown Markovian (predictable) processes in sequences of displays (see Hailman, 1977, for a discussion of Markovian processes and their application to visual communication; see also Altmann 1965; Moynihan 1966; Smith et al. 1978; and Robinson 1979).

We have studied two types of syntactic systems in our animals. The first is a rule system for the sequencing of calls in the cotton-top tamarin and the second is a syntactic order for dialogues between individuals vocalizing during the antiphonal calling of pygmy marmosets.

Sequences within animal calls

As mentioned earlier, the calls of cotton-top tamarins consist of two basic elements that are found in several definable variations (Cleveland and Snowdon 1982). These units (chirps and whistles) can appear as individual units unrelated to other calls. However, a call may also consist of several syllables in a sequence. The simplest type of sequence is the rapid repetition of an individual unit. In most cases these repeated units function to intensify the meaning of a single unit. When chirps and whistles are combined, chirp elements always precede whistle segments and there is a decreasing center frequency across successive elements. Thus far, these are the most complex rules for the construction of sequences that we have found. The syntactic system is relatively simple.

Recently Marler (1977) differentiated two types of syntax: "phonetic" and "lexical." "Phonetic" syntax is analogous to the formation of different words through the rearrangement of phonemes. The resulting word has a different meaning and function than do the individual components. "Lexical" syntax is analogous to the combining of individual words into phrases so that the resulting structure is the sum of the meanings of the individual components. Marler argued that most syntax to be found in animals would be phonetic, and that lexical syntax would be rather rare.

In our study of the cotton-top tamarin (Cleveland and Snowdon 1982) we found that most call sequences could be best described as phonetic syntax. However, the meaning of two frequently occurring phrases appeared to be the sum of the individual units. The first phrase consisted of a type E (alarm) chirp and a low arousal alerting call. This combination was always given after initial type E (alarm) chirps had aroused the animals. When this combination phrase occurred, the animals remained stationary and scanned the environment with reduced arousal. A short time later they began to move about, giving only the low-arousal alerting call.

In the other phrase the type F chirp (given in response to hearing a distant group) was combined with one whistle from the intergroup-spacing long call (normally a series of whistles). The type F chirp+whistle combi-

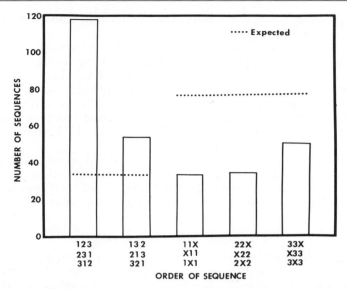

Figure 9.5. Sequences of calls between three pygmy marmosets. Dotted line indicates number of sequences of each type expected by chance. Histograms show observed number of sequences of each type.

nation appeared after the initial type F chirps and prior to the onset of long calls. The sequence appropriately combined the chirp's function of orientation to a strange group's call and the intergroup-spacing function of the long call.

The cotton-top tamarins, like many other animals, use sequences of calls. The sequences can be described by simple rule systems. And some of the combinations of the cotton-tops suggest a type of lexical syntax.

Exchanges between animals

Antiphonal calling, which is often found between paired animals (Thorpe 1972; Deputte, chap. 4, this volume), represents one form of rule system for between-animal communication. Animal B calls after animal A and animal A calls after animal B. We became interested in the antiphonal calling of our pygmy marmosets and focused on the order of callers in a group of three animals (Cleveland and Snowdon, in preparation). We recorded a total of 297 call exchanges, or dialogues, varying in length from 3 to 27 calls. Figure 9.5 presents the results of the analysis of the initial 3 calls of each exchange. There are 27 possible sequences of calling orders, given three animals. There are two basic patterns in which all three animals could each be represented within any 3-call exchange (animal 1, animal 2, animal 3, with permutations 2,3,1 and 3,1,2 representing the same order of individuals with different starting points, i.e., initiating caller, and 1,3,2 with 2,1,3 and 3,2,1). There are also seven possible orders in which one individual could be represented twice or three times within the three-call exchange. As the figure shows, orders of vocal exchanges with each of the three animals calling occurred much more often than would be predicted by chance, whereas exchanges in which one animal repeated itself before both of the others called occurred less often than would be predicted. Thus, animals generally did not call again until each of the other animals had called. Furthermore, the order of individual callers – 1,2,3 with its permutations occurred much more frequently than the order 1,3,2 with its permutations – indicates that the animals called in a specific pattern; they did not call equally in both patterns nor did they exchange calls at random. These results indicate that an ordered system of antiphonal calling exists among animals in a group.

The simplicity of these primate examples of syntax should be stressed. In no way do they approach the complexity of human rules for sequencing words and sentences or determining conversational interactions; they do indicate that rule-governed communication systems per se are not unique to humans. One can find simple rule systems both within primate calls and between primate callers. The use of rule systems for vocal communication is not limited to human beings.

ONTOGENY

The development of communication in primates has received surprisingly little attention. Although studies involving deafening and tutoring to indicate degrees of plasticity in song learning in birds have been undertaken (see Marler 1970b; Nottebohm 1970, for review), the ontogeny of communication in monkeys has been the subject of only a few studies, notably the work of Chevalier-Skolnikoff (1974 and chap. 13, this volume) on the ontogeny of visual communication in monkeys and great apes and the work of Newman and Symmes (chap. 11, this volume) on the ontogeny of vocal communication in squirrel monkeys.

Some studies demonstrated impairment of call production following social deprivation or deafening (Talmage-Riggs, Winter, Ploog, and Mayer 1972; Newman and Symmes 1974) and a few successfully conditioned vocalizations (Larson, Sutton, Taylor, and Lindeman 1973; Wilson 1975). Our approaches to ontogeny contrast with these studies.

First, the naturalistic description of vocal communication has been the primary step in the understanding of the development of speech and language in human beings. However, aside from the work of Chevalier-Skolnikoff (1974), Bertrand (1969), and Symmes et al. (1979), we have no natural histories of monkey vocal ontogeny. Currently we are gathering vocal samples from each of our infant cotton-top tamarins. We record approximately 3 hours each week from each infant during the first 20 weeks of life. Although we have not completed the overall analyses, there are a few interesting preliminary results.

The infants go through an extensive period of "babbling" that is analogous to that of human infants. When the monkey infants are separated from their parents during the first two months, they emit a constant stream of vocalizations in highly variable forms (compared with the acoustic structure of adult calls) and in inappropriate juxtaposition with one another. Over time these vocalizations become more stereotyped and closely modeled to adult forms, and they also appear in more appropriate contexts. The callitrichids we have studied are much more similar to human beings in the extent and complexity of their early babbling than any other monkeys of which we are aware.

We observed a Piagetian-like progression of vocal development that seems to parallel the nature of social relationships between infant and parent. This is best illustrated through two examples of developmental retardation. In one group of pygmy marmosets twins were born to parents who died soon after. Other pygmy marmoset infants progressed from infant babbling through juvenile stages to essentially adultlike vocal behavior within five months, when their parents would produce new infants. In contrast, the parentless twins displayed a much slower rate of development, not reaching the juvenile stage until the age of 1, and throughout their lives, they never attained a fully adult vocal repertoire (Pola and Snowdon 1975).

In the stump-tailed macaque, Lillehei and Snowdon (1978) found that the older infants in a group displayed complex, individual coo vocalizations, whereas the youngest animal had a simple call structure. Fortuitously, for the purposes of the study, the mother of an older infant died, and the infant regressed to an earlier stage of social dependency and began nursing from an older sister. In addition to the regression in social behavior, the infant experienced a regression toward a simpler structure in its contact calls. Both examples seem to confirm that progression toward an adult repertoire, which would include a greater complexity of call structure, is related to the degree of social independence from its parents that a young animal has attained. Stage of independence rather than mere chronological age is the main determinant of adult vocal competence.

If callithrichids are to be useful models in the study of nonhuman–human vocal communication, a mechanism for the acquisition of vocal signals similar to that found in human infants must be demonstrated.

CONCLUSIONS

In this chapter I have argued that primate communication is considerably more complex than was previously thought and that traditional ethological models for studying communication are inadequate for the study of primates. Models and paradigms from linguistics and psycholinguistics have been introduced and applied to the analysis of various problems of primate vocal communication. The technique of distinctive-features analysis has been used to demonstrate subtle contextual, or phoneticlike, variations within classes of calls that formerly had been considered unitary. Successfully synthesized monkey calls have been used in playback studies to determine the capability of monkeys to categorize the phonetic-like differences in their calls. This apparent categorization is evidently a function of the type of paradigm and the type of stimuli used in asking the question. When nonsynthesized calls of familiar individuals are used, the responses indicate individual differentiation; that is, they are no longer categorical.

The use of quantitative-analytic techniques has demonstrated the existence of subtle and specific variants within calls. The types of variability include: individual, which allows individual recognition by fellow group members; populational, which distinguishes populations – at least at the level of subspecies, and localization, which allows the caller to compensate for distance from the recipient.

The sequences of animal calls are definable and predictable. Simple grammars can be constructed to describe sequences, and Markovian analyses have shown sequential dependencies in several species. In addition, pygmy marmosets seem to employ a rule system to govern antiphonal exchanges, for individuals called in sequence and certain sequences were more likely to occur than others.

The ontogeny of vocal communication in monkeys needs extensive study. Only a few studies indicte any role for auditory experience and learning mechanisms in the acquisition of calls. Immature marmosets and tamarins babble extensively, much more than any other nonhuman primate, and thus provide an interesting characteristic for the pursuit of ontogenetic questions. The achievement of a given level of social maturity appears to be closely correlated with a given level of vocal maturity. Preliminary evidence suggests that long-call structure in saddlebacked tamarins may result from learning experiences.

Although similar study techniques are being used with both monkeys and human beings, and although some analogous results have been obtained, it must be stressed that there is no identity between the natural communication system of any primate and the natural communication system (including language) of human beings. Nor is there any biological reason to assume an identity of communication systems between species. However, a comprehensive study of primate communication can illuminate exactly which aspects of human language are species-unique and which have some rudimentary origins in the natural communication systems of other species. Nonhuman species able to mimic human speech and language disorders may provide better models for the study of those disorders. The more we know about the communication systems of a greater variety of nonhuman primates, the greater will be our understanding of our own unique communication system.

REFERENCES

Altmann, S. A. 1965. Sociobiology of rhesus monkeys. II. Stochastics of social communication. *Journal of Theoretical Biology* 8:490–522.

Beecher, M. D., Petersen, M. R., Zoloth, S. R., Moody, D. B., and Stebbins, W. C. 1979. Perception of conspecific vocalizations by Japanese macaques. *Brain, Behavior and Evolution* 16:443–60.

Bertrand, M. 1969. The behavioral repertoire of the stumptail macaque: a descriptive and comparative study. *Bibliotheca Primatologica* 11:1–273.
Brown, C. H., Beecher, M. D., Moody, D. B., and Stebbins, W. C. 1979. Locatability of vocal signals in Old World monkeys: design features for the communication of position. *Journal of Comparative and Physiological Psychology* 93:806–19.
Chevalier-Skolnikoff, S. 1974. The ontogeny of communication in the stumptail macaque (*Macaca arctoides*). *Contributions to Primatology* 2:1–174.
Chomsky, N. 1957. *Syntactic structures*. The Hague: Mouton.
Cleveland, J., and Snowdon, C. T. 1982. The complex vocal repertoire of the adult cotton-top tamarin, *Saguinus oedipus oedipus. Zeitschrift für Tierpsychologie* 58:231–270.
    In preparation. Predictable sequences of antiphonal calling in pygmy marmosets.
Eimas, P., Siqueland, E., Jusczyk, P., and Vigorito, J. 1971. Speech perception in infants. *Science* 171:303–6.
Eisenberg, J. F. 1976. Communication mechanisms and social integration in the black spider monkey (*Ateles fusciceps robustus*) and related species. *Smithsonian Contributions to Zoology* 213:1–108.
Fodor, J. A., Bever, T. G., and Garrett, M. F. 1974. *The psychology of language*. New York: McGraw-Hill.
French, J. A., and Snowdon, C. T. 1981. Sexual dimorphism in responses to unfamiliar intruders in the tamarin (*Saguinus oedipus*). *Animal Behaviour* 29: 822–9.
Gallup, G. G. 1977. Self-recognition in primates: a comparative approach to the bidirectional properties of consciousness. *American Psychologist* 32:329–38.
Gautier, J.-P. 1974. Field and laboratory studies of the vocalizations of talapoin monkeys (*Miopithecus talapoin*). *Behaviour* 51:209–73.
Green, S. 1975a. Variations of vocal pattern with social situation in the Japanese monkey (*Macaca fuscata*): a field study. In *Primate behavior:* Vol. 4, *Developments in field and laboratory research,* ed. L. A. Rosenblum, pp. 1–102. New York: Academic Press.
    1975b. Dialects in Japanese monkeys: vocal learning and cultural transmission of locale specific behavior? *Zeitschrift für Tierpsychologie* 38:304–14.
Green, S., and Marler, P. 1979. The analysis of animal communication. In *Handbook of behavioral neurobiology*, vol. 3, *Social behavior and communication,* ed. P. Marler and J. G. Vandenbergh, pp. 73–158. New York: Plenum.
Hailman, J. P. 1977. *Optical signals*. Bloomington: Indiana University Press.
Hodun, A., Snowdon, C. T., and Soini, P. 1981. Subspecific variation in the long calls of the tamarin *Saguinus fuscicollis. Zeitschrift für Tierpsychologie* 57: 97–110.
Krebs, J. R., and Kroodsma, D. E. 1980. Repertoires and geographical variation in birdsong. In *Advances in the study of behavior,* Vol. 11, ed. J. S. Rosenblatt, R. A. Hinde, C. Beer, and M.-C. Busnell, pp. 143–77. New York: Academic Press.
Kuhl, P. K., and Miller, J. D. 1975. Speech perception by the chinchilla: voiced–voiceless distinction in alveolar plosive consonants. *Science* 190:70–72.
Larson, C. R., Sutton, D., Taylor, E. M., and Lindeman, R. 1973. Sound spectral properties of conditioned vocalizations in monkeys. *Phonetica* 27:100–10.
Liberman, A. 1970. The grammars of speech and language. *Cognitive Psychology* 1:301–23.
Liberman, A. M., Cooper, F. S., Shankweiler, D. P., and Studdert-Kennedy, M. 1967. Perception of the speech code. *Psychological Review* 74:431–61.
Lieberman, P. 1975. *On the origins of language*. New York: Macmillan.

Lillehei, R. A., and Snowdon, C. T. 1978. Individual and situational differences in the vocalization of young stumptail macaques (*Macaca arctoides*). *Behaviour* 65:270–81.

Marler, P. 1970a. A comparative approach to vocal learning: song development in white-crowned sparrows. *Journal of Comparative and Physiological Psychology, Monograph Supplement* 71:1–25.

1970b. Birdsong and speech development: could there be parallels? *American Scientist* 58:669–73.

1977. The structure of animal communication sounds. *Recognition of complex acoustic signals,* ed. T. H. Bullock, pp. 17–35. Berlin: Dahlem Konferenzen.

Marler, P., and Hobbett, L. 1975. Individuality in the long range vocalizations of wild chimpanzees. *Zeitschrift für Tierpsychologie* 38:97–109.

Marshall, J. T., and Marshall, E. R. 1976. Gibbons and their territorial songs. *Science* 193:235–7.

Miller, C. A., and Morse, P. A. 1976. The "heart" of categorical speech discrimination in young infants. *Journal of Speech and Hearing Research* 19:578–89.

Miller, G. A. 1951. *Language and communication.* New York: McGraw-Hill.

Molfese, D. 1978. Electrophysiological correlates of categorical speech perception in adults. *Brain and Language* 5:25–35.

Morse, P. A., and Snowdon, C. T. 1975. An investigation of categorical speech discrimination by rhesus monkeys. *Perception and Psychophysics* 17:9–16.

Moynihan, M. 1966. Communication in the titi monkey, *Callicebus. Journal of Zoology* 150:77–127.

1970a. The control, suppression, decay, disappearance and replacement of displays. *Journal of Theoretical Biology* 29:85–112.

1970b. Some behavior patterns of platyrrhine monkeys. II. Saguinus goeffroyi and some other tamarins. *Smithsonian Contributions to Zoology* 28:1–77.

Newman, J. D., and Symmes, D. 1974. Vocal pathology in socially-deprived monkeys. *Developmental Psychobiology* 7:351–8.

Nottebohm, F. 1970. The ontogeny of birdsong. *Science* 167:950–6.

Petrinovich, L., and Peeke, H. D. 1973. Habituation to territorial song in the white-crowned sparrow (*Zonotrichia leucophrys*). *Behavioral Biology* 8:743–8.

Pisoni, D. 1977. Identification and discrimination of the relative onset time of two component tones: implications for voicing perception in stops. *Journal of the Acoustical Society of America* 61:1352–61.

Pisoni, D. R., and Lazarus, J. H. 1974. Categorical and non-categorical modes of speech perception along the voicing continuum. *Journal of the Acoustical Society of America* 55:328–33.

Pola, Y. V., and Snowdon, C. T. 1975. The vocalizations of pygmy marmosets (*Cebuella pygmaea*). *Animal Behaviour* 26:192–206.

Robinson, J. G. 1979. An analysis of the organization of vocal communication in the titi monkey, *Callicebus moloch. Zeitschrift für Tierpsychologie* 49:381–405.

Rowell, T. A., and Hinde, R. 1962. Vocal communication in the rhesus monkey (*Macaca mulatta*). *Proceedings of the Zoological Society of London* 138:279–94.

Simmons, J. 1973. The resolution of target range by echolocating bats. *Journal of the Acoustical Society of America* 54:157–73.

Sinnott, J. M., Beecher, M. D., Moody, D. B., and Stebbins, W. C. 1976. Speech sound discrimination by monkeys and humans. *Journal of the Acoustical Society of America* 60:687–95.

Smith, W. J. 1965. Message, meaning and context in ethology. *American Naturalist* 99:405–9.

1969. Messages of vertebrate communication. *Science* 165:145–50.

1977. *The behavior of communicating*. Cambridge: Harvard University Press.

Smith, W. J., Pawlukiewicz, J., and Smith, S. T. 1978. Kinds of activities correlated with singing patterns of the yellow-throated vireo. *Animal Behaviour* 26:862–84.

Snowdon, C. T. 1979. Response of non-human animals to speech and to species-specific sounds. *Brain, Behavior and Evolution* 16:409–29.

Snowdon, C. T., and Cleveland, J. 1980. Individual recognition of contact calls by pygmy marmosets. *Animal Behaviour* 28:717–27.

Snowdon, C. T., Cleveland, J., and French, J. A. In press. Responses to context- and individual-specific cues in cotton-top tamarin long calls. *Animal Behaviour*.

Snowdon, C. T., and Hodun, A. 1981. Acoustic adaptations in pygmy marmoset contact calls: locational cues vary with distances between conspecific. *Behavioral Ecology and Sociobiology* 9:295–300.

In preparation. Troop responses to long calls of isolated tamarins, *Saguinus mystax*.

Snowdon, C. T., and Pola, Y. V. 1978. Interspecific and intraspecific responses to synthesized marmoset vocalizations. *Animal Behaviour* 26:192–206.

Symmes, D., Newman, J. D., Talmage-Riggs, G., and Lieblich, A. K. 1979. Individuality and stability of isolation peeps in squirrel monkeys. *Animal Behaviour* 27:1142–52.

Talmage-Riggs, G., Winter, P., Ploog, D., and Mayer, W. 1972. Effects of deafening on the vocal behavior of the squirrel monkey (*Saimiri sciureus*). *Folia Primatologica* 17:404–20.

Thorpe, W. H. 1958. The learning of song patterns in birds with special reference to the song of the chaffinch, *Fringilla coelebs*. *Ibis* 100:535–70.

1972. Duetting and antiphonal singing in birds: its extent and significance. *Behaviour, Supplement* 18:1–197.

Thürwächter, W. 1980. Einzel-und Gruppenverhalten desk Krallenäffchens Saguinus oedipus oedipus (Wollkopftamarin) in Gefangenschaft unter besonder Berüchsichtigung der Lautäusserung "Long Call." Zulassungsarbeit, Facultät Biologie, Universität Freiburg im Breisgau.

Umiker-Sebeok, J., and Sebeok, T. A. 1980. Questioning apes. In *Speaking of apes*, ed. T. Sebeok and J. Umiker-Sebeok, pp. 1–59. New York: Plenum.

Verner, J. A., and Milligan, M. M. 1971. Responses of male white-crowned sparrows to playback of recorded song. *Condor* 73:56–64.

Waser, P. M. 1977. Individual recognition, intragroup cohesion and intergroup spacing: evidence from sound playback to forest monkeys. *Behaviour* 60: 28–74.

Waser, P. M., and Waser, M. S. 1977. Experimental studies of primate vocalization: specializations for long-distance propagation. *Zeitschrift für Tierpsychologie* 43:239–63.

Waters, R. S., and Wilson, W. A. 1976. Speech perception by rhesus monkeys: the voicing distinction in synthesized labial and velar stop consonants. *Perception and Psychophysics* 19:285–9.

Wiley, R. H., and Richards, D. G. 1978. Physical constraints on acoustic communication in the atmosphere: implications for the evolution of animal vocalizations. *Behavioral Ecology and Sociobiology* 3:69–94.

Wilson, W. A. 1975. Discriminative conditioning of vocalizations in *Lemur catta*. *Animal Behaviour* 23:432–6.

Winter, P. 1969. Dialects in squirrel monkeys: vocalizations of the Roman arch type. *Folia Primatologica* 10:216–29.

10 · How monkeys see the world: A review of recent research on East African vervet monkeys

ROBERT M. SEYFARTH and
DOROTHY L. CHENEY

Although laboratory studies have demonstrated that apes can be taught to use symbols to represent objects in the external world (Gardner and Gardner 1969; Premack 1976; Rumbaugh 1977; Patterson 1978), it generally has been assumed that the vocalizations of monkeys and apes under natural conditions convey information only about the motivational state of the signaler (e.g., Smith 1977). This dichotomy between the "semantic" signaling of apes in captivity and the "affective" signaling of the same and related species in the wild has tended to divert attention from two crucial questions: First, what selective forces might have given rise to the potential for representational signaling in nonhuman primates? Second, do we know enough about the signals of monkeys and apes to deny the possibility that such communication occurs under natural conditions? Similarly, there have been almost no comparative studies of communicative development in human and nonhuman species, largely because research on monkeys and apes and research on the language of children have been based on fundamentally different assumptions.

This chapter attempts to bring together research on communication in captive and free-ranging primates and to offer some analogies between nonhuman primate communication and the development of human language. We begin by describing the predator alarm calls of East African vervet monkeys and we discuss how the use of such calls develops during ontogeny. We then use this relatively simple communication system as a model to examine more complex forms of communication, arguing that research on the vocalizations of monkeys and apes, like research on grooming, alliance formation, and other social interactions, should focus on the question of how each communication system functions under natu-

We thank the government of Kenya for permission to conduct research in Amboseli National Park, and we are grateful to C. Brown, M. Petersen, and C. Snowdon for taking the time to organize the symposium for which this chapter was prepared. Our current research is supported by the Harry Frank Guggenheim Foundation and NSF Grant BNS 8008946. Between 1977 and 1980 work was supported by NIMH Postdoctoral Fellowship MH07446 to Robert M. Seyfarth, a grant from the Wenner–Gren Foundation and a NSF postdoctoral fellowship to Dorothy L. Cheney, and a National Geographic Society Grant and NSF Grant BNS 16894 to Peter Marler.

ral conditions. The long-term objective of our research is therefore to determine both what monkeys and apes *need* to signal about in their social interactions with each other and how individuals capable of representational signaling might enjoy an evolutionary advantage over others.

Because this chapter attempts to bring together research from a number of fields, we have deliberately borrowed technical terms from human psychology, linguistics, and child development with the intention of applying them to nonhuman primates. In doing so, one must remember that terms such as "semanticity" and "categorization," when applied to humans, carry implications of conscious intent that may not apply to nonhuman species. Thus, when we say, for example, that vervet monkeys have different alarm calls for different predators, we have no evidence that monkeys consciously select different vocalizations in precisely the same way that humans select different words to refer to different objects. Instead, our aim is to emphasize the functional parallels, however rudimentary, between the way monkeys use vocalizations and the way humans use words.

VERVET MONKEY ALARM CALLS

Do nonhuman primates in their natural habitat signal about objects or events in the world around them? A methodological problem confronting any observer who attempts to answer this question is that an animal cannot be interviewed. Instead, the observer must try to arrive at the content of each signal by studying the responses it evokes in other individuals. If the call, for example, conveys subtle information, such responses may not be immediate or obvious (Marler 1961; see also Snowdon, chap. 9, this volume). To draw a parallel with human language, it is as if an observer were attempting to discover the meaning of a conversation by studying only the overt responses of those listening to it. When investigating the possible semantic content of nonhuman primate vocalizations, we therefore began with a subset of vocalizations that might be expected to evoke noticeable and measurably different responses: the predator alarm calls of vervet monkeys. If semantic signaling could be demonstrated in this relatively simple case, vervet alarm calls might serve as a model for research on the more complex and subtle vocalizations used by monkeys in their social interactions with each other.

Vervet monkey alarm calls were first studied by Struhsaker (1967) in Amboseli National Park, Kenya. Struhsaker suggested that acoustically different alarm calls were given by the monkeys in response to different predators and that each call evoked a different and seemingly adaptive response. Our preliminary observations and recordings, made in Amboseli National Park between 1977 and 1978, supported Struhsaker's findings and indicated that adult vervets gave acoustically different alarm calls to at least three classes of predators: large mammalian carnivores, eagles, and snakes. Within each class, alarm calls were largely restricted to a single species: Thus large mammalian carnivore alarms (hereafter called leopard alarms) were given primarily to leopards (*Panthera pardus*); eagle

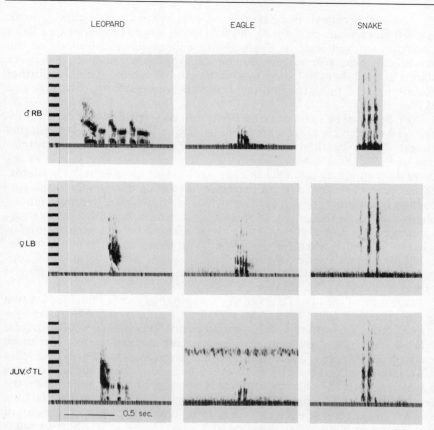

Figure 10.1. Examples of the alarm calls given by three individuals to leopards, martial eagles, and pythons. Top row shows alarms by adult male RB (group C), middle row shows alarms by adult female LB (group A), and bottom row shows alarms by 2½-year-old juvenile male TL (group C). The *x*-axis shows time; the *y*-axis shows frequency in units of 1 kHz. In eagle alarm of juvenile male TL, noise around 5 kHz is insects. (From Seyfarth, Cheney, and Marler 1980b. Reprinted with permission.)

alarms were given primarily to martial eagles (*Polemaetus bellicosus*); and snake alarms were given primarily to pythons (*Python sebae*). The acoustical features of each alarm-call type (Figure 10.1) were such that (1) alarm calls could be assigned unambiguously to one type, both by sound spectrography in the laboratory and by ear in the field, and (2) alarm calls could be distinguished from the nonalarm vocalizations they most closely resembled (Seyfarth, Cheney, and Marler 1980b).

In addition to being acoustically distinct, each alarm-call type was also associated with a different set of responses, each of which seemed to represent an adaptive escape strategy for coping with the hunting behavior of the predator involved. For example, when monkeys were on the ground, leopard alarms caused them to run into trees, where they appeared to be safe from leopard attacks. Eagle alarms caused them to look up or run into bushes, where they seemed to be safe from an eagle stoop, and snake alarms caused them to look down on the ground around them (for further descriptions of predator hunting behavior, see Seyfarth, Cheney, and Marler 1980b).

Both Struhsaker's and our own observations suggested that each alarm-call type effectively represented, or signified, a different class of external danger. However, there remained potential ambiguities in the interpretation of these results. Most serious was the possibility that monkeys apparently responding to an alarm call might in fact have seen the predator. Differences in response might therefore have been due simply to the perception of different predators, in which case acoustical differences among alarms would be irrelevant. Moreover, it was also possible that our perception of the monkeys' responses was affected by our expectations of what a "correct" response should be. With these points in mind, we decided to conduct a series of experiments in which tape recordings of leopard, eagle, or snake alarms given by known individuals would be played to the monkeys in the absence of actual predators.

Playbacks of the alarm calls of vervet monkeys were conducted on two groups between July 1977 and May 1978. The calls used had been tape-recorded from known individuals during actual encounters with leopards, martial eagles, and pythons. Playbacks were divided among the alarm calls of adult males, adult females, and juveniles. As shown in Figure 10.1, leopard alarm calls by adult males were longer and contained more acoustical units than alarms to other types of predators. Anticipating the possibility that call length might influence responses, we constructed long and short versions of each call type. Long versions contained a mean of five units and had a mean duration of 3.7 sec ($SD = 2.5$). Short versions contained a single unit and had a mean duration of 0.3 sec ($SD = .2$). To control for possible effects of amplitude, calls used in some trials differed naturally in amplitude across alarm types, with leopard alarms being louder than eagle alarms, which, in turn, were louder than snake alarms. In a second subset of trials, calls did not differ significantly in amplitude across alarm types (Seyfarth, Cheney, and Marler 1980b).

In each trial, an alarm call was played to monkeys (at least one adult male, two adult females, and two immatures) from a previously concealed speaker. Subjects were filmed using a sound movie camera for 10 sec preceding and 10 sec following each call playback. Subjects experienced 50 trials on the ground and 38 in the trees. No two playbacks to the same group of monkeys were conducted within 24 hr of each other, nor was any trial run within 15 min of alarming by nearby vervets or by the subjects' own group. Call-type order and speaker position relative to the subjects were varied systematically.

Alarm-call playbacks produced two kinds of response. Subjects in all age–sex classes looked toward the speaker and scanned the surrounding area, more in the 10 sec after playback than in the 10 sec before. They be-

Table 10.1. *Responses of monkeys to playbacks of leopard, eagle, and snake alarms*

Responses of monkeys	Alarm type					
	Leopard ($N = 19$)	Eagle ($N = 14$)	Snake ($N = 19$)	Leopard ($N = 10$)	Eagle ($N = 17$)	Snake ($N = 9$)
On ground						
Run into tree	8*	2	2			
Run into cover	2	6**	2			
Look up	4	7**	2			
Look down	1	4	14*			
In tree						
Run higher in tree				4	4	2
Run out of tree				0	5**	0
Look up				3	11	5
Look down				4	12	9*

Note: Entries show the number of trials in which at least one subject showed a given response that was longer in the 10 sec after playback than in the 10 sec before. Asterisks denote cases in which a particular response to a given call type occurred in significantly more trials than with either one or both of the other alarm types.
* $p < .01$. ** $p < .05$.

haved as if searching for additional cues, both from the alarmist and elsewhere. In addition, each alarm-call type elicited a distinct set of responses (Table 10.1). When subjects were on the ground, leopard alarms were more likely than other alarm types to cause them to run into trees. Eagle alarms made them look up and/or run into cover, and snake alarms caused them to look down. When subjects were in trees, eagle alarms were more likely than other alarm types to evoke looking up and/or running out of the tree, whereas snake alarms were more likely to cause subjects to look down. The monkeys behaved as though each alarm-call type designated a different external object or event.

This view of vervet alarm calls as rudimentary semantic signals contrasts with many earlier interpretations, which have tended to regard animal signals simply as manifestations of different levels of arousal, lacking clearly defined external referents (e.g., Smith 1977). If such was the case, we might expect that responses to alarms would have differed in relation to call features that mirror arousal levels, such as call length or amplitude. To test this hypothesis, we compared responses to long and short versions of each call type (see earlier discussion). Responses to leopard, eagle, and snake alarm playbacks were also compared (1) when amplitude differed systematically across alarm type and (2) when amplitude was controlled (see earlier discussion). Results indicated that variation in call length and equation of amplitude, as well as variation in the age–sex class of both alarmists and responders, failed to blur distinctions among major response categories. Variation in the acoustical structure of different call types was the only feature both necessary and sufficient to explain differences in response (Seyfarth, Cheney, and Marler 1980b).

It has also been argued that animal signals do not refer to specific, narrowly defined external referents but that each signal instead encodes one of a small number of very generalized "messages" such as "attack," "escape," or "frustration" (Smith 1977). A few such general signals are thought to be capable of eliciting a wide variety of responses because they are given in different contexts. By this interpretation, the "meaning" of each signal is highly context-dependent. In our experiments, however, context was not a systematic determinant of the responses of vervets to alarm calls. Different alarms evoked different responses in the same context, and responses to some alarms remained constant despite contextual variation. The most parsimonious interpretation is that each alarm represented a certain class of danger and that monkeys responded according to their vulnerability to that danger at the time.

HOW MONKEYS SEE THE WORLD

The relation between semantic signals and natural categories

The preceding discussion of vervet alarm calls raises the question of how monkeys classify objects in the world around them. If an organism is to use different signals to represent the virtually infinite number of objects in its environment, it must either employ an infinite number of signals or sort objects into groups. Present indications are that the repertoires of natural, meaningfully distinct signals are limited in animals. Thus any study of representational communication must inevitably consider how a given species "categorizes" objects in its natural habitat. Although research on the "natural categories" of nonhuman species recently has received considerable attention (e.g., Herrnstein, Loveland, and Cable 1976; Cerella 1979; Sandell, Gross, and Bornstein 1979), few studies have gone beyond the phenomenon as it is expressed in the laboratory to consider how such cognitive abilities may have evolved under natural conditions. Category formation is also well documented in humans (e.g., Berlin and Kay 1969; Rosch 1973, 1977), where the ontogeny of word–object associations has been well studied (e.g., Clark 1973; Nelson 1973; Anglin 1977; see also later discussion).

By giving alarm calls to some species but not to others, and by giving acoustically different alarms to different predators, vervet monkeys effectively categorized objects in the world around them. When giving alarm calls, adults were most selective, giving leopard alarms primarily to leopards, eagle alarms primarily to martial eagles, and snake alarms primarily to pythons. In marked contrast to adults, infants (monkeys under 1 year of age) and juveniles (monkeys older than 1 year but not yet adult size) gave alarm calls to a significantly wider variety of species and were significantly more likely to give alarms to nonpredators like warthogs, pigeons,

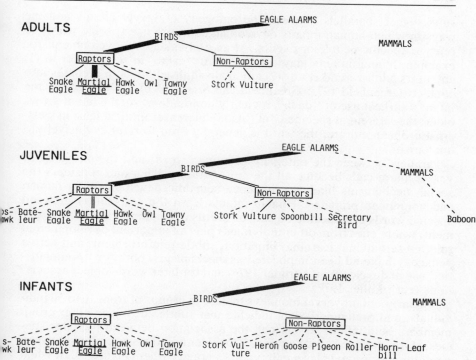

Figure 10.2. Number of eagle alarms given by adult, juvenile, and infant monkeys to different species or objects. Broken line, 1–5 alarms; single line, 6–10 alarms; double line, 11–15 alarms, dark line, >15 alarms. Data collected over a 14-month period from 31 adults, 16 juveniles, and 17 infants. (From Seyfarth and Cheney 1980. Reprinted with permission.)

and falling leaves that posed no danger to them (Seyfarth and Cheney 1980; Seyfarth, Cheney, and Marler 1980a). An example of these age-related differences in alarm-calling behavior is shown in Figure 10.2.

Intriguingly, however, although infants gave alarm calls to a wider variety of species than did adults, infant alarm-calling behavior was not entirely random. Infants gave leopard alarms primarily to terrestrial mammals, eagle alarms to birds, and snake alarms to snakes or long thin objects (Seyfarth and Cheney 1980). In other words, from a very early age infants distinguished between general predator *classes* (e.g., terrestrial mammal vs. flying bird), whereas adults distinguished between particular predator *species* within such classes (e.g., leopards vs. other terrestrial mammals and martial eagles vs. other birds).

Development in a social setting

These data on vervet alarm calls suggest some ways in which nonhuman primate semantic signals, and the categorization they imply, develop under natural conditions in the absence of human training. Moreover, results suggest parallels between the ontogeny of vervet alarms and the ways in which human infants develop an ability to use specific words to refer to specific people or events. At an early stage in development, for example, human infants may use the word "dada" to refer either to (1) any of its adult caretakers or (2) any male adult with whom it comes into contact (Greenfield 1973; Greenfield and Smith 1976). In both cases, the infant's earliest use of "dada" is clearly nonrandom. As the infant grows older, the referential specificity of "dada" increases until, at least in western European culture, the word is generally used to refer exclusively to one person.

Parallels between the ontogeny of human words and the ontogeny of vervet alarm calls become all the more striking when one considers the social mechanisms likely to influence semantic development in human and nonhuman primates. Human infants learn the association between specific words and specific objects or events in a complex social environment, where cues from other individuals play an important developmental role. Observational learning, imitation, subtle reinforcement, and active pedagogy have all been implicated as mechanisms, both for "prelinguistic" social development (Bruner 1976) and for later word–object associations (e.g., Miller 1977).

Similarly, infant vervet monkeys mature in groups of from 10 to 50 individuals, and preliminary data indicate that subtle cues from other group members help infants to sharpen the association between predator species, alarm-call type, and response. Consider first the association between predator species and alarm-call type. As noted earlier (Figure 10.2), when infants first began to given alarm calls each alarm type was restricted only to a certain broad class of objects (e.g., flying bird vs. terrestrial mammal), whereas the alarms of older individuals were limited to a particular species within each class (e.g., martial eagle vs. other flying bird). Fieldwork currently in progress suggests that subtle reinforcement may be one mechanism by which infants restrict the number of species to which they give alarms. For example, if an infant is the first to alarm at a martial eagle – one of the monkeys' main predators – it is virtually certain that other group members will also give alarms. In contrast, if an infant alarms at a species that does not attack monkeys, the probability of subsequent alarms by adults is much lower.

In an analogous manner, there is evidence for observational learning and/or imitation in the development of alarm-specific escape responses. As noted earlier, adults not only gave different alarm calls to different predators, they also responded differently to each alarm type. In contrast, playback experiments showed that infant responses were more generalized than those of adults and that infants were significantly more likely than adults to respond in ways that were potentially maladaptive (Seyfarth and Cheney 1980). Given that adultlike, alarm-specific responses develop only after experience, it is of interest that infants differed from adults in two further respects. First, infants in playback experiments were significantly more likely than adults to respond in a given way only after first looking at another animal who had already begun that same response;

and second, infants near their mothers were more likely to show adultlike responses than infants whose mothers had temporarily wandered more than 5 m away. Adults showed no such differences in response depending on the presence of other individuals.

In summary, it seems clear that predator classification by young monkeys improves with age and experience and that the monkeys' social environment plays an important role in this development. Although results thus far suggest intriguing parallels with human language development, further research is needed to determine the precise nature of the process of perceptual categorization that occurs among vervets and the exact roles of early predisposition, experience, and adult tutelage in such development.

How monkeys classify each other

Although vervet monkeys clearly discriminated among a variety of animal species and divided them into groups, we found no conclusive evidence either for or against the hypothesis that vervets can create a true, hierarchical taxonomy with all the formal relations this implies (Kay 1971). Premack (1976), for example, was able to teach the chimpanzee Sarah not only different signs for blue, green, and red but also the relation "blue is a type of color," "red is a type of color," and so on. Although the data shown in Figure 10.2 are suggestive, they do not prove that vervet monkeys, as they mature, come to understand the relation between specific objects (e.g., martial eagle, tawny eagle) and the higher-order groups to which they belong (e.g., raptor, bird, etc.).

Under what conditions might natural selection confer a reproductive advantage on individuals capable of creating hierarchical taxonomies, and how might such behavior function in the everyday social interactions of group-living monkeys? In an attempt to answer such questions, we conducted a series of experiments that examined how monkeys classify what is perhaps the most important part of their environment: each other.

Perhaps the most ubiquitous examples of hierarchical taxonomy in human societies are found in systems of kinship. Members of a given group are distinguished on one level as individuals and at the same time are grouped together at higher levels, such as the family, clan, and so on. Such classification is not only hierarchical (e.g., Fox 1967; Kay 1971), it also goes beyond a purely "egocentric" view of one's group: Each individual regards others not only in terms of his own relationship with them but also in terms of the relations other individuals have with each other. As applied to nonhuman primates, this view of recognition and social organization suggests the following question: If a group-living monkey, an adult female, for example, can "recognize" her offspring, does this mean she simply divides individuals into two groups (own offspring vs. others), or is there evidence of more subtle individual recognition on the basis of relations that animals other than the adult female are seen to have with each other? Observational data suggest that such complex recognition does indeed occur. Among both baboons and vervet monkeys, for example, individuals not only interact at high rates with members of their own immediate family but also compete with each other to establish bonds with members of high-ranking families (Cheney 1977, 1978; Seyfarth 1980). Individuals seem to "know" not only who ranks higher than they do but also which animals should be grouped together as members of the same kin group (e.g., Kurland 1977).

Experiments that attempt to document the extent to which monkeys recognize the relations that other group members have with each other share at least one methodological problem with experiments on semantic signaling: How can we induce animals to tell us what they know about the identity of another group member? As a first step in this investigation, we conducted a series of playback experiments on maternal recognition of the screams of their 2-year-old offspring. Adult female vervet monkeys and their offspring were selected for study because previous research had indicated that mothers intervened on behalf of their offspring in a large proportion of the offsprings' disputes. We therefore reasoned that playback of juvenile screams could evoke responses from mothers that were both strong and easily distinguishable from the responses of other adults.

Playback experiments were conducted in two groups of monkeys, each of which contained at least seven juveniles between 1 and 3 years of age. We selected for experimentation three to four typical scream bouts from each of four juveniles (two from each of two groups) matched for age, sex, and mother's dominance rank. When conducting an experiment, we first waited until the mother of one of the experimental juveniles was out of sight of her offspring and in proximity to two other "control" females who also had offspring in the group. Control females were also always out of sight of their 2-year-old offspring. A speaker was then concealed in a bush approximately 7 to 15 m from the females. Among trials for each mother, we systematically varied the speaker's position relative to the mother's orientation, so that no mother received all screams from the same direction. Filming of all females began 10 sec before each scream was played and continued for 45 sec after the scream bout had ended. We therefore were able to obtain an accurate record to estimate the probability that each female would look toward the speaker (or show some other response) in the first 10 sec before and after each scream, as well as the latency and duration (up to 45 sec) of each female's response.

Observational data collected before experimentation suggested that the screams of a juvenile often initially attracted wide attention in the group; we therefore predicted that all females would show some response to the playbacks but that mothers' responses would be stronger than controls'. In comparing the behavior of each female before and after screams, we found that playbacks significantly increased the probability that both mothers and controls would look toward the speaker (Cheney and Seyfarth 1980). In general, however, mothers showed both a shorter latency and a longer duration of looking than did controls. In addition, mothers were significantly more likely than controls to move toward the speaker (Cheney and Seyfarth 1980).

The responses of mothers demonstrate that adult female vervets are able to classify juveniles into at least two categories (offspring vs. others) on the basis of voice alone. Are they capable of further discrimination, among the offspring of others? When the responses of control females were compared to their behavior before each experiment, we found that playbacks significantly increased the probability that controls would look at the mother. In contrast, there was no change in the probability that one control would look at another.

There are two possible explanations for the behavior of control females. First, controls may have looked at mothers simply because the stronger response of mothers caused controls to orient toward them. Second, controls may have looked at mothers because they were able to discriminate among the screams of particular juveniles and to associate each juvenile with a particular adult female. To test between these two hypotheses, we examined separately those trials in which mothers did not approach the speaker. Results from these experiments still indicated a significant increase in the probability that controls would look at the mother (Cheney and Seyfarth 1980). Moreover, analysis of the gaze and position of control females in these trials suggested that it is unlikely that the controls' behavior was caused simply by cues from the mother.

The results of these experiments strongly suggest that, in recognizing other group members, individual vervet monkeys go beyond simple discriminations such as "close relatives versus others" and create more complex, hierarchical categories in which individuals are distinguished both on one level as individuals and on higher levels as members of particular kin groups. The ability of monkeys to classify individuals in this manner is almost certainly the result of natural selection acting on animals within a complex social framework, where detailed knowledge of relations among all group members is essential. Indeed, we may speculate that the ability of monkeys to classify predators (e.g., Figure 10.2) and their own vocalizations (e.g., Snowdon 1979; Zoloth, Petersen, Beecher, Green, Marler, Moody, and Stebbins 1979), as well as the ability of captive chimpanzees to understand the relation "X is different from Y but both are members of the set A" (Premack 1976), first emerged in the primates' need to classify each other in an analogous manner.

DISCUSSION: FROM SIMPLE TO COMPLEX FORMS OF COMMUNICATION

Language allows us to represent objects in the external world by means of relatively arbitrary, context-independent labels and therefore also reflects the way in which we classify objects for the purpose of representation. In at least a rudimentary sense, certain vocalizations of nonhuman primates function in an analogous manner. The alarm calls discussed in this chapter suggest that, in at least one aspect of their vocal repertoire, free-ranging monkeys are capable of signaling about objects. Moreover, data on both the ontogeny of alarm calls and individual vocal recognition indicate that complex classifications of objects and fellow group members may occur on the basis of vocal cues alone.

There is no reason to assume that such complex signaling abilities are restricted to one relatively simple and narrow subset of the monkeys' vocalizations. Indeed, when thinking about the evolution of language, we frequently assume that the first appearance of rudimentary representational signals brought a considerable selective advantage to its users. If such an assumption is correct, it seems unlikely that animals capable of semantic signaling under a specialized set of conditions (e.g., encounters with predators) will not make use of this ability during more frequently occurring social interactions.

By far the most commonly used vocalizations among vervet monkeys are an apparently graded series of grunts, uttered during a variety of social situations (Struhsaker 1967). Such grunts evoke few immediately obvious responses and at first appear to be simply manifestations of different levels of arousal. Research on the coo vocalizations of Japanese macaques, however, has suggested that subtle variations within the monkeys' graded vocal repertoire are specific to particular social interactions (Green 1975). The same vocalizations also appear to be perceived in a categorical manner (Zoloth et al. 1979). Snowdon and Pola (1978) previously demonstrated that pygmy marmosets categorically responded to synthesized continua of their trills. Moreover, field experiments on the grunts of vervet monkeys now in progress indicate that such calls are not only context-specific but also individually distinctive and in some cases clearly serve to designate different external events (Cheney and Seyfarth, in press). Such results both suggest that we are only beginning to understand how much information is conveyed in animal signals and emphasize the limited usefulness of any strict dichotomy between "semantic" human language and "affective" animal communication.

Premack (1975), for example, has argued that an "affective" signaling system that conveys information only about the signaler's motivational state can become functionally semantic if vocalizations are individually distinctive and if all members of a group "agree" about the level of arousal of different events (see also Marler 1977; Green and Marler 1979). Thus, if a listener can recognize the voice of another individual and also knows how that individual responds to different events, the listener will be able to determine without further cues what that individual is doing, or to what it is reacting, when it vocalizes.

All the assumptions on which Premack's argument is based are satisfied in groups of free-ranging primates. As we have suggested, primate groups are composed of individuals who appear to recognize each other by sight, and, in the absence of visual cues, there is also evidence that monkeys can recognize the vocalizations of others. Even if we accept the suggestion that, for example, different alarm calls are simply manifestations of different levels of arousal, group members seem to "agree" on which predator should be associated with each arousal level.

Premack's argument emphasizes that the apparent dichotomy between "affective" and "semantic" signaling breaks down when we begin to examine how signals function under natural conditions. Whatever the evolutionary history of primate vocalizations, at least some such vocalizations can now be said to function in a semantic manner. As research on the social signals of monkeys and apes continues, we think it is important to remember that animals will use complex communicative signals only when it becomes selectively advantageous to do so. Thus, a central question for future research would appear to be: How might the use of semantic signals benefit those who employ them? Research on behavior such as grooming and alliance formation suggest that these interactions are best understood as means by which animals form those social bonds that subsequently bring benefit to them. Similarly, only by studying vocalizations within the broader context of other forms of social interaction and by considering what animals *need* to signal about will we begin to understand the evolutionary history of representational signaling and the selective forces that may have favored one type of communication over others.

REFERENCES

Anglin, J. M. 1977. *Word, object and conceptual development.* New York: Norton.

Berlin, B., and Kay, P. 1969. *Basic color terms: their universality and evolution.* Berkeley: University of California Press.

Bruner, J. S. 1976. From communication to language – a psychological perspective. *Cognition* 3:255–87.

Cerella, J. 1979. Visual classes and natural categories in the pigeon. *Journal of Experimental Psychology: Human Perception and Performance* 5:68–77.

Cheney, D. L. 1977. The acquisition of rank and the formation of reciprocal alliances in free-ranging baboons. *Behavioral Ecology and Sociobiology* 2: 303–18.

1978. Interactions of male and female baboons with adult females. *Animal Behaviour* 26:389–408.

Cheney, D. L., and Seyfarth, R. M. 1980. Vocal recognition in free-ranging vervet monkeys. *Animal Behaviour* 28:362–7.

In press. How vervet monkeys perceive their grunts: field playback experiments. *Animal Behaviour.*

Clark, E. 1973. What's in a word? On the child's acquisition of semantics in his first language. In *Cognitive development and the acquisition of language,* ed. T. E. Moore, pp. 65–110. New York: Academic Press.

Fox, R. 1967. *Kinship and marriage.* Harmondsworth, Eng.: Penguin.

Gardner, R. A., and Gardner, B. T. 1969. Teaching sign language to a chimpanzee. *Science* 165:664–72.

Green, S. 1975. Communication by a graded vocal system in Japanese monkeys. In *Primate behavior, Vol. 4: Developments in field and laboratory research,* ed. L. A. Rosenblum, pp. 1–102. New York: Academic Press.

Green, S., and Marler, P. 1979. The analysis of animal communication. In *Handbook of behavioral neurobiology. Vol. 3: Social behavior and communication,* ed. P. Marler and J. G. Vandenbergh, pp. 73–158. New York: Plenum.

Greenfield, P. M. 1973. Who is "dada"? Some aspects of the semantic and phonological development of a child's first words. *Language and Speech* 16:34–43.

Greenfield, P. M., and Smith, J. H. 1976. *The structure of communication in early language development.* New York: Academic Press.

Herrnstein, R. J., Loveland, D. H., and Cable, C. 1976. Natural concepts in pigeons. *Journal of Experimental Psychology: Animal Behavior Processes,* 2:285–302.
Kay, P. 1971. Taxonomy and semantic contrast. *Language* 47:866–87.
Kurland, J. 1977. Kin selection in the Japanese monkey. *Contributions to Primatology* 12:1–145.
Marler, P. 1961. The logical analysis of animal communication. *Journal of Theoretical Biology* 1:295–317.
    1977. Primate vocalizations: affective or symbolic? In *Progress in ape research,* ed. G. H. Bourne, pp. 85–96. New York: Academic Press.
Miller, G. A. 1977. *Spontaneous apprentices.* New York: Seabury Press.
Nelson, K. 1973. Some evidence for the cognitive primacy of categorization and its functional basis. *Merrill-Palmer Quarterly of Behavior and Development* 19:21–40.
Patterson, F. G. 1978. The gestures of a gorilla: language acquisition in another pongid. *Brain & Language* 5:72–97.
Premack, D. 1975. On the origins of language. In *Handbook of psychobiology,* ed. M. S. Gazzaniga and C. B. Blakemore, pp. 591–605. New York: Academic Press.
    1976. *Intelligence in apes and man.* Hillsdale, N.J.: Erlbaum.
Rosch, E. H. 1973. On the internal structure of perceptual and semantic categories. In *Cognitive development and the acquisition of language,* ed. T. E. Moore, pp. 111–44. New York: Academic.
    1977. Classification of real-world objects: origins and representations in cognition. In *Thinking: readings in cognitive science,* ed. P. N. Johnson-Laird and P. C. Wason, pp. 212–22. Cambridge University Press.
Rumbaugh, D., ed. 1977. *Language learning by a chimpanzee: the Lana Project.* New York: Academic Press.
Sandell, J. H., Gross, C. G., and Bornstein, M. C. 1979. Color categories in macaques. *Journal of Comparative Physiology and Psychology,* 93:626–35.
Seyfarth, R. M. 1980. The distribution of grooming and related behaviours among adult female vervet monkeys. *Animal Behaviour* 28:798–813.
Seyfarth, R. M., and Cheney, D. L. 1980. The ontogeny of vervet monkeys alarm calling behavior: a preliminary report. *Zeitschrift für Tierpsychologie* 54:37–56.
Seyfarth, R. M., Cheney, D. L., and Marler, P. 1980a. Monkey response to three different alarm calls: Evidence of predator classification and semantic communication. *Science* 210:801–3.
    1980b. Vervet monkey alarm calls: semantic communication in a free-ranging primate. *Animal Behaviour* 28:1070–94.
Smith, W. J. 1977. *The behavior of communicating: an ethological approach.* Cambridge, Mass.: Harvard University Press.
Snowdon, C. T. 1979. Response of non-human animals to speech and to species-specific sounds. *Brain, Behavior and Evolution* 16:409–29.
Snowdon, C. T., and Pola, Y. V. 1978. Interspecific and intraspecific responses to synthesized pygmy marmoset vocalizations. *Animal Behaviour* 26:192–206.
Struhsaker, T. T. 1967. Auditory communication among vervet monkeys (*Cercopithecus aethiops*). In *Social communication among primates,* ed. S. A. Altmann, pp. 281–324. Chicago: University of Chicago Press.
Zoloth, S. R., Petersen, M. R., Beecher, M. D., Green, S., Marler, P., Moody, D. B., and Stebbins, W. 1979. Species-specific perceptual processing of vocal sounds by Old World monkeys. *Science* 204:870–3.

PART IV

Ontogeny and primate communication

Questions about the mechanisms of ontogeny in primate communication have received surprisingly little emphasis. The pioneering studies of Thorpe (1958), Marler (1970), and their students have given us a picture of bird song development that, at least in some species, incorporates a combination of genetic and experiential components. As yet we have no comparable demonstration in primates that learning is a significant component in vocal or any other modality of communication. Attempts to parallel birdsong research have produced some communication deficits in deafened and socially isolated monkeys. Nevertheless, despite profound deprivations in monkeys, the bulk of their communicative repertoire remains unaltered, and, indeed, data presented by Newman and Symmes indicate a fixity of call structure from the first days after birth. However, before we dismiss the importance of learning in communication, we must consider the evidence we have from general observations on primates.

Several preceding chapters have mentioned data pertaining to primate ontogeny. Gautier and Gautier-Hion, for example, presented data that infants called more frequently, but with less diversity of call structure, than did adults. Snowdon reported that marmoset and tamarin infants often vocalized by "babbling," that is, by giving imperfect forms of adult calls in inappropriate situations. Seyfarth and Cheney described the response of young vervet monkeys to classes of animals that could constitute predators. To the young monkey, predator classes were much broader than those of adults, although they did use the appropriate alarm call with the appropriate general class of predator. These instances of developmental changes suggest that at least some communication is learned. However, the content of that learning cannot be simply characterized and manipulated.

The chapters in Part IV, which represent three different approaches, deal more directly with problems of ontogeny. Although each approach emerges from a disparate tradition, each chapter holds considerable promise for future research.

Following in the tradition of birdsong research, Newman and Symmes examine the contributions of genetic and experiential mechanisms to the development of vocal communication in monkeys. They review primate studies analogous to those previously successful with songbirds that look at the effects of restricting acoustic experience through social isolation and deafening. They conclude that under certain circumstances the morphology of some vocal structures can be altered but that the necessity of auditory feedback and imitation for development of normal vocal behavior in primates has not been adequately demonstrated as yet.

Based on their own extensive experience with squirrel monkeys, Newman and Symmes review longitudinal studies of the isolation-peep vocalization. Individual specific features of the call remain constant over the animals' first two years, though there are slight changes in the duration of the call. Since squirrel monkeys have two distinct pelage populations – the Roman-arch-eyebrow type and the Gothic-arch-eyebrow type – and since each population is associated with a distinctively different structure of isolation-peep call, the effects of hybridization between the two morphs offer an appropriate arena for the study of genetic versus learned components. Hybrids have the morphological features of their mothers, that is, a Roman-arch mother produces a Roman-arch infant. Likewise, the structure of the isolation peeps of most hybrid infants corresponds to the isolation-peep structure of their mothers. There appears to be a close association between the inheritance of morphology and the structure of the isolation peep.

Data such as these support the notion that primate calls are inherited, that they develop without much auditory experience, and that they cannot be modified by auditory experience. However, it should be cautioned that too few primate species and too few vocalization types have been studied as yet. It should be noted that bird song, so often shown to be malleable by experience, is modifiable only in a subset of bird species and, furthermore, is only one of several vocalizations given by birds. Demonstrations of vocal learning in primates may have been somewhat less than successful because few studies have focused on a call that is a functional analog of birdsong. Newman and Symmes suggest that calls involved in the establishment of intragroup social relationships might be more amenable to demonstrations of vocal learning.

Another reason for the failure of studies of primate vocal learning may be the lack of an equivalent of the white-crowned sparrow (Marler 1970). To date, deafening and isolation techniques have been used with only two species, the squirrel monkey and the rhesus macaque. No cross-fostering studies have been done. Species that display sufficient variability in the structure and usage of adult vocalizations over different populations – the primate equivalent of the white-crowned sparrow – may be waiting to be discovered. Any parallels developed between primate communication and human communication will never be truly impressive until mechanisms for vocal acquisition similar to those of human beings can be demonstrated in primates.

Epple, Alveario, and Katz provide another perspective on ontogeny. Their chapter demonstrates the separation of the hormonal from the social determinants of communication. The authors worked with tamarin monkeys, which differ from many other primate species by being monogamous and highly territorial. Tamarins exhibit considerable intrasexual aggression, especially directed toward unfamiliar intruders. These animals also have highly developed scent glands, which are used in both sexual and aggressive contexts. Because such signal systems involving reproduction and aggression are thought to be under direct control of gonadal hormones, one can ask whether scent-gland morphology and scent-marking behavior as well as other aggressive activities are under strict gonadal control.

Gonadal hormones are theorized to have two types of effects on behavior. The first, an organizing effect, is the more or less permanent impact of early hormonal experience and is best tested by manipulating hormonal state early in development. The second, an activational effect, concerns the immediate influence of hormones on a given behavior. It is best tested by manipulating hormonal state after development is completed. To distinguish between these two effects, Epple, Alveario, and Katz looked at gonadectomized and sham-gonadectomized animals. Some animals experienced gonadectomy prior to puberty when gonadal hormones would be expected to have an organizing effect, and some animals experienced gonadectomy after puberty when the organizational effect would be minimal. To test the effect of social determinants, the experimenters placed animals in various social groups – situations in which they were dominant or subordinate, with an experienced mate or a mate as inexperienced as the test animal.

Their results indicated that gonadal hormones prior to and during puberty are necessary for the full elaboration of glandular fields, whereas the absence of gonads in fully developed adults produced little change in glandular structure. Similarly, adult gonadectomy produced little effect on scent marking and other aggressive activities. Prepubertal gonadectomy also produced only a transient suppression of scent-marking behavior and did not impair other aggressive activities. The social experience of an animal appears to be much more important in the expression of aggressive behavior. When an animal was dominant or with an older, more experienced mate, it showed more aggressive responses to intruders than it did when it was subordinate or when both pair members were inexperienced.

Thus Epple and her colleagues show that even though the glandular structures for scent marking are under gonadal control, the behavioral activities of marking and other aggressive actions are not under gonadal control. The role of social experience is more important than the organizing or activating effects of gonadal hormones in the expression of aggressive communication in tamarin monkeys.

No psychologist could review the ontogeny of human communication without mentioning the tradition of cognitive development that originated with Jean Piaget. However, this tradition has rarely been applied to the study of primate communication. Chevalier-Skolnikoff offers us insight into the differing abilities of primate species to develop communication skills. The Piagetian tradition, basically an observational approach to the longitudinal study of development, is rigorously guided by a detailed and sequential theory of intellectual development. Recognizing the complexity of the primate, Chevalier-Skolnikoff employs complex models to study their communication.

Piaget's theory has been applied previously to other aspects of primate behavior. Jolly (1972) used it as a framework for integrating the large and diverse literature on primate learning. Chevalier-Skolnikoff (1977) used the Piagetian model for describing social development in several primate species. Parker (1977) used it as a model for comparing intelligence and nonstereotyped behavior in different species. Chapter 13, however, is the first explicit application of the theory to the study of communication.

The advantages of using the Piagetian framework are several. First, as a theory explicitly modeled after human development, it articulates the various cognitive and communicative abilities of human infants and suggests a normative rate and sequence for human infant development. This provides a standard for comparison with other species. Insofar as the various behaviors indicative of a particular stage of development can be successfully translated into behavioral acts observable in other species, a variety of primate species can be directly compared. For example, Chevalier-Skolnikoff shows that during an early stage of development, the sensorimotor phase, apes display the same levels of accomplishment as human infants, whereas monkeys demonstrate only five of the first six levels. Thus for the study of communication, the Piagetian framework can suggest a metric for comparison between species and can be extended to a comparison of individuals.

A second advantage of this theoretical system is as an organizational structure for the presentation of a vast amount of longitudinal observational data. The end result is not just a descriptive study but a highly organized and logical presentation of observational data. All too frequently the lack of a clear organizational framework for observational data is an inhibition to the gathering of such data. Yet without the basic natural history of communication ontogeny, we cannot hope to progress to more analytical and experimental studies.

One criticism of Piagetian studies has been the degree to which the classification of activities into the appropriate stage is both reliable and, when applied to nonhuman species, valid. To meet the test of reliability, Chevalier-Skolnikoff has had her observations scored by others, and to meet the test of validity as well, she has translated the tasks from human infants to monkeys and apes. As a way of comparing ontogeny in a wide variety of species, the Piagetian approach holds much promise.

REFERENCES

Chevalier-Skolnikoff, S. 1977. A Piagetian model for describing and comparing socialization in monkey, ape and human infants. In *Primate biosocial development,* ed. S. Chevalier-Skolnikoff and F. E. Poirier, pp. 159–87. New York: Garland.
Jolly, A. 1972 *The evolution of primate behavior.* New York: Macmillan.

Marler, P. 1970. A comparative approach to vocal learning: Some developments in white-crowned sparrows. *Journal of Comparative and Physiological Psychology, Supplement* 71:1–25.

Parker, S. T. 1977. Piaget's sensorimotor series in an infant macaque: A model for comparing unstereotyped behavior and intelligence in human and nonhuman primates. In *Primate biosocial development,* ed. S. Chevalier-Skolnikoff and F. E. Poirier, pp. 43–112. New York: Garland.

Thorpe, W. H. 1958. The learning of song patterns by birds, with special reference to the song of the chaffinch, *Fringilla coelebs. Ibis* 100:535–70.

11 · Inheritance and experience in the acquisition of primate acoustic behavior

JOHN D. NEWMAN and DAVID SYMMES

The interaction of genetic mechanisms with environmental influences in organizing the neural networks underlying behavior remains one of the elusive problem areas of modern biology. In primates this question has not yet been directed toward specific behavior patterns, partly because the limits of influence on behavior exerted by inheritance and experience have not been clearly defined. Within the context of communicative behavior, progress in identifying these limits has been made for the acoustic behavior of other animals. In this chapter, we consider evidence pertaining to the roles of inheritance and experience in primate acoustic behavior. We also summarize the kinds of vocalizations exhibited by infant primates and their resemblance to adult call morphology. Finally, we discuss acoustic behavioral traits for which patterns of inheritance are emerging, including some new data on the inheritance of vocalizations in squirrel monkeys.

ROLE OF EXPERIENCE IN THE NORMAL DEVELOPMENT OF PRIMATE VOCALIZATIONS

Vocal learning is a conspicuous attribute in the development of social sounds in many birds (e.g., Marler and Mundinger 1971) and, of course, in human speech. As Thorpe (1967) points out, vocal learning in tropical birds is related to maintaining pair and family bonds. If this were also true in primates, one would expect vocal learning to be widespread, since both a tropical ecology and a cohesive family characterize many primate spe-

A number of individuals have provided assistance of various kinds. Dr. P. D. MacLean, Laboratory of Brain Evolution and Behavior, NIMH, provided access to a hybrid squirrel monkey born in his laboratory. Dr. T. C. Jones, New England Regional Primate Research Center, assisted in acquiring two other hybrids. Dr. B. White, Laboratory of Experimental Pathology, National Institute of Allergy and Infectious Diseases, and Dr. N. Ma, New England Regional Primate Research Center, performed karyotype analyses. The staff of the Primate Project, Iquitos, Peru, particularly Dr. M. Moro, Mr. R. Kingston, and Sr. F. Encarnacion, identified and trapped groups of squirrel monkeys from specified geographic locations. Dr. H. J. Smith, Ms. K. Johnson, and Mrs. D. Bernhards assisted with sound recording and analyses in our laboratory.

cies. However, a number of authorities (e.g., Nottebohm 1972) have concluded that vocal learning plays an insignificant role in the development of primate vocalizations. Nottebohm (1972) views the evolution of vocal learning as an efficient mechanism for preventing saltatory gene dispersal. He argues that this adaptation fits the distribution pattern of migratory birds better than that of primates, since the former can more readily disperse over large geographic distances. However, this role for vocal learning does not fit as well in tropical birds nor in the closely spaced dialect system of the New Zealand saddleback (Jenkins 1978). Consequently, there is no satisfactory theoretical basis for explaining why vocal learning is present in some animal groups and not in others.

It would be a strange paradox to confirm that experience plays only an insignificant role in nonhuman primate vocal behavior. Given the long period of social development in many primates, the opportunity to acquire vocal traits through experience and practice are manifold. The absence of learned vocal behavior is not due to inherently poor learning skills in primates; the literature contains many references to learned behaviors under both laboratory (e.g., Rumbaugh 1971) and natural (e.g., Imanishi 1957) conditions. Nor is it due to an inherent lack of dexterity in vocal performance, given the wide range of sounds in primate vocal repertoires and the elaborate vocal performances of gibbons and titi monkeys (see the following discussion and Deputte, chap. 4, this volume). An inability of nonhuman primates to utter *human* speech (Lieberman 1968) is, of course, irrelevant to this point. Yet, the generally, almost casually, accepted view is that nonhuman primate vocalizations develop normally without the influence of learning. The evidence supporting this view deserves review and analysis.

Effects of restricting acoustic experience

Several authors have made reference to the normal vocal behavior of primates reared in the absence of conspecific vocalizations. Boutan (1913, cited in Marler 1963) claimed that all the calls of a hand-reared gibbon (*Hylobates leucogenys*) developed normally. However, as Marler points out, this claim was made without a clear indication of how well the author knew gibbon vocalizations. In any case, the claim preceded the availability of the sound spectrograph. Marler (1969) commented that a hand-reared chimpanzee produced rough grunts whose basic patterning and spacing were like those of wild chimps (although no spectrographic evidence was presented). Gautier (1974) stated that a newborn talapoin monkey (*Miopithecus talapoin*) reared without contact with conspecifics gave almost the entire vocal repertoire: distance–cohesion–isolation calls, flight calls, alarm calls. Similarly, Winter, Handley, Ploog, and Schott (1973) reported that squirrel monkeys (*Saimiri sciureus*) produce nearly all the species' vocal repertoire within the first week of life (the only exception being twitters) and that calls of infants reared with muted mothers were within the normal range for frequency and duration. Eisenberg (1976) analyzed the vocalizations of a black spider monkey (*Ateles fusci-*

ceps robustus) separated from its mother at 8 hours postpartum and raised apart from adult conspecifics. Although most adult vocalizations were not recognizable until 5 to 7 months of age, the author concluded that the infantile vocalizations were very similar to adult call morphology. In contrast to these findings of vocal normality, our work on rhesus monkeys (*Macaca mulatta*) reared in partial social isolation demonstrated that the development of normal coo calls is dependent on both hearing conspecifics and maternal interaction (Newman and Symmes 1974).

Auditory feedback and imitation in primate vocal development

A number of reports suggest a possible role of vocal "practicing" in shaping the final morphology of vocalizations.

Infant squirrel monkeys reared with conspecific animals produce a range of calls when separated from their mothers. Most infants we have studied in captive groups produce a variety of structurally diverse sounds, ranging from tonal calls resembling the isolation peep (IP) but with several slope reversals ("baby-piepen" of Ploog, Hopf, and Winter 1967) to calls with tonal and noisy segments to calls composed of continuous wideband noise superimposed upon a tonal element (Lieblich, Symmes, Newman, and Shapiro 1980). These non-IP isolation calls are entirely absent in older infants and in some of our animals were never recorded. By 1 year of age, these variable isolation calls no longer occur. Further, at about this same age, the IP reaches its adult form, subsequent samples showing little or no change in structure (Lieblich et al. 1980). In as much as the IPs of a particular adult squirrel monkey vary little over a time span of years (Symmes, Newman, Talmage-Riggs, and Lieblich 1979), the stabilization of IP form around 1 year of age imparts a "vocal signature" that, insofar as we can tell, remains for the life of the individual. Winter et al. (1973) demonstrated that the IPs of an infant squirrel monkey deafened soon after birth were within the normal range for frequency and duration. However, since these authors did not quantify variability in IP structure and since this infant died at 9 months of age, there was no demonstration of a stable adult IP in that animal. The only published study of squirrel monkeys deafened as adults did not include the IP in their analysis of vocalizations (Talmage-Riggs, Winter, Ploog, and Mayer 1972). Thus, a role for auditory feedback in acquiring and maintaining an individually distinctive stable IP in adults cannot be ruled out.

The territorial-defense songs of gibbons differ from those of other primates in that in most species the female utters the more elaborate vocalization, or great call (see Tembrock 1974; Marshall and Marshall 1976; Tenaza 1976; Marler and Tenaza 1977; and Deputte, chap. 4, this volume). Juvenile females have been observed singing along with their mothers, their contributions being quite simple compared with the mothers'. Whether these juvenile efforts represent a form of imitative practicing, necessary for correct performance of the elaborate adult vocalization, is presently unknown.

Troop-specific dialects have been discovered in certain calls of provi-

sioned Japanese macaques (*M. fuscata*) by Green (1975b). This author
suggests that the imitative learning of calls uttered in the past during early
provisioning experiences has resulted, by a form of cultural transmission,
in the maintenance of these dialects. Studies of undisturbed primate populations, including ours on squirrel monkeys in Peru (see later discussion),
have so far failed to find convincing evidence for troop-specific dialects.

One group of workers has aimed directly at shaping the structure of vocalizations by differential reinforcement (Larson, Sutton, Taylor, and
Lindeman 1973; Sutton, Larson, Taylor, and Lindeman 1973). In those
studies, coo calls of rhesus macaques gradually increased in duration and
stabilized in structure (harsh noise and inflections disappearing) over the
10-week conditioning period. This is perhaps the best documentation of
the voluntary alteration in structure of a primate vocalization.[1] However,
as the authors point out, there is some room for doubt as to the degree of
voluntary control. Since the arousal level of the subjects dropped with increased time in the conditioning task, a change in the quality of their vocalizations also may have occurred. The changes noted by these authors
in the quality of the calls over time are consistent with this notion, in that
several authors have linked a calm demeanor with low-pitched, unin-
flected tonality and an absence of noise (see Green 1975a).

Taken together, the evidence cited in this section does not preclude a
role of vocal learning in primate vocal development. That vocal morphology can change under specifiable conditions appears clear. However,
whether auditory feedback and imitation are *essential* for normal vocal
maturation remains to be demonstrated.

ONTOGENY OF PRIMATE ACOUSTIC BEHAVIOR

From the standpoint of understanding the influence of inheritance and experience on the development of primate vocalizations, one would like to
know how the various sounds of the adult repertoire mature and under
what conditions these maturational stages can be disrupted. Some calls
are not readily followed developmentally, notably the long calls of adult
males of numerous species, which are generally used in intertroop spacing. Such species typically have a single adult male, who is both dominant
and the sole producer of the long call for its troop (see Gautier and Gautier
1977). Since, in these same species, supralaryngeal accessory organs or
vocal sacs are prominent in adult males, the ability to produce fully developed long calls may be dependent on the maturation of these vocal accessory structures. There is ample opportunity for other males to hear
long calls but no clear documentation of such individuals practicing these

---

[1] The Lombard reflex, an increase in vocal amplitude in response to an increase in the level
of the background noise, also has been demonstrated in monkeys in a vocal-conditioning
paradigm. Monkeys, like humans, elevate vocal amplitude about 2 dB for every 10 dB increase in the background noise level. This observation suggests that monkeys monitor
their own voice and adjust their vocal amplitude to compensate for the masking effect of
noise. Thus both vocal duration and amplitude appear subject to some degree of modification in monkeys (Sinnott, Stebbins, and Moody 1975).

sounds. Thus, on the basis of existing evidence, the ability to produce male long calls is dependent only on sexual maturation, with social factors possibly controlling which males actually utter the calls.

The typical infant vocal repertoire in primates is rich in sounds related to establishing or maintaining physical contact with the mother. Calls given by infants briefly separated from their mothers are long in duration and tonal in quality, generally without abrupt frequency shifts. These are variously referred to as "isolation," "lost," or "separation" calls (the same terms being applied to similar sounds emitted by isolated adults). Longer separation results in calls exhibiting an increase in noisy, or frequency-shifted, components. Similar calls are given as part of "weaning tantrums" in older infants. Upon regaining contact with the mother, softer, briefer sounds are emitted, as is the case during nursing or affinitive maternal contact. Table 11.1 lists examples of infant calls from a variety of nonhuman primates.

As has been described for several other primate species, the squirrel monkey isolation call is especially significant for an infant who accidentally becomes separated from its mother. Baldwin (1967) described a case in the semi-free-ranging colony at Monkey Jungle in Goulds, Florida, in which a baby who fell off a female "aunt" and began emitting isolation calls was quickly retrieved by its mother. Baldwin (1967) experimentally tested the significance of the isolation call by administering a tranquilizer to a baby being carried by its mother, causing the baby to drop off the mother as it became sedated. The baby was largely ignored by the mother until it began to move and utter isolation calls, whereupon the mother exhibited retrieval behavior (approaching the infant and extending a shoulder toward it).

A study in our laboratory (Lieblich et al. 1980) traced the maturation of squirrel monkey isolation calls over the first two years of life. Although differing in detail between individuals, a change in call structure characteristic of all animals can be followed (Figure 11.1). The most obvious change involves a gradual lengthening of the call with age. In Roman-arch individuals (MacLean 1964), a significant decrease in the slope of the tonal fundamental also occurs, so that change in frequency over the middle of the call becomes less rapid with increasing age. However, from the standpoint of the adult isolation call, it is clear that its characteristic features are present in the calls of even a 1-day-old infant (Figure 11.1). This is true of both Gothic-arch and Roman-arch infants.

Development of intragroup vocalizations

Implicit in considerations of primate vocal development is the importance of adequate social experience, a factor that is dismissed as inconsequential in many studies of the acquisition of vocal traits in nonmammalian species. The primate worker wishing to control access to conspecific sounds while providing, at the same time, an adequate social environment is faced with a formidable task. Nevertheless, some significant observa-

Table 11.1. *Major classes of infant vocalizations by species*

Species	Call	Main context	Call structure
Mountain gorilla[a]	Cry	Separated or in difficulty	Tonal, superimposed noise; 0.05–0.15 sec
	Shriek	"Temper tantrum"; stress	Noisy
	Chuckle	Play	Raspy expiration
Chimpanzee[b,c,d]	Scream	Separation	Tonal, noisy overlay; 0.25 sec or more
	Hoo-whimper	Separation; temper tantrum	Soft, low-pitched tone; 0.01 sec
	Laughter	Play (tickling)	Variable pulsed noise
Orangutan[e]	Scream	Separated, hurt	
	Hoot	Frightened infant	
	Whimper	Frightened infant	
	Fear squeak	Frightened older infant	
Kloss's gibbon[f]	Loud squeal	Separated juvenile	
White-handed gibbon[g]	Hoo-sigh	Separated immature juvenile	
Siamang[h]	Bleating gurgle	Isolated infant; intense play	
Lion-tailed macaque[i]	Whistly call	Separation	Resemble adult spacing calls
Japanese macaque[j]	Smooth early high (coo)	Separation	Coo variant; early frequency peak
	Stop	Retrieval by mother	Long tone, periodic amplitude drop
	Whistly	Rejection by mother	Frequency modulated
	Squeal	Weaning tantrum	
	Whine	Maternal acceptance following weaning tantrum	
Stump-tailed macaque[k]	Trilled whistle	Separation	
	Infantile whistle	Weaning tantrum	Like trilled whistle, but interrupted
Pig-tailed macaque[l]	Long cry	Separation	Harsh
	Squeak	Rough handling by mother	
	Emotive coo	Separated older infant	Tonal, "mournful"
Black and white colobus[m]	Scream	Abandoned newborn	Strong fundamental at 1–2 kHz
	Cawing	Abandoned infant	
	Squeak	Rejection	

Table 11.1 (*cont.*)

Species	Call	Main context	Call structure
Howler monkey[n]	Caw	Separation	
	Squeak	Fear of abandonment	
	Whimper	After retrieval	
Rufous-naped tamarin[o]	Infantile squeak	Distress; separated	
	Long whistle	Separation persists	
Black spider monkey[p]	Eee-awk	Separation	
	Squawk	Rough handling	
Night monkey[q]	Squeak	Separation	
	Scream	Separation	
	High trill	Separation	
Pygmy marmoset[r]	Tsik-phee	Separation	Sequence
	Infant J-call	Separated older infant	Series of rapidly ascending tones
Common marmoset[s]	Phee	Separation; rejection	Mainly above 8 kHz
	Tsik	Separation; rejection	Rapidly descending
Titi monkey[t]	Distress call	Separation	
Squirrel monkey[u,v,w]	Isolation peep	Separation	Sustained tone
	Location trill	Reunion with mother	Series of FM notes
	Scream	First sound at birth; separation; rejection	Quavering tone with some noise
	Play peep	Play	Rapidly ascending

Sources: [a] Fossey 1972; [b] Van Lawick-Goodall 1968; [c] Van Lawick-Goodall 1969; [d] Marler and Tenaza 1977; [e] MacKinnon 1974; [f] Tenaza 1976; [g] Ellefson, cited in Marler and Tenaza 1977; [h] Chivers 1976; [i] Green 1981; [j] Green 1975a; [k] Chevalier-Skolnikoff 1974; [l] Grimm 1967; [m] Marler 1972; [n] Baldwin and Baldwin 1976; [o] Moynihan 1970; [p] Eisenberg 1976; [q] Moynihan 1964; [r] Pola and Snowdon 1975; [s] Epple 1968; [t] Robinson 1979; [u] Winter 1968; [v] Winter et al. 1973; [w] Lieblich et al. 1980.

tions have been made in both the laboratory and the field pertaining to the development of intragroup vocal communication.

The most thorough analysis has come from the study of several representatives of the genera *Macaca* and *Cercopithecus* (see Gautier and Gautier 1977 for a review). In the macaques, tonal sounds designated as coos ("clear calls" of Rowell and Hinde 1962; "type A" calls of Itani 1963) are common in generally peaceful intragroup situations. Coos occur in a variety of structural forms, and relevant to our topic here is the evidence for age-related differences in the occurrence of these variants. In the most complete study of this class of sounds, Green (1975a) noted seven coo

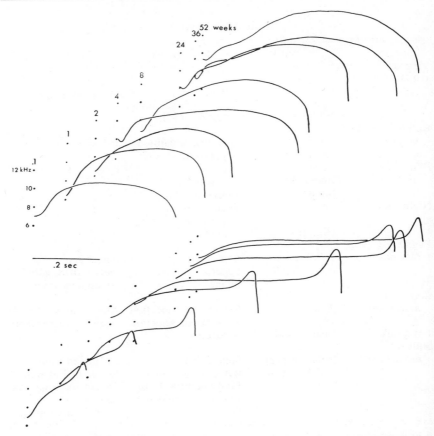

Figure 11.1. Maturation of the isolation peep from one day to one year of age in a typical Gothic-arch (above) and Roman-arch (below) squirrel monkey. Shown are computer-generated plots of 16 calls at each age ("averaged pitch profiles"; see Symmes et al. 1979).

types in the repertoire of the Japanese macaque, *M. fuscata*. Two of these, the smooth early high (SEH) and dip early high (DEH), were given by young animals when alone. The SEH also was given by older individuals when the troop was dispersed, occurring as coordinated vocal sequences between troop members. A third variant, smooth late high (SLH), was uttered primarily by females in estrus, but cases of young calling to their mothers nearby with this call were also recorded. Coo variants of the SEH and SLH morphology have also been recorded from young stump-tailed macaques (*M. arctoides;* Lillehei and Snowdon 1978). Of interest for cross-species comparison is the fact that in the stump-tails, also, SEH variants were given by isolated individuals and SLH variants by animals attempting to maintain close contact with their mothers. Several authors have made passing reference to the gradual transition of infantile macaque vocalizations to adult morphology. However, there has been no systematic tracking of the maturation of coos (or any other macaque call). Hence, we do not know whether SLH variants occur in separated newborn Japanese or stump-tailed macaques, or if they do not, what factors are associated with their appearance. Without this basic information, it is impossible at this writing to evaluate the role of experience in the expression of this or any other coo variant.

INHERITANCE AND EXPERIENCE IN THE PERCEPTION OF PRIMATE VOCALIZATIONS

Although most of our discussion has been directed toward the physical attributes of primate vocalizations and their ontogeny, attention also must be given to relevant perceptual mechanisms. In the absence of documented vocal imitation in primates, unlike the evidence available for many birds, where auditory discriminative capabilities can be measured by the imitative reproduction of model playback sounds, other procedures are required to test the biological significance of various acoustic features perceived by a species member.

The evidence cited for Japanese macaques using SLH and SEH coos in different contexts formed the basis for psychoacoustic experiments that tested possible perceptual correlates (see Beecher, Petersen, Zoloth, Moody, and Stebbins 1979; Petersen, chap. 8, this volume). The results of these experiments led the authors to believe that they had uncovered a species-specific perceptual mechanism, since their *M. fuscata* subjects labeled tape-recorded examples of that species' coo calls as SEH or SLH (although the tokens overlapped in average pitch and other attributes) and the control species (*M. nemestrina, M. radiata, C. aethiops*) did not. It is not known whether the other macaque species use coos with peak frequency differences equivalent to SEH and SLH in different contexts, although available data indicate that *M. nemestrina* produces coos with these features (Grimm 1967). Since stump-tailed macaques (*M. arctoides*) use SLH and SEH variants in the same contexts as *M. fuscata* (see earlier discussion), it would be interesting to perform similar psychoacoustic tests in this species. The prediction would be that stump-tails label SLH and SEH tokens as well as do Japanese macaques. More important, however, is how individual macaques acquire this labeling tendency. Is this an inherited mechanism, exhibited by individuals regardless of their auditory experience, or must a Japanese macaque grow up with the social and acoustic experience acquired from intragroup interactions before this perceptual mechanism is functional?

Evidence for a role for experiential factors in primate auditory perception is only inferential. For example, juvenile rhesus macaques recognize the calls of their mothers (Hansen 1976), and mother squirrel monkeys recognize the calls of their infants (Kaplan, Winship-Ball, and Sim 1978). Analysis of the behavior of gray-cheeked mangabey troops to natural calls and their tape-recorded playback indicates that troop members can differentiate between the whoopgobble of different males (Waser 1977). The behavior of vervet monkeys to three types of tape-recorded alarm calls led Seyfarth, Cheney, and Marler (1980; Seyfarth and Cheney, chap. 10, this volume) to conclude that adults of this species can communicate the presence of different predators by these calls but that infants are much less specific in their interpretation of the meaning of the sounds.

The findings cited in the preceding paragraph suggest that experience can play a significant role in perceptual adaptations. In addition to extending and confirming these findings, the more difficult task of establishing the *limits* of experiential control over vocal perception needs to be undertaken.

INHERITANCE OF VOCAL TRAITS

To this point, we have indicated some clues as to a possible role for experience in primate vocal behavior. These clues should not overshadow the fact that genetic mechanisms clearly influence the development of primate vocalizations. Among the fundamental issues are questions pertaining to *modes* of inheritance and precisely *what* attributes are inherited. Conventional approaches to these questions include the crossing of highly inbred strains and the analysis of phenotypic introgression in natural populations.

Not surprisingly, in view of the long generation times involved, no strains of primates have been selectively bred for their vocal characteristics, as has been done in birds. Although tracing the inheritance patterns of *any* characteristic would be a slow process in primates, it would be worthwhile if it aided in understanding the genetic mechanisms controlling species-specific differences in vocalizations. However, another approach to the analysis of genetic control of vocal behavior has been through the crossbreeding of closely related species. Work on the transmission of vocal traits in interspecific hybrids has been conducted in several animal groups.

Evidence for the control of male calling-song patterns by a polygenic system has been forwarded for *Teleogryllus* crickets (Bentley and Hoy 1974). Male F_1 hybrids produce songs intermediate in pulse duration to those of the parent species, whereas the backcross of a hybrid to one of the parent species results in offspring whose song is like the parental type. A similar result was achieved by crossing two species of *Chorthippus* grasshoppers (Helversen and Helversen 1975), in that F_1 hybrids were intermediate in song pattern to the two parental types. The hybrids also showed greater intraindividual variability in their song patterns, although a stronger expression of the maternal trait was evident. Natural hybrids have been described for several species of frogs. The hybrid individuals are found in regions where the ranges of two species overlap and are recognizable by their distinctive body coloration patterns. Mating calls of these hybrids are intermediate to the parental types and highly variable in pulse frequency (e.g., Gerhardt 1974; Littlejohn and Watson 1976). As for birds, the inheritance of vocal traits has been little studied, perhaps because the demonstrated prominence of song learning in some species requires that the acoustic experience of the species under study be very carefully controlled. However, one study (Lade and Thorpe 1964) examined the characteristics of hybrid mating calls in doves and found inheritance patterns suggestive of those in orthopteran insects and frogs; the temporal pattern of calling in a dove hybrid was often intermediate to that of the parents.

Study of hybrid primate vocalizations is only beginning. Gautier and Gautier (1977) illustrated two calls, given under undisclosed circumstances, of a 1-month-old female hybrid whose parents were *Cercopithecus ascanius* (mother) and *C. pogonias*. The illustrated calls contain structural attributes similar to those of both parents. The songs of two hybrid gibbons are described by Marler and Tenaza (1977). The mother of these hybrids was an *Hylobates muelleri* and the father was an *H. lar*. The hybrids, a male and female, behaved as adults as a mated pair, each singing their sex-specific part in compound songs. The songs of the female were structurally intermediate to those of each parent, and the male hybrid's songs resembled those of his father. The song of another female hybrid gibbon born in captivity was analyzed by J. T. Marshall and W. Y. Brockelman (personal communication). This individual, Barbara, was the offspring of a female *H. agilis* and a male *H. pileatus*. Barbara's songs, recorded at age 35, combined elements of the song of her mother with elements from *other* gibbon species. A similar finding has been discovered by Marshall and Brockelman in natural hybrids between *H. lar* and *H. pileatus*, in that hybrid and backcross offspring may produce songs resembling those of other hylobatids not involved in the interspecific matings.

Patterns of isolation peep inheritance in squirrel monkeys

Questions about the inheritance of vocal traits relate to various levels of social organization, from species-specific characteristics through troop-specific dialects to parental traits. Given the uncertain taxonomic status of *Saimiri* varieties (Hill 1960; Napier and Napier 1967), questions about species-specificity (vs. subspecific variation) must remain somewhat tentative. However, descriptions of *Saimiri* in Hill (1960) suggest that his *S. sciurea* corresponds to the Gothic-arch variety described by MacLean (1964; see also Cooper 1968) and that *S. voliviensis* corresponds to the Roman-arch variety. Our studies have found that these two types of *Saimiri* also differ consistently in the structure of their isolation peeps. It would therefore appear that IPs can be useful as a phenotypic marker at the species or subspecies level. Winter (1969) was the first to report that Roman and Gothic IPs differ in structure. He identified the terminal part of the call as containing the key differentiating feature, in that it sweeps upward in Romans and downward in Gothics. Having examined an extensive sample of IPs, we find this observation to hold for most animals. However, there are exceptions. In some monkeys clearly Gothic arch in

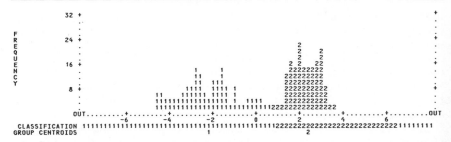

Figure 11.2. Distribution of discriminant scores for 150 Gothic and 150 Roman squirrel monkey isolation peeps with three descriptors. Scores are based on linear discriminant-function analysis using peak frequency location, tail slope, and average frequency as variables. (1) Gothic, (2) Roman.

facial characteristics, the IP terminates in a slight upsweep; the converse is also true, some Roman-arch monkeys ending their IPs with an abrupt downsweep. These observations led us to search for more universally diagnostic features separating Roman and Gothic IPs. We have found that the key differentiating attribute is the temporal location within a call at which the frequency of the tonal fundamental reaches its highest value (Symmes et al. 1979). This peak frequency location (PFL) is expressed as percent of total call duration. Roman IPs invariably have a PFL in the last 10% of the call (PFL = 90–100%). Gothic IPs are more variable in PFL, but there is virtually no overlap with Roman IP values, since Gothic IP PFL rarely exceeds 80%. Using linear discriminant-function analysis, Roman and Gothic IPs can be classified with an accuracy of 93% using only PFL as the discriminative variable. To achieve 100% correct classification, two additional variables are required, the slope of the final 50 msec (tail slope) and the average frequency along the midportion of the call. This set of three descriptors was used to derive discriminant functions for 300 IPs taken from animals with clear Gothic-arch and Roman-arch facial features and diverse geographic origins (Romans from Bolivia and Peru; Gothics from Colombia, the Guyanas, and Peru) and from adults reared in our laboratory. If one looks at the spread of the Roman and Gothic discriminant scores (Figure 11.2), it is apparent that the Gothic scores are more dispersed around their group mean (centroid). This means that Gothic IPs are more variable, as a group, in their species-specific attributes than are the Roman IPs. The discriminant scores from the two groups do not overlap, resulting in 100% correct classification as to the Roman-arch or Gothic-arch source of the calls. The discriminant functions of these adult IPs can be used to determine the extent to which infant IPs are classifiable as to species identity (i.e., Roman or Gothic). Such an analysis was performed on the IPs of 1- and 2-day-old infants, four born to Roman-arch parents and four to Gothic-arch parents. Of the 160 IPs analyzed, 99% were correctly classified. Thus, even at this early age, species-specific attributes are present in the IP, strong evidence against any role for auditory experience in shaping species-specificity in IP structure.

Table 11.2. *Classification of Roman ×*
Gothic hybrid isolation peeps by linear
discriminant-function analysis

Hybrids	Isolation peeps	
	Gothic	Roman
Gothic-arch[a]		
I 4	16	24
N A	36	4
763	3	37
Roman-arch[b]		
H 23	0	40
728	0	40
G 31	0	40
G 32	0	40
H 3	5	35

[a] Mean percent correctly classified: 46.
[b] Mean percent correctly classified: 98.

IP inheritance patterns in Roman × Gothic hybrids

Given the well-defined differences between Roman and Gothic IPs present from birth, working out the inheritance patterns for these differences through crossbreeding becomes a logical next step. Viable hybrid offspring from Roman × Gothic crosses have been known to exist for several years (Jones, Thorington, Hu, Adams, and Cooper 1973), along with a distinctive hybrid karyotype. We have recorded IPs from a total of eight adult Roman × Gothic hybrids, five of them resulting from crossbreeding known Roman-arch and Gothic-arch parents, the other three identified as hybrids by the hybrid karyotype of 11 acrocentric chromosomes. Three of the hybrids studied by us had a recognizably Gothic-arch facial pattern (two of whom were known to have Gothic-arch mothers). The other five had the Roman-arch facial pattern and, in the three with known parentage, the mother also had the Roman-arch phenotype. Thus, our analyses of hybrid IPs examined the expression of Roman-specific IP characteristics in the five Roman-arch hybrids and the expression of the Gothic-specific IP characteristics in the three Gothic-arch hybrids. These analyses were performed with the same IP descriptors used on the Roman-arch and Gothic-arch adults and infants. The results of this analysis, summarized in Table 11.2, show that the Roman-specific IP phenotype is almost as clear in Roman-arch hybrids as in the Roman-arch parental species, whereas Gothic-arch hybrids show greater variability in the expression of the Gothic-specific IP trait.[2] This can also be seen by inspection of individual hybrid sound spectrograms (Figure 11.3). Examination of the distribution of discriminant scores of the hybrid IPs indicates that Roman-arch hy-

---

[2] Isolation peeps of newborn hybrids with Roman-arch mothers are also strongly Roman.

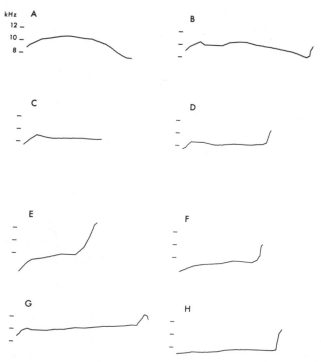

Figure 11.3. Tracings of individual sound spectrograms of isolation peeps from adult purebred and hybrid squirrel monkeys: (A) purebred Gothic female, (B–D) hybrid females with Gothic mothers and Roman fathers, (E) purebred Roman female, (F–H) hybrid females with Roman mothers and Gothic fathers.

brids are nearly as compact in their score range as are purebred Roman-arch monkeys. It is worth noting that the incorrectly classified IPs for both Roman and Gothic hybrids fall into the zone where discriminant scores overlap instead of being strongly Gothic or Roman. Thus, these calls may be said to be "intermediate" relative to the parental types. However, both Roman- and Gothic-arch hybrids produce IPs that would be indistinguishable from IPs from purebred individuals.

Geographic diversity in IPs of natural squirrel monkey populations

Nothing is known about the extent to which Gothic-arch and Roman-arch troops interact or interbreed in the wild. There appear to be no major natural barriers to prevent them from intermixing. The distinctively different IPs in these forms of *Saimiri* may play a role in behaviorally segregating them, thus reducing the likelihood of interbreeding in natural populations. Our knowledge of the degree to which squirrel monkey troops retain their integrity, the degree to which interbreeding occurs between neighboring troops, and, particularly, the distribution patterns of IP phenotypes in

Table 11.3. *Classification accuracy of isolation peeps by geographic origin*

Actual group	Percent of actual group's isolation peeps				
	Group A	Group B	Group C	Group D	Group E
Tahuayo (group A)	80	15	0	2	3
San Joaquin (group B)	30	36	4	8	22
Yucuruchi (group C)	5	15	35	45	0
Sarapanga (group D)	1	19	3	69	9
Palizada (group E)	5	2	0	28	65

Note: Five groups of Roman-arch squirrel monkeys, each group captured together at sites along the Amazon River in Peru between the towns of Iquitos and Requena. The capture sites were located in an approximate line from northeast (group A) to southwest (group E), with about 90 km separating sites A and E.

wild populations is extremely limited. The available evidence suggests that some degree of troop integrity is maintained (Thorington 1968) but that new individuals can interact sexually with troop members (Baldwin 1967). We have made a start at sampling IP phenotypes in a geographically delimited region. With the cooperation of the Pan-American Health Organization Primate Project in Iquitos, Peru, we recorded IPs from representatives of five separate Roman-arch troops, identified as to capture site. Analysis of IP structural characteristics, using eight descriptors (see Symmes et al. 1979), failed to identify site-specific phenotypes (dialects) for each of the five troops. Nevertheless, the accuracy of classifying the IPs by troop reached 80% in a troop captured about 40 km south of Iquitos (the Tahuayo troop), and the rest of the troops achieved better than chance correct classification (Table 11.3). These data suggest that panmixia over this geographical area probably does not exist but that some gene exchange takes place between neighboring populations.

Primate vocalizations and the genetics of behavior

Intraspecific consistency of structure over wide geographic areas has been noted for a number of primate vocalizations. Consequently, these sounds serve as species-specific phenotypic markers (see Struhsaker 1970; Marshall and Marshall 1976; Gautier and Gautier 1977; Oppenheimer 1977 for examples). However, little work has come forth concerning interdemic differences in vocal phenotype. Our analysis of intertroop differences in Peruvian squirrel monkeys is an effort in that direction. Such work ideally should be matched with an analysis of biochemical and cytogenetic differences in these same populations, following the direction of some avian studies (Nottebohm and Selander 1972).

The reports of naturally occurring interspecific primate hybrids (e.g., Bernstein 1966; Struhsaker 1970) suggest that reproductive barriers occasionally break down between sympatric primate species, a fact that is well established in other animal groups. The role of acoustic behavior in maintaining reproductive isolation in natural primate populations is still poorly understood. However, a larger issue is the extent to which reproductive isolation has failed. This is not so great a problem in F_1 hybridization, since such individuals may have either distinctive coloration (e.g., gibbons: Marshall and Brockelman, in preparation) or a distinctive karyotype (e.g., squirrel monkeys: Jones et al. 1973). However, where hybrids can interbreed with the parental species and with each other, considerable phenotypic introgression may occur (see Short 1969 for an avian example). In such cases, having a priori knowledge of karyotypic and biochemical markers in identified backcross and F_2 populations may help in establishing the genetic makeup of natural populations and in establishing the inheritance patterns of their vocalizations. Progress in identifying such markers is under way in macaques (Avise and Duvall 1977) and in squirrel monkeys in our laboratory.

DISCUSSION AND CONCLUSIONS

Returning to the proposition stated at the outset of this chapter, that primate vocalizations develop without auditory experience, let us recapitulate the evidence. The data in support of this argument are, in fact, rather meager. The conclusion that vocal traits are inherited is based mainly on impressions of the similarity between calls of infants reared apart from congeners and adult calls. With the exception of studies of the squirrel monkey isolation peep (Winter et al. 1973; Lieblich et al. 1980; this chapter), there has been no attempt to demonstrate quantitatively those structural attributes of specific calls that are shared by infants and adults. Further, in every case so far reported (with the questionable exception of Boutan, cited in Marler 1963), only *some* of the calls of the adult repertoire have been reported in these restrictively reared infants. If experience *was* to play a role in primate vocal development, it most likely would be concentrated on those vocalizations important in the establishment of intragroup social partnerships. It is therefore significant that such calls show a dependence on normal social experience (Newman and Symmes 1974).

Some authors (e.g., Marler 1976) argue that primates have moved phylogenetically from discrete to intergraded vocal repertoires and that this increased emphasis on graded sounds, most conspicuous in chimpanzees and gorillas, was a major step toward human speech. Determining the proportion of graded sounds in a particular species has been a persistent problem in primate bioacoustics, but most authorities agree that developing individuals utter more variable and intergraded sounds even in species with relatively discrete adult calls. This being the case, one wonders why discrete calls would be adaptive in the adult repertoire of many primates but not in chimpanzees and gorillas. The answer, perhaps all too obvious, may be that the vocal repertoire of *adult* apes is no more graded than that of other primate adults. A recent analysis of the chimpanzee vocal repertoire (Marler and Tenaza 1977) showed that the most common calls of older adults, the pant-hoot and rough grunt, only rarely intergraded with other call types. The most variable calls were either infant calls or those used in agonistic encounters, as is the case with other primates. Perhaps the longer maturational time of the great apes has disproportionately weighted the calls of developing individuals in assessments of variability in these species. If the antecedents to human speech are to be found in the graded calls of nonhuman primates, we must look to the sounds used by the younger, still-maturing subadults of a species, for freed from the roles of sexual partner, troop defender, and status seeker, they have the greatest opportunity for learning and friendly social interactions. The adult members of a primate group, on the other hand, serve as teachers by providing, in the consistency of their vocalizations and their behavioral responses to the calls of others, a set of "rules" for subadults to learn (see Seyfarth and Cheney, chap. 10, this volume). Thus, the primary role of learning in the acoustic behavior of primates, including humans, may be to gradually restrict the variability of vocal utterances and behavioral responsiveness with age. By this process, the individual would acquire, at maturity, a considerable degree of communicative predictability. This, in turn, would provide an opportunity for a primitive lexical syntax to evolve. However, the variability of vocal performance in subadults would also provide a rich source of new sounds, which, if coupled with some external reinforcement, could enlarge the adult repertoire of the individuals involved.

REFERENCES

Avise, J. C., and Duvall, S. W. 1977. Allelic expression and genetic distance in hybrid macaque monkeys. *Journal of Heredity* 68:23–30.

Baldwin, J. D. 1967. A study of the social behavior of a semifree-ranging colony of squirrel monkeys (*Saimiri sciureus*). Ph. D. dissertation, Johns Hopkins University.

Baldwin, J. D., and Baldwin, J. I. 1976. Vocalizations of howler monkeys (*Alouatta palliata*) in southwestern Panama. *Folia Primatologica* 26:81–108.

Beecher, M. D., Petersen, M. R., Zoloth, S. R., Moody, D. B., and Stebbins, W. C. 1979. Perception of conspecific vocalizations by Japanese macaques. *Brain Behavior and Evolution* 16:443–60.

Bentley, D., and Hoy, R. R. 1974. The neurobiology of cricket song. *Scientific American* 231:34–44.

Bernstein, I. S. 1966. Naturally occurring primate hybrid. *Science* 154:1559–60.

Chevalier-Skolnikoff, S. 1974. The ontogeny of communication in the stumptail macaque (*Macaca arctoides*). *Contributions to Primatology* 2:1–174.

Chivers, D. J. 1976. Communication within and between family groups of siamang (*Symphalangus syndactylus*). *Behaviour* 57:116–35.

Cooper, R. W. 1968. Squirrel monkey taxonomy and supply. In *The squirrel monkey,* ed. L. A. Rosenblum and R. W. Cooper, pp. 1–29. New York: Academic Press.

Eisenberg, J. F. 1976. Communication mechanisms and social integration in the black spider monkey, *Ateles fusciceps robustus,* and related species. *Smithsonian Contributions to Zoology* 213:1–108.

Epple, G. 1968. Comparative studies on vocalization in marmoset monkeys (*Hapalidae*). *Folia Primatologica* 8:1–40.

Fossey, D. 1972. Vocalizations of the mountain gorilla (*Gorilla gorilla beringei*). *Animal Behaviour* 20:36–53.

Gautier, J.-P. 1974. Field and laboratory studies of the vocalization of talapoin monkeys (*Miopithecus talapoin*). *Behaviour* 51:209–73.

Gautier, J.-P., and Gautier, A. 1977. Communication in Old World monkeys. In *How animals communicate,* ed. T. A. Sebeok, pp. 890–964. Bloomington: Indiana University Press.

Gerhardt, H. C. 1974. The vocalizations of some hybrid treefrogs: acoustic and behavioral analyses. *Behaviour* 59:130–51.

Green, S. 1975a. Variation of vocal pattern with social situation in the Japanese monkey (*Macaca fuscata*): a field study. In *Primate behavior,* Vol. 4: *Developments in field and laboratory research,* ed. L. A. Rosenblum, pp. 1–102. New York: Academic Press.

1975b. Dialects in Japanese monkeys: vocal learning and cultural transmission of locale-specific behavior? *Zeitschrift für Tierpsychologie* 38:304–14.

1981. Sex differences and age gradations in vocalizations of Japanese and liontailed monkeys. *American Zoologist* 21:165–83.

Grimm, R. J. 1967. Catalogue of sounds of the pig-tailed macaque (*Macaca nemestrina*). *Journal of Zoology* 152:361–73.

Hansen, E. W. 1976. Selective responding by recently separated juvenile rhesus monkeys to the calls of their mothers. *Developmental Psychobiology* 9:83–8.

Helversen, D. von, and Helversen, O. von. 1975. Verhaltensgenetische Untersuchungen am akustischen Kommunikationssystem der Felheuschrecken (Orthoptera, Acrididae). I. Der Gesang von Artbastarden zwischen *Chorthippus biguttulus* und *Chorthippus mollis. Journal of Comparative Physiology: Sensory, Neural, and Behavioral Physiology* 104:273–99.

Hill, W. C. O. 1960. *Primates: comparative anatomy and taxonomy.* Vol. 4: *Cebidae,* Pt. A. New York: Interscience.

Imanishi, K. 1957. Social behavior in Japanese monkeys, *Macaca fuscata. Psychologia* 1:47–54.

Itani, J. 1963. Vocal communication of the wild Japanese monkey. *Primates* 4:9–66.

Jenkins, P. F. 1978. Cultural transmission of song patterns and dialect development in a free-living bird population. *Animal Behaviour* 26:50–78.

Jones, T. C., Thorington, R. W., Hu, M. M., Adams, E., and Cooper, R. W. 1973. Karyotypes of squirrel monkeys (*Saimiri sciureus*) from different geographic regions. *American Journal of Physical Anthropology* 38:269–78.

Kaplan, J. N., Winship-Ball, A., and Sim, L. 1978. Maternal discrimination of infant vocalizations in squirrel monkeys. *Primates* 19:187–93.

Lade, B. I., and Thorpe, W. H. 1964. Dove songs as innately coded patterns of specific behavior. *Nature* 202:366–8.

Larson, C. R., Sutton, D., Taylor, E. M., and Lindeman, R. 1973. Sound spectral properties of conditioned vocalizations in monkeys. *Phonetica* 27:100–10.

Lieberman, P. 1968. Primate vocalizations and human linguistic ability. *Journal of the Acoustical Society of America* 44:1574–84.

Lieblich, A. K., Symmes, D., Newman, J. D., and Shapiro, M. 1980. Development of the isolation peep in laboratory-bred squirrel monkeys. *Animal Behaviour* 28:1–9.

Lillehei, R. A., and Snowdon, C. T. 1978. Individual and situational differences in the vocalizations of young stumptail macaques (*Macaca arctoides*). *Behaviour* 45:270–81.

Littlejohn, M. J., and Watson, G. F. 1976. Mating-call structure in a hybrid population of the *Geocrinia laevis* complex (Anura: Leptodactylidae) over a seven-year period. *Evolution* 30:848–50.

MacKinnon, J. 1974. The behaviour and ecology of wild orangutans (*Pongo pygmaeus*). *Animal Behaviour* 22:3–74.

MacLean, P. D. 1964. Mirror display in the squirrel monkey, *Saimiri sciureus*. *Science* 146:950–2.

Marler, P. 1963. Inheritance and learning in the development of animal vocalizations. In *Acoustic behaviour of animals,* ed. R.-G. Busnel, pp. 228–43. Amsterdam: Elsevier.

  1969. Vocalizations of wild chimpanzees: an introduction. In *Recent Advances in Primatology*, Vol. 1: *Behavior,* ed. C. R. Carpenter, pp. 94–100. Basel: Karger.

  1972. Vocalizations of East African monkeys. II. Black and white colobus. *Behaviour* 42:175–97.

  1976. An ethological theory of the origin of vocal learning. In *Origins and evolution of language and speech,* ed. S. R. Harnad, H. D. Steklis, and J. Lancaster. *Annals of the New York Academy of Sciences* 280:386–95.

Marler, P., and Mundinger, P. 1971. Vocal learning in birds. In *Ontogeny of vertebrate behavior,* ed. H. Moltz, pp. 389–450. New York: Academic Press.

Marler, P., and Tenaza, R. 1977. Signaling behavior of apes with special reference to vocalization. In *How animals communicate,* ed. T. A. Sebeok, pp. 965–1033. Bloomington: Indiana University Press.

Marshall, J. T., Jr., and Marshall, E. R. 1976. Gibbons and their territorial songs. *Science* 193:235–7.

Moynihan, M. 1964. Some behavior patterns of platyrrhine monkeys. I. The night monkey (*Aotus trivirgatus*). *Smithsonian Miscellaneous Collections* 146:1–84.

  1970. Some behavior patterns of platyrrhine monkeys. II. *Saguinus geoffroyi* and some other tamarins. *Smithsonian Contributions to Zoology* No. 28:1–77.

Napier, J. R., and Napier, P. H. 1967. *A handbook of living primates.* London: Academic Press.

Newman, J. D., and Symmes, D. 1974. Vocal pathology in socially deprived monkeys. *Developmental Psychobiology* 7:351–8.

Nottebohm, F. 1972. The origins of vocal learning. *American Naturalist* 106:116–40.

Nottebohm, F., and Selander, R. K. 1972. Vocal dialects and gene frequencies in the Chingolo sparrow (*Zonotrichia capensis*). *Condor* 74:137–43.

Oppenheimer, J. R. 1977. Communication in New World monkeys. In *How animals communicate,* ed. T. A. Sebeok, pp. 851–89. Bloomington: Indiana University Press.

Ploog, D., Hopf, S., and Winter, P. 1967. Ontogenese des Verhaltens von Totenkopf-Affen (*Saimiri sciureus*). *Psychologische Forschung* 31:1–41.

Pola, Y. V., and Snowdon, C. T. 1975. The vocalizations of pygmy marmosets (*Cebuella pygmaea*). *Animal Behaviour* 23:826–42.

Robinson, J. G. 1979. An analysis of the organization of vocal communication in the titi monkey, *Callicebus moloch. Zeitschrift für Tierpsychologie* 49:381–405.

Rowell, T. E., and Hinde, R. A. 1962. Vocal communication by the rhesus monkey (*Macaca mulatta*). *Proceedings of the Zoological Society of London* 138:279–94.

Rumbaugh, D. M. 1971. Evidence of qualitative differences in learning processes among primates. *Journal of Comparative Physiological Psychology* 76:250–5.

Seyfarth, R. M., Cheney, D. L., and Marler, P. 1980. Monkey responses to three different alarm calls: evidence of predator classification and semantic communication. *Science* 210:801–3.

Short, L. L., Jr. 1969. "Isolating mechanisms" in the blue-winged warbler–golden-winged warbler complex. *Evolution* 23:355–6.

Sinnott, J., Stebbins, W., and Moody, M. I. 1975. *Journal of the Acoustical Society of America* 58:412.

Struhsaker, T. T. 1970. Phylogenetic implications of some vocalizations of *Cercopithecus* monkeys. In *Old World monkeys: evolution, systematics, and behavior,* ed. J. R. Napier and P. H. Napier, pp. 365–444. New York: Academic Press.

Sutton, D., Larson, C., Taylor, E. M., and Lindeman, R. 1973. Vocalization in rhesus monkeys: conditionability. *Brain Research* 52:225–31.

Symmes, D., Newman, J. D., Talmage-Riggs, G., and Lieblich, A. K. 1979. Individuality and stability of isolation peeps in squirrel monkeys. *Animal Behaviour* 27:1142–52.

Talmage-Riggs, G., Winter, P., Ploog, D., and Mayer, W. 1972. Effect of deafening on the vocal behavior of the squirrel monkey (*Saimiri sciureus*). *Folia Primatologica* 17:404–20.

Tembrock, G. 1974. Sound production of *Hylobates* and *Symphalangus*. In *Gibbon and Siamang,* ed. D. M. Rumbaugh, 3:196–205. Basel: Karger.

Tenaza, R. R. 1976. Songs, choruses and countersinging of Kloss' gibbons (*Hylobates klossii*) in Siberut Island, Indonesia. *Zeitschrift für Tierpsychologie* 40:37–52.

Thorington, R. W., Jr. 1968. Observations of squirrel monkeys in a Colombian forest. In *The squirrel monkey,* ed. L. A. Rosenblum and R. W. Cooper, pp. 69–85. New York: Academic Press.

Thorpe, W. H. 1967. Vocal imitation and antiphonal song and its implications. *Proceedings of the 14th International Ornithological Congress,* pp. 245–63.

van Lawick-Goodall, J. 1968. A preliminary report on expressive movements and communication in the Gombe Stream chimpanzees. In *Primates: studies in adaptation and variability,* ed. P. C. Jay, pp. 313–74. New York: Holt, Rinehart and Winston.

1969. Mother–offspring relationships in free-ranging chimpanzees. In: *Primate ethology,* ed. D. Morris, pp. 365–436. Garden City, N.Y.: Doubleday.

Waser, P. M. 1977. Individual recognition, intragroup cohesion and intergroup spacing: evidence from sound playback to forest monkeys. *Behaviour* 60:28–74.

Winter, P. 1968. Social communication in the squirrel monkey. In *The squirrel monkey,* ed. L. A. Rosenblum and R. W. Cooper, pp. 235–53. New York: Academic Press.

1969. Dialects in squirrel monkeys: vocalizations of the Roman arch type. *Folia Primatologica* 10:216–29.

Winter, P., Handley, P., Ploog, D., and Schott, D. 1973. Ontogeny of squirrel monkey calls under normal conditions and under acoustic isolation. *Behaviour* 47:230–9.

12 · The role of chemical communication in aggressive behavior and its gonadal control in the tamarin (*Saguinus fuscicollis*)

GISELA EPPLE, MARY CATHERINE ALVEARIO, and YAIR KATZ

Chemical signals derived from such sources as feces, urine, saliva, genital discharge, and specialized skin glands form part of the communicatory repertoire of a great number of mammals, including some primates (Mykytowycz 1970; Epple 1974a; Cheal 1975; Keverne, chap. 15, and Goldfoot, chap. 16, this volume). In some of these species, chemical signals are important in the control of aggressive behavior and social dominance (cf., Ralls 1971; Epple 1974b; Thiessen 1976). Aggression and the ability to establish and maintain social dominance are also influenced by an individual's endocrine condition and, in turn, affect its endocrinology. The relationship between aggressive behavior, gonadal hormones and the communicatory functions of scents has been studied in detail in rodents, where gonadal steroids are usually necessary for the ontogenetic development and adult maintenance of scent glands, marking behavior, and aggression (cf., Bronson and Desjardins 1971; Hart 1974; Leshner 1975; Thiessen 1976; Quadagno, Briscoe, and Quadagno 1977). Our understanding of the hormonal control of aggression and social dominance, and their relation to chemical communication, is more incomplete in primates than in rodents. However, it has been shown that the gonads are of some importance in controlling primate aggression and chemical communication in studies of scent marking that coincides with gonadal activation during the breeding season (Jolly 1967; Schilling 1974; Harrington 1975; Mazur 1976), in studies of the influence of androgens on the fetal nervous system (cf. Eaton, Goy, and Phoenix 1973; Rose, Bernstein, Gordon, and Catlin 1974), and in studies on the effects of gonadectomy and hormone therapy (cf. Mazur 1976; Dixson 1980).

This chapter reviews some of our studies on the relationship between the gonads, aggressive behavior, and chemical communication in the saddleback tamarin, a New World monkey.

These studies were supported by Grants BNS 76-06838 and BNS 78-06172 from the National Science Foundation and by Research Career Development Award K04 HD 70575 from the National Institutes of Health.

BASIC SOCIAL STRUCTURE OF SADDLEBACK
TAMARIN GROUPS

Saguinus fuscicollis, a member of the family Callitrichidae, inhabits the primary and secondary forests of the upper Amazonian region (Hershkovitz 1977). Little is known about the social behavior of this species in the wild. However, field studies by Castro and Soini (1978) and J. Terborgh (personal communication) suggest that the social organization of saddleback tamarins may be similar to that of *S. oedipus* and *S. nigricollis,* which were studied in some detail in their natural habitats (Dawson 1978; Izawa 1978; Neyman 1978). These species form relatively small groups, each containing a stable nucleus consisting of a breeding female, her dependent young, and a male with whom she probably has a monogamous pair bond. There appears to be only one breeding female in each group. Maturing offspring and nonbreeding adults apparently show a tendency to leave their families and immigrate into other conspecific groups. These transient animals may change group affiliations repeatedly (Dawson 1978; Izawa 1978; Neyman 1978).

Tamarins may inhabit overlapping home ranges and occasionally form temporary feeding aggregations consisting of more than one group (Thorington 1968; Moynihan 1970; Castro and Soini 1978; Izawa 1978). However, the occupation and defense of territories in the strictest sense also has been reported, and the spacing behavior adopted appears to depend on a number of ecological factors (Thorington 1968; Moynihan 1970; Dawson 1978; Neyman 1978). The *S. fuscicollis* studied in Peru by Terborgh (personal communication) is strictly territorial, and, in addition to a number of displays, aggression forms part of their territorial defense.

In the laboratory, saddleback tamarins, like other callitrichids, establish monogamous pair bonds and are characterized by high levels of aggression directed at nonrelated conspecific adults (Kleiman 1977; Epple 1978). In laboratory families developing naturally from an adult pair and its progeny, as well as in groups containing several nonrelated males and females, only one female ever bears offspring (Epple 1975). In families, the mother remains the only reproductively active female, and little sexual activity occurs among her adult offspring (Epple, unpublished data). In groups containing several nonrelated adults, one female usually establishes dominance over all other females, pair-bonds to a dominant male, and appears to inhibit pair bonding in the other females, although these may engage in sexual activities (Epple 1972, 1978a).

The establishment of a pair bond may be an important factor in determining the social status and reproductive success of males and particularly of females. Our understanding of the mechanisms involved in the establishment and maintenance of the pair bond in callitrichids, however, is very incomplete. Preference for the mate, expressed in spatial proximity, a number of "affectionate behaviors," and a close sociosexual association between mated partners may be involved in these processes (Epple 1978a; Kleiman 1978; Poole 1978). In addition to these associative behaviors, saddleback tamarins apparently utilize aggressive rejection of potential challengers to a bond to enforce monogamy (Epple 1978a). Pair-bonded adults violently attack nonrelated adult group members relatively frequently, necessitating the removal of the recipient of such aggression from the group. Overt aggression also is used to prevent friendly social contact with conspecific adults not belonging to the same social group (Epple 1978a,b). Females tend to be more aggressive than males, and the predominant targets of their aggression are other females (Epple 1978a, b). It is obvious that overt aggression may be more frequent in the laboratory, where the animal eliciting such behavior has no chance to avoid it by retreating.

SCENT GLANDS, SCENT-MARKING BEHAVIOR, AND THEIR COMMUNICATIVE FUNCTIONS

Males and females of all species of callitrichids studied so far possess specialized scent glands in the circumgenital area and on the midchest above the sternum. The morphological appearance and histological composition of these glands and the occurrence of sexual dimorphisms in the structures vary among species (Wislocki 1930; Perkins 1966, 1968, 1969a, b, c; Epple and Lorenz 1967; Starck 1969; Sutcliffe and Poole 1978). In *S. fuscicollis,* the sternal gland consists of an accumulation of large apocrine glands (Perkins 1966) and is only recognizable as a morphological specialization on close inspection of the animal. The circumgenital-suprapubic gland is a very conspicuous structure, consisting of an accumulation of large sebaceous and apocrine glands (Perkins 1966). It involves the skin of the external genitalia (scrotum and labia majora) and extends rostrally across the symphysis pubis as a thick, largely hairless, and darkly pigmented glandular pad. Although there is no dramatic sexual dimorphism of the gland, the female gland tends to be longer than the male gland. The length of the midline of the suprapubic part of the gland from the cranial edge of the glandular differentiation to the base of the penis or clitoris was measured in 24 adult males and 15 adult females under sedation. The average length of the male gland was 27.8 ± 0.9 mm, that of the female gland 33.6 ± 1.4 mm. These measurements suggest that females possess larger gland than males. The shape of the glands and their thickness, however, vary among individuals. Since exact measurement of these parameters and of the size of the external genitalia are difficult to obtain with living animals, the extent to which the glands are sexually dimorphic awaits histological analysis.

The scent glands develop as macroscopically recognizable differentiations when the animals are 7 to 10 months old and grow until they reach their adult size at the time the animals are 2 years old or even older. Scent marking is performed with the sternal as well as with the circumgenital-suprapubic gland. Sternal marking, however, is a relatively infrequent behavior, most often shown in situations of high social arousal. It involves rubbing the sternal area against environmental items. The motor pattern is variable and adapted to the nature and position of the item to be marked

(Epple 1975). Marking with the circumgenital-suprapubic gland is a very frequent behavior, regularly shown by socially dominant adults. Items in the environment and, occasionally, conspecifics are marked with these glands. The glandular areas are pressed against the substratum and their secretions, a few drops of urine and perhaps some genital discharge, are deposited. The motor patterns involved in marking are variable, depending both on the nature and position of the substratum, the parts of the glands involved, and the level of arousal of the animal. The lowest-intensity marking is performed in a sitting position. The animal sits, presses the circumgenital and circumanal areas against the substrate, and rubs back and forth and/or from side to side ("sit rubbing," Moynihan 1970). Secretions from the suprapubic part of the gland are applied when the animal assumes a prone position, pressing the lower abdomen and the suprapubic gland to the substrate and either pulling itself forward with its hands and/or pushing the body with the feet ("pull rubbing," Moynihan 1970). The suprapubic glands also may be rubbed against small protuberances while in a sitting position.

"Sit rubbing" and "pull rubbing" are quite distinct. Each may be performed by itself, but frequently both are given in succession. Although in cotton-top tamarins sit rubbing and pull rubbing may serve different functions (French and Snowdon 1981), we have no evidence that this is the case in *S. fuscicollis*. Saddleback tamarins seem to sit-rub at low levels of arousal but pull-rub or show both patterns as arousal increases. At very high levels of arousal, suprapubic marking and sternal marking also are combined into one pattern, during which the animal rubs the entire ventral surface over the substrate. Since sternal marking is so infrequent, all the following experimental data exclude sternal marking and pertain only to marking with the circumgenital-suprapubic gland.

The well-developed scent glands and the frequent use of the circumgenital-suprapubic gland suggest that the chemical signals deposited this way are of considerable importance as a means for social communication. In a series of experiments published previously we studied the communicatory contents of the circumgenital-suprapubic marks and urine. All these experiments employed 5 min choice tests in which the subjects were simultaneously presented with two different scent samples. Fresh pieces of white pine or clean aluminum plates that had been marked by donor monkeys or treated with a predetermined amount of stimulus fluid served as stimulus objects. The occurrence of a statistically significant discriminatory response in contacting, sniffing, and scent-marking the stimulus objects was interpreted as proof that the animals could discriminate between the stimuli. The analysis of the communicatory content of natural scent marks and urine has established that the tamarins prefer[1] homospecific scent marks and urine over marks and urine of other marmoset species. Moreover, *S. f. fuscicollis* prefers its own marks over those of *S. f. illigeri,* showing that the scent is also subspecies-specific. It is not clear,

[1] The term "preference" in this context signifies only a quantitative difference in response; it does not imply a specific motivation.

however, whether this specificity is based on qualitative or quantitative differences. To the human observer, *S. f. fuscicollis* usually has a stronger body scent than *S. f. illigeri*. Interestingly, *S. f. illigeri* shows a tendency to prefer the scent of the opposite subspecies over its own, but hybrids between both subspecies show no preference (Epple, Golob, and Smith 1979). Male conspecific marks and urine are preferred over female conspecific odors when tested under a variety of conditions in which the amount of scent from each gender was varied systematically. Since a preference for male scent over female scent is shown regardless of the amount of scent and since male marks are recognized even in a mixture of male and female scents, we believe that gender is encoded qualitatively (Epple 1973, 1974a, 1978c). Intact male marks are preferred over castrated male marks, and the scent marks of dominant males are preferred over those of submissive males (Epple 1973, 1979). Discrimination between individual marks but not urine samples is shown when the monkeys are presented with two scent samples from familiar individuals, both of the same sex. If one of the scent odors has had a recent aggressive encounter with the subject, the subject prefers the opponent's odor over that of the neutral donor. The subjects were usually dominant in these encounters.[2] They showed preference for their opponent's scent even days after the encounter, which suggests that their response is not caused by a change of the opponent's odor as a result of the encounter (Epple 1973). In a choice test to determine the ability of the tamarins to discriminate between the scent marks of two unfamiliar, same-sexed individuals following a period of habitation to one of the stimulus odors, a preference is shown for the novel odor (Epple, Golob, and Smith 1979).

Concurrently with these behavioral studies we conducted work on the chemical analysis of the major volatile constituents of the scent marks. Combined gas-chromatographic–mass-spectrometric analysis of hexane-soluble material from either male or female scent marks indicated the common but variable presence of 16 major components (squalene and 15 esters of *n*-butyric acid), representing 96% (by weight) of the total volatile material (Figure 12.1; Smith, Yarger, and Epple 1976; Yarger, Smith, Preti, and Epple 1977). In addition to squalene and the 15 esters, a large number of highly volatile compounds are present in low concentrations. Their identification is presently in progress. Preliminary data suggest that the relative concentrations of the esters and squalene, probably in synergy with the more volatile compounds, may encode some of the communicatory signals described above (Epple, Golob, and Smith 1979).

Marking with the circumgenital-suprapubic glands occurs in a large number of situations. Marking activity varies not only among individuals but also among behavioral contexts. It is easily evoked by a wide variety of arousing stimuli as long as these do not elicit fear responses. It is therefore likely that marking behavior and the chemical signals deposited in the environment are important in controlling a number of behavioral interactions. The actual behavioral functions of the scent marks, as they are pro-

---

[2] The animals who were submissive in these encounters were not tested.

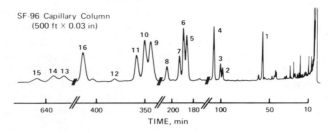

Peak	Molecular Formula	Identification
		SATURATED
1	$C_{20}H_{40}O_2$	
4	$C_{22}H_{44}O_2$	
8	$C_{24}H_{48}O_2$	
12	$C_{26}H_{52}O_2$	
		MONOENES—TYPE A
2	$C_{22}H_{42}O_2$	
6	$C_{24}H_{46}O_2$	
10	$C_{26}H_{50}O_2$	
14	$C_{28}H_{54}O_2$	
		MONOENES—TYPE B
3	$C_{22}H_{42}O_2$	
7	$C_{24}H_{46}O_2$	
11	$C_{26}H_{50}O_2$	
15	$C_{28}H_{52}O_2$	
		DIENES
5	$C_{24}H_{44}O_2$	
9	$C_{26}H_{48}O_2$	
13	$C_{28}H_{52}O_2$	
		SQUALENE
16	$C_{30}H_{50}$	

Figure 12.1. The major volatile constituents of the scent marks of saddleback tamarins.

duced, perceived, and utilized in specific social situations, probably are dependent upon the recipient's experience, the social and environmental context at the time, and the impact of other contiguous signals (e.g., calls or visual displays). For example, chemical information on the sex of an individual might function as an attractant for an animal of the opposite sex but as a repellent or aggression-eliciting signal for an animal of the same sex, particulary when combined with information on reproductive state. Similarly, individual odors that may serve as social or sexual attractants to recipients who have experienced friendly interactions with the sender may be regarded as threat signals by recipients who have experienced agonistic interactions with the sender. Thus, scent marks of largely similar chemical composition are adaptable to a number of different functional roles. Behavioral processes such as group cohesion, spacing and territoriality, intergroup and intragroup aggression, parent–infant interactions, recognition of, and attraction to, a sexual partner at the appropriate stage of reproductive activity, sexual arousal, temporary or permanent male–female pair bonding, reproductive synchrony, and so on may, in part, be controlled by chemical messages.

One area in which scent and scent-marking behavior appear to be important is in the demonstration of aggressive motivation and the establishment and maintenance of high social rank. Aggressive dominance interactions among group members are usually accompanied by a strong increase in the marking activity of the dominant animal and a decrease in that of the submissive animal. Aggressive interactions with conspecific animals not belonging to the same social group result in a dramatic increase in the marking activity of dominant adults as compared to their marking activities in undisturbed situations, but may result in a complete suppression of marking in adults introduced into established groups (Epple 1978a). These results suggest that one function of marking is the communication of aggressive motivation. In this context, the scent may actually serve as an aggressive threat signal. The high marking activity of an aggressive animal results in a large amount of this individual's odor in the environment. Since tamarins sniff conspecific marks frequently, a subdominant animal must be aware of the identity and amount of the dominant's odor. This awareness, combined with the experience of previous defeat or with other aggressive signals received from a social partner, may effectively control the recipient's behavior.

GONADAL CONTROL OF THE SCENT GLANDS, AGGRESSIVE BEHAVIOR, AND SCENT MARKING

Scent-marking activity, aggression, and dominance are associated in many mammalian species (cf. Ralls 1971). With a few exceptions, aggression, the activity of the scent glands, and marking behavior are dependent on gonadal hormones in most nonprimate species (Thiessen 1976), whereas the aggressive behavior of simian primates is strongly influenced by social experience (e.g., Rose et al. 1974; Mazur 1976; Dixson 1980). Therefore, we studied the relative importance of the gonads and a number of social factors influencing these characteristics in the tamarins. The social interactions of established, permanently cohabiting pairs and trios with strange conspecific adults were studied in situations analogous to territorial defense in the wild. Three experiments were performed, each testing the aggressive interactions of the tamarins with strangers under identical conditions (Epple 1978a,b, 1980, 1981a,b).

Methods

Experiment 1: Effects of gonadectomy in adulthood. The effects of gonadectomy in adulthood on aggressive behavior, the scent glands, and scent marking were studied in 14 permanently cohabiting, adult male–female pairs (Epple 1978b, 1980). Following presurgical studies of their behavior, the males of 7 pairs and the females of 2 pairs were surgically gonadectomized while their partners remained intact. The males of the remaining 5 pairs underwent sham castration, replicating all surgical procedures except removal of the testes.

Each pair lived in a testing room (3 × 3 × 2.5 m), where they were the only residents. Each test room was equipped with a home cage, which contained natural branches and a sleeping box. Each pair remained in the test room throughout the period of presurgical testing, but the tamarins were not confined to their cage. After gonadectomy and control surgery, they were placed in a cage (0.61 × 1.22 × 1.83 m) in a colony room. Seven months after surgery, each pair was returned to a test room and remained there for all postsurgical testing.

The behavior of each pair was tested in the following manner before and after surgery. Every pair was given 20 social encounters with 20 different intact conspecific adults both before and after surgery. Once before and once after surgery, 10 males and 10 females belonging to other groups were each introduced into the subject's room for 10 min. The "intruder" was wheeled into the test room in a wire mesh cage (0.61 × 0.91 × 1.22 m) and remained confined to the cage during the entire encounter. Subjects and intruder could interact through the ¼-in. wire mesh, which allowed a limited amount of contact but prevented serious injuries. The subjects were never tested more than once a day and usually only three encounters per week were given.

Each 10-min test was divided into 40 intervals of 15 sec each, indicated by an audible signal. Two observers, each focusing on one subject, recorded the interactions between subjects and intruder. The tests were scored in the following way. A score of 1 per interval was given for performing each of the following behavior patterns, described in detail by Epple (1978b), regardless of their actual frequencies of occurrence: scent marking, arching of the back, threat-facing, full ruffle (piloerection of the whole body), and attacking, chasing, and fighting. For final analysis of the raw aggression data, an accumulative score for injurious aggression was computed by summing the scores per interval of attacking the intruder by violent grabbing and/or scratching it, fighting with it through the mesh of the transport cage, and chasing it along the cage wall. Average scores for each individual before and after surgery were computed and all data were analyzed by nonparametric statistics following Siegel (1956).

In addition to assessing marking behavior during aggressive encounters both before and seven months after surgery, long-term records on the marking activity of the subject males on novel objects were kept from the time of gonadectomy up to a maximum of 4½ years (Epple and Cerny 1979). The marking activity of these males was assessed in series of 5-min tests, during which the animals received an aluminum bar (61 × 5 × 0.6 cm) that had been marked several times by an intact adult male immediately before the test. The 5-min test was divided into 60 intervals of 5 sec each. For each interval the subject received a score of 1 if it scent-marked the stimulus bar or any other place in the cage. During marking tests, the subjects were confined to one-half of their two-compartment home cage while their mates remained in the adjacent compartment.

Three of the seven castrates were tested as intacts (64 tests), and all seven subjects were tested between 2 weeks and 12 months after castration (442 tests). Testing of two of the males was extended until 21 and 25 months after surgery. During these months, all seven males cohabited with their original females. Four of the subjects were removed from their females 13, 13, 14, and 25 months after surgery and each was placed into an experimental trio consisting of an intact male, the castrated male, and a female, all nonrelated adults. The four castrates were given 184 marking tests while living in the trios. At 26, 38, and 51 months after surgery, three castrates living in a trio were again placed alone with a permanent female partner, either by removing the intact male from the group or by removing the castrate and pairing him with another female. These males received 140 marking tests after being returned to a pair situation. The change in the social setting provided an opportunity to assess the relative importance of gonadal hormones and social conditions in influencing scent-marking behavior. The marking activity of the sham-castrated males who remained with their original partners was tested before (94 tests) and after surgery (99 tests) in the same manner as that of the castrates.

Experiment 2: Effects of prepubertal gonadectomy. Nine males, six of whom were twins, served as subjects. All were born in the laboratory and lived with their parents and siblings until they were 5½ months old. At that age, prior to any macroscopically recognizable glandular differentiation, five males were castrated, and four were sham-castrated, assigning the twin brothers to different conditions. After surgery, the males were allowed 2 weeks with their families to recover. At the age of 6 months, each male was removed from his family and paired with an adult, sexually experienced, nonpregnant female. Eight males cohabited with their female partners until they were past 2 years of age. The partner of one sham castrate died after 4 months of cohabitation and was replaced by another adult female before behavioral tests began.

When the males were between 10 and 12 months old, their social interactions with strange adult conspecifics and their scent-marking behavior were tested. Each pair received seven social encounters with seven different intact adult female intruders and seven encounters with seven intact adult male intruders. The encounters were conducted and evaluated in a way identical to that used for males castrated and sham-castrated in adulthood. Details are reported by Epple (1981a).

Following completion of all social encounter tests, the subjects and their partners were housed in a colony room until the males were 2 years old. At that time, all castrates and three control males each received five social encounters with five strange males and five social encounters with five strange females. The same intruders who had encountered the 1-year-old subjects were used. The methods of conducting, recording, and analyzing the encounters were identical to those used during encounters given at 1 year of age with one exception. A sufficient number of test rooms to house all subject pairs was no longer available. Therefore, the following procedure was adopted. All subjects lived in a colony room. Each pair inhabited a home cage that was visually screened from other monkeys. Prior to every encounter the subject pair, inside its home cage, was wheeled into the test room, released from its cage, and allowed to run free in the room for 2 hours. This habituation period was followed by a 10-min social encounter test. Fifteen minutes after the encounter was over the pair was returned to its home cage and transferred to the colony room.

The scent-marking activity of the castrates and sham castrates on novel objects was also evaluated. The marking responses to clean aluminum plates, to plates scent-marked by an adult intact male immediately before the test, and to plates scent-marked by an adult intact female immediately before the test were studied under conditions identical to those used for the males castrated as adults. Between the ages of 10 and 11 months, the subjects received 10 tests of 5-min duration with clean, male-marked and female-marked plates in the presence of their female partners. This was done to habituate the subjects to the testing procedure and to reinforce, by the presence of their mates, their spontaneous tendency to investigate conspecific odors. Following the habituation tests, all castrates and three sham castrates were tested alone, receiving 20 tests with a clean stimulus plate, 20 tests with a male-marked plate, and 20 tests with a female-marked plate. At 2 years of age, all castrates and two of the control males received 10 tests each with clean, male-scented and female-scented plates.

Experiment 3: The effects of social experience. This experiment investigated the effect of maturation under different social conditions on the ontogenetic development of aggression and scent marking in gonadally intact animals. The subjects, 20 males and 20 females, lived in their families until they were 6 months of age. At that time, each subject was removed from its family and permanently grouped with one or two unrelated partners of the opposite sex. Ten 6-month-old females were each paired with one 6-month-old male. Four 6-month-old females were each paired with one sexually experienced adult male. Six young females were placed in trios, each consisting of the juvenile female, one sexually experienced adult male, and one 6-month-old male. Four 6-month-old males were each paired with one nonpregnant adult female.

When the subjects were between 10 and 12 months old, each pair and trio was placed in a test room of its own. Their social interactions with strange adult conspecifics were studied using the same procedures as described in Experiment 1 for the tamarins gonadectomized in adulthood. Each pair and trio encountered five adult male intruders and five adult female intruders (Epple 1981b).

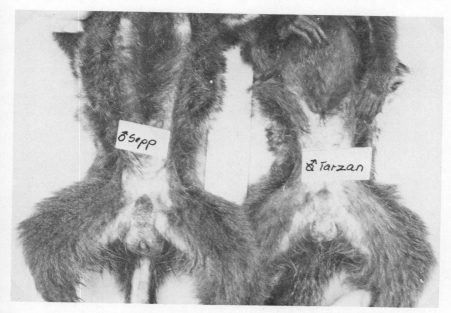

Figure 12.2. Circumgenital-suprapubic scent gland of a male castrated in adulthood (right) and an intact male (left). (From Epple 1980. Reprinted with permission.)

Results

The effect of gonadectomy on the scent glands. Gonadectomy of adult males resulted in atrophy of the scrotum, which forms part of the gland, thereby decreasing its surface area. However, the appearance of the suprapubic part of the gland was little changed by castration, and its average length of 26.9 ± 2.3 mm was in the range of that of intact males (Figure 12.2). Figure 12.3 shows that the morphology of female scent glands also does not seem to be strongly effected by ovariectomy in adulthood. The average length of the glands of three ovariectomized females (33.7 ± 3.3 mm) was also within the range of that of intact females.

Castration prior to puberty inhibited the development of the scrotum, which developed normally in the control males. Moreover, it permanently retarded, but did not inhibit completely, the development of the suprapubic part of the scent gland. At 1 year of age, the scent glands of the control males had developed normally, although they had not yet reached adult size. The suprapubic part of the pads showed an average length of 21.5 ± 1.8 mm and the scrota an average width of 16.1 ± 0.1 mm. In contrast to the sham castrates, 1-year-old castrates showed unpigmented, flat patches of glandular skin of an average length of 4.6 ± 0.7 mm in the area above the symphysis pubis. No glandular differentiation was noticeable

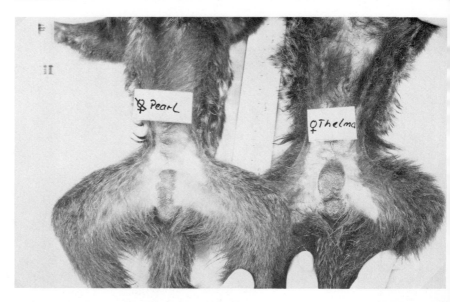

Figure 12.3. Circumgenital-suprapubic scent gland of a female ovariectomized in adulthood (left) and an intact female (right). (From Epple 1980. Reprinted with permission.)

on the underdeveloped scrota. By the time the control males were 2 years old, their scrota (average width 20.3 ± 0.9 mm) and suprapubic scent glands had reached adult size. The suprapubic pads (average length of 30.0 ± 2.9 mm) were fully pigmented and raised well above the surrounding skin. The suprapubic scent glands of the castrates had also increased in size by the time these males were 2 years old. They now formed a largely unpigmented, only slightly raised narrow strip of glandular skin with an average length of 17.6 ± 1.4 mm. The scrota also showed some glandular differentiation (Figure 12.4). Moreover, the thickness of the suprapubic pads and their pigmentation appeared to increase somewhat in 4 of the 5 castrates during their third and fourth years of life. The difference between the glands of castrated and control males persisted to their present age of 4 years and is illustrated by Figure 12.4.

The effect of gonadectomy on injurious aggression, displays, and scent marking. Table 12.1 shows the mean behavior score obtained by males and females gonadectomized as adults during their interactions with the intruders. When the scores obtained before surgery are compared with the postsurgical scores, it becomes obvious that gonadectomy and sham castration of sexually experienced adult subjects had no statistically significant effect on their display of injurious aggression, scent marking, and visual displays. Although the postsurgical aggression scores and the marking scores of the adult males tended to be higher than their presurgical scores, particularly in the castrates (Table 12.1), the increase was not

Figure 12.4. Circumgenital-suprapubic scent glands of 4-year-old twin brothers castrated (right) and sham castrated (left) at 5½ months of age. Note the lack of pigment in the castrate. (From Epple 1981a. Reprinted with permission.)

Table 12.1. *Mean behavior scores (± standard errors) of males and females gonadectomized in adulthood and adult sham castrates before and after surgery*

Behaviors	Mean behavior scores		
	Sham-castrated males[a]	Castrated males[a]	Ovariectomized females[b]
Threat facing			
Before surgery	1.0 ± .6	3.4 ± .9	2.3 ± .03
After surgery	3.5 ± .8	3.2 ± .9	1.8 ± 1.0
Arching			
Before surgery	.6 ± .2	.2 ± .1	3.9 ± 3.7
After surgery	.6 ± .4	.6 ± .2	1.9 ± 1.8
Ruffling			
Before surgery	8.4 ± 4.0	6.6 ± 1.8	11.3 ± 4.2
After surgery	8.0 ± 4.7	5.8 ± 1.6	6.6 ± .4
Scent marking			
Before surgery	8.0 ± 1.9	8.0 ± 1.4	13.0 ± .6
After surgery	9.4 ± 1.2	9.2 ± 2.7	13.5 ± 1.7
Accumulative aggression			
Before surgery	7.4 ± 3.1	8.7 ± 3.3	3.3 ± 2.4
After surgery	10.9 ± 3.1	15.1 ± 4.7	8.4 ± 6.0

[a] No significant statistical difference by Walsh test. [b] No statistical analysis was performed on the scores of the two ovariectomized females.

Table 12.2. *Mean behavior scores (± standard errors) of prepubertally castrated and sham-castrated males tested at 1 and 2 years of age*

	Mean behavior scores	
Behaviors	1-year-olds	2-year-olds
Threat facing		
Sham-castrated	9.0 ± .4	8.0 ± 1.5
Castrated	.7 ± .3[a]	3.2 ± .7[b]
Arching		
Sham-castrated	2.7 ± 1.2	2.4 ± 1.7
Castrated	.2 ± .1[c]	1.3 ± .5[b]
Ruffling		
Sham-castrated	4.8 ± 1.4	9.4 ± 3.3
Castrated	1.4 ± .4[d]	4.8 ± 1.0[b]
Scent marking		
Sham-castrated	10.8 ± 1.2	11.6 ± 1.1
Castrated	.04 ± .02[c]	5.3 ± 1.6[e]
Accumulative aggression		
Sham-castrated	2.7 ± 2.0	19.2 ± 7.8
Castrated	1.8 ± .9[b]	9.7 ± 3.8[b]

Mann–Whitney U test: [a] $p = .016$. [b] ns. [c] $p = .008$. [d] $p = .032$. [e] $p = .018$.

statistically significant. The two ovariectomized females also showed an increase in their aggression scores. However, no statistical analysis was done on the scores of only two subjects.

The behavior scores of prepubertally castrated and sham-castrated males are shown in Table 12.2. At 1 year of age, both groups of males showed considerably less aggression than males castrated in adulthood. The sham-castrated juvenile males tended to have higher accumulative aggression scores than the castrates. However, the average scores of the subject groups were not statistically different. On the other hand, sham castrates scent-marked, threat-faced, ruffled, and arched significantly more frequently than castrates (Table 12.2). When tested at 2 years of age, most differences between the behavior scores of castrates and sham castrates had disappeared. However, the castrates still scent-marked less frequently than the sham castrates (Table 12.2).

In experiments 1 and 2, there was a significant correlation between the accumulative scores for injurious aggression received by the subjects and the corresponding scores received by their bonded partners. The scores of males castrated and sham castrated as adults were correlated with those of their females both before ($r_s = .9$, $p < .01$, all correlations by

Table 12.3. *Pre- and postsurgical scent-marking scores (± standard errors) on male-scented aluminum plates*

Condition	Mean scent-marking scores	
	Sham castrates	Castrates
Before surgery	8.0 ± 1.6	7.9 ± 2.7
After surgery		
Total		7.3 ± 4.0
Pair	6.4 ± 2.7	8.8 ± 2.2
Trio		2.4 ± 1.3
Pair		15.0 ± 5.2[a]

[a] Kruskal–Wallis one-way analysis of variance: $p < .05$ for differences between pair and trio conditions.

Table 12.4. *Mean scent-marking scores (± standard errors) of males prepubertally sham-castrated and castrated, tested at 1 and 2 years of age in the presence of their females and alone*

Males	Mean scent-marking score	
	1-year-olds	2-year-olds
With mate		
Sham castrated	10.2 ± 1.3[a]	
Castrated	6.4 ± 2.2[a]	
Alone		
Sham castrated	10.0 ± 2.7[a]	11.5 ± 5.9[a]
Castrated	8.4 ± 2.2[a]	6.2 ± 1.9[a]

Note: Clean, male- and female-scented plates were pooled.
[a] No significant statistical difference by Mann–Whitney U test.

Spearman Rank correlation coefficient) and after surgery ($r_S = .853$, $p < .01$). Likewise, the scores of prepubertal castrates and sham castrates were correlated with those of their females when the males were 1 year old as well as at 2 years of age ($r_S = .7$, $p < .05$).

The scent-marking activity of males castrated and sham castrated as adults and of prepubertal castrates and sham castrates on differentially scented aluminum plates is illustrated in Tables 12.3 and 12.4. Long-term studies of the marking behavior of the males castrated as adults on male-scented plates showed no changes in marking activity attributable to gonadectomy. The total overall average marking scores of castrates and sham castrates before and after surgery did not differ (Table 12.3). The total score of the adult castrates includes data obtained in the course of the 4½-year study during which the males lived in three different social settings. When the marking scores obtained by the castrates under each social condition are considered separately, it becomes clear that marking activity was significantly different under the three conditions. Whereas the marking activity of the castrates during cohabitation with their original females was in the range of their overall total scores, the scent-marking behavior of each of the four males studied when living in a trio with an intact male and an intact female decreased. The decline in the marking activity of the castrates during life in trios was completely reversible when their social situation changed. After placing two of them with a single female partner again, their marking activity increased strongly. Another castrate killed the intact male living in his group during a dominance fight. Thereafter, he remained with the adult female and her offspring. Although his scent-marking activity also recovered significantly after the death of the intact male, it did not increase to its original level. The average score of these three males, who had been castrated 2, 4, and 4½ years before, was now above the overall average for all castrates (Table 12.3).

The marking activity of prepubertally castrated and sham castrated 1-year- and 2-year-old males on scented plates is shown in Table 12.4. All these subjects lived with one adult female during the time of testing. As the table shows, the marking activity of castrates and sham castrates did not differ under these conditions, and some scores received by these males tended to be slightly, but not significantly, higher than the overall scores of the adult males shown in Table 12.3. Although the prepubertal sham castrates tended to scent mark more frequently than prepubertal castrates, the standard errors of their mean scores overlap under most conditions. Unscented stimulus plates and plates carrying conspecific odors elicited different levels of scent marking ($p = .05$ at 1 year; $p = .001$ at 2 years, Kruskal–Wallis one-way analysis of variance). Plates carrying male odor were scent-marked by 1- and 2-year-old males more frequently than clean plates ($p = .016$, Walsh test) and plates carrying female odor ($p = .008$ at 1 year; $p = .016$ at 2 years). The gonadal condition of the males did not influence this behavior, which is also shown by intact males and females when they are given a simultaneous choice between male and female odors (Epple 1974a,b).

The effects of social experience on injurious aggression, displays, and scent marking of tamarins. The behavior scores of the young intact males and females maturing under different social conditions in experiment 3 are shown in Table 12.5. As the table shows, there were significant differences in the aggression scores among the three groups of male subjects as well as among the three groups of female subjects. Males living with adult females were more aggressive than those paired with same-aged partners ($p = .02$; all two-sample comparisons by Mann–Whitney U-test) or those living in trios ($p = .005$). Females living with either an adult male or an

Table 12.5. Mean behavior scores (± standard errors) of 1-year-old intact males and female subject animals cohabiting either with a 1-year-old mate, an adult mate, or in a trio

| | Mean behavior scores | | | | | |
| | Young male and | | | Young female and | | |
Behavior	Young female	Adult female	Young female, adult male	Young male	Adult male	Young male, adult male
Threat facing	.7 ± .3	6.1 ± 2.2[a]	.5 ± .2	2.1 ± .6	3.8 ± 1.1	3.6 ± 1.4
Arching	.2 ± .2	.6 ± .4	.1 ± .1	3.6 ± 1.5	7.5 ± 2.7	1.7 ± .8
Ruffling	2.2 ± .5	4.0 ± 1.4	4.8 ± 2.2	7.5 ± 1.9	10.2 ± 4.4	7.8 ± 2.6
Scent marking	5.6 ± 1.7	13.4 ± 1.7[a]	.9 ± .3	3.0 ± 1.0	10.6 ± 3.1[a]	9.6 ± 2.5
Accumulative aggression	.9 ± .4	11.0 ± 3.7[a]	.7 ± .3	1.2 ± .5	11.8 ± 5.3[a]	10.4 ± 3.7

[a] Kruskal–Wallis one-way analysis of variance tested differences among the scores obtained by same-sexed animals living under different social situations: $p = .01$.

adult and a young male were more aggressive than those cohabiting only with a young partner ($p = .02$). Except for scent marking and male facial threat, there were no significant differences in the display scores among subject groups (Table 12.5). Males ($p < .05$) and females ($p < .02$) mated to adult partners scent marked more frequently than subjects mated to young partners and young males living in trios ($p < .005$). As in experiments 1 and 2, the accumulative aggression scores of the young subjects and their mates were correlated. This was evident when the scores of all 1-year-old subjects mated to 1-year-old partners and to adult partners were analyzed as a group ($r_S = .43, p < .1$). In this analysis, the young females in the trios were considered to be mated to the adult males; the scores of the young males were not included.

SUMMARY

Experiments 1 and 2 show that gonadal hormones are necessary for the full morphogenetic development of the circumgenital-suprapubic scent glands. However, prepubertal castration did not prevent glandular development completely. As in intact males, the glands of prepubertal castrates grew larger as the animals aged beyond 1 year. Gonadectomy of adult tamarins, on the other hand, had little effect on the morphological appearance of the scent glands of males and females. It appears, therefore, that gonadal hormones are essential for the full development of the glands. However, once these structures have developed fully, they may become largely hormone independent or may be maintained by extragonadal hormones. Adrenal androgens, which provide a source of extragonadal steroids, may possibly support the glands after gonadectomy in adulthood and also allow a certain amount of development of the glands in prepubertal castrates. This hypothesis, of course, awaits histological confirmation and further experimental study.

Gonadectomy had no detectable influence on aggressive behavior, displays, and scent marking when performed on socially experienced adult males and females. The scent-marking behavior of these castrated males appeared to be much more sensitive to social stimuli than to gonadal hormones. This is clearly demonstrated by the dramatic decrease in scent marking shown by castrated males when placed in trios with an intact male and female and the complete reversal of this decrease when the social situation changed.

Although gonadectomy in adulthood had no effect on the behaviors under study, prepubertal castration significantly influenced the social interactions of the males with strange conspecifics. However, these effects were not permanent and have to be evaluated in the context of social experience. At 1 year of age, castrates directed fewer aggressive displays against intruders than did control males. Moreover, they showed almost no scent marking during encounters with strangers. However, the amount of injurious contact aggression showed by castrates did not differ from that shown by control males. Both groups of 1-year-old males showed less injurious aggression than adults and less aggression than the intact 1-year-old males cohabiting with adult females in Experiment 3. The low marking activity of the castrates was specific to social encounters. The castrates, whose marking in the presence of intruders was negligible, marked at approximately the same level as the control males when tested with novel objects in their home cages. Moreover, even at the age of 1 year the marking scores of castrates and control males on novel objects was already in the range of the adult males of experiment 1. By the time they were 2 years old, both the castrates and control males directed as much injurious aggression against conspecific intruders as adult intact males and males castrated in adulthood when tested under identical conditions. Moreover, most differences in the display activities of prepubertal castrates and sham castrates had vanished by that time, although the castrates still marked less frequently than the control males.

The results of Experiments 1 and 2 show that although gonadal hormones influence the development of the behavior patterns under investigation, they are not the only determinants of these behaviors, that is, there is no simple causal relationship between the gonads and various manifestations of aggression and scent marking. Social factors appear to be as important as hormonal factors in determining the development of these behaviors. Among these, the behavior and motivations of the mate may be important. In all three experiments, a correlation between the accumulative aggression scores of the bonded partners constituting the subject pairs was evident. This suggests that aggression displayed by one subject is not independent of that displayed by its mate and shows that the behavior of the mate is another stimulus that shapes social responses. That the aggression scores of 1-year-old intact males living with adult females in Experiment 3 were so much higher than the scores of 1-year-old sham castrates and castrates may be due to the influence of their females. There was, indeed, a significant difference between the aggression scores of the females of Experiment 3 and the two groups of females living with castrates and sham castrates in Experiment 2 ($p = .049$, Kruskal–Wallis analysis of variance).

The importance of social experience in determining the behavioral development of these primates is further documented by the results of Experiment 3, which evaluates the influence of some social experiences on the ontogenetic manifestation of several behaviors in intact males and females. Since these animals had the same social history and received social encounter tests with strange conspecific adults at the same age and under identical conditions as the prepubertal castrates and sham castrates, the scores of the males in both experiments at 1 year of age are directly comparable. As mentioned above, the accumulative aggression scores of the four intact males cohabiting with adult females were higher than those obtained by the prepubertal castrates and sham castrates, and the visual-display and scent-marking scores were in the same range. The accumulative-aggression scores and most of the display scores of the 10 males cohabiting with a female of *their own age* were significantly lower than those of the males living with adult females and were in the same range as those of the prepubertal castrates. When the scores of intact young males and females living with adult partners are compared with those of the socially experienced adults of Experiment 1, it becomes obvious that most of their behavior scores are at, or even above, the level of sexually and socially experienced adults.

Studies on the sociosexual behavior within the pairs and trios of Experiment 3 have shown that the interaction profiles of young subjects and their adult mates resemble those of monogamously mated adults more closely than do those of the subjects with same-aged partners in pairs as well as in trios (Epple and Katz 1980). Moreover, young females cohabiting with adult males and young males cohabiting with adult females also reproduced at a significantly earlier age than the subjects living with young partners, and a higher percentage of pregnancies progressed to term in females cohabiting with adult males (Epple and Katz 1980). These results suggest that young males and young females are capable of fulfilling the behavioral and reproductive roles of adult breeders at an early age when given a chance to form a pair bond with an adult. Under natural conditions this situation may not be uncommon, and the ability to respond to it either behaviorally and/or physiologically may be an important part of the reproductive strategy of these small primates. In the wild, the population of transients, many of whom appear to be late juveniles, forms a reservoir from which potential breeding females and dominant males can be recruited should a group lose one or both of the bonded nuclear pair. Our laboratory observations suggest that many saddleback tamarin groups observe a behavioral incest tabu. If this occurs in the wild, it is likely that the breeding position of a group suffering a loss will be occupied by a stranger associated with the group rather than a family member. This behavioral mechanism, of course, greatly increases the gene flow across the population.

The results of all the studies reviewed in this chapter clearly demonstrate that *both* hormonal condition and a variety of social factors influence the complex behavior of the tamarins. Previous social experience can influence the behavior of these primates as much, or even more, than their gonadal condition. In this respect, the tamarins resemble other simian primates. Although gonadal hormones appear to affect and to reflect the dominance status and the aggressiveness of an individual in several of the higher primate species (Clark and Birch 1945; Birch and Clark 1946; Joslyn 1973; Rose et al. 1974; Mazur 1976; Cochran and Perachio 1977; Dixson and Herbert 1977), a wide variety of social factors may completely override the action of hormones on primate behavior (Mirsky 1955; Wilson and Vessey 1968; Green, Whalen, Rutley, and Battie 1972; Dixon and Herbert 1977; Bernstein, Gordon, and Peterson 1979).

REFERENCES

Bernstein, I. S., Gordon, T. P., and Peterson, M. 1979. Role behavior of an agonadal alpha male rhesus monkey in a heterosexual group. *Folia Primatologica* 32:263–7.

Birch, H. G., and Clark, G. 1946. Hormonal modification of social behavior: II. The effects of sex-hormone administration on the social dominance status of the female-castrate chimpanzee. *Psychosomatic Medicine* 8:320–31.

Bronson, F. H., and Desjardins, C. 1971. Steroid hormones and aggressive behavior in mammals. In *The physiology of aggression and defeat*, ed. B. E. Eleftheriou and J. P. Scott, pp. 43–63. New York: Plenum.

Castro, R., and Soini, P. 1978. Field studies on *Saguinus mystax* and other callitrichids in Amazonian Peru. In *The biology and conservation of the Callitrichidae*, ed. D. G. Kleiman, pp. 73–8. Washington, D.C.: Smithsonian Institution Press.

Cheal, M. L. 1975. Social olfaction: a review of the ontogeny of olfactory influences on vertebrate behavior. *Behavioral Biology* 15:1–25.

Clark, G., and Birch, H. G. 1945. Hormonal modification of social behavior: I. The effect of sex hormone administration on the social status of a male-castrate chimpanzee. *Psychosomatic Medicine* 7:321–9.

Cochran, C. A., and Perachio, A. A. 1977. Dihydrotestosterone proprionate effects on dominance and sexual behaviors in gonadectomized male and female rhesus monkeys. *Hormones and Behavior* 8:175–87.

Dawson, G. 1978. Composition and stability of social groups of the tamarin *Saguinus oedipus geoffroyi* in Panama: ecological and behavioral implications. In *The biology and conservation of the Callitrichidae*, ed. D. G. Kleiman, pp. 23–37. Washington, D.C.: Smithsonian Institution Press.

Dixson, A. F. 1980. Androgens and aggressive behavior in primates: a review. *Aggressive Behavior* 6:37–67.

Dixson, A. F., and Herbert, J. 1977. Testosterone, aggressive behavior and dominance rank in captive adult male talapoin monkeys (*Miopithecus talapoin*). *Physiology and Behavior* 18:539–43.

Eaton, G. G., Goy, R. W., and Phoenix, C. H. 1973. Effects of testosterone treatment in adulthood on sexual behaviour of female pseudohermaphrodite rhesus monkeys. *Nature* 242:119–20.

Epple, G. 1972. Social behavior of laboratory groups of *Saguinus fuscicollis*. In *Saving the lion marmoset. Proceedings of the Wild Animal Preservation Trust Golden Lion Marmoset Conference*, ed. D. D. Bridgewater, pp. 50–8. Oglebay Park, Wheeling, W. Va.:WAPT.

  1973. The role of pheromones in the social communication of marmoset monkeys (Callitrichidae). *Journal of Reproduction and Fertility, Supplement* 19:447–54.

  1974a. Primate pheromones. In *Pheromones*, ed. M. C. Birch, pp. 366–85. Amsterdam: North-Holland.

  1974b. Olfactory communication in South American primates. *Annals of the New York Academy of Sciences* 237:261–78.

  1975. The behavior of marmoset monkeys (Callithricidae). In *Primate Behavior: developments in field and laboratory research*, Vol. 4: ed. L. A. Rosenblum, pp. 195–239. New York: Academic Press.

  1978a. Notes on the establishment and maintenance of the pair bond in *Saguinus fuscicollis*. In *The biology and conservation of the Callitrichidae*, ed. D. G. Kleiman, pp. 231–7. Washington, D.C.: Smithsonian Institution Press.

  1978b. Lack of effects of castration on scent marking, displays and aggression in a South American primate (*Saguinus fuscicollis*). *Hormones and Behavior* 11:139–50.

  1978c. Studies on the nature of chemical signals in scent marks and urine of *Saguinus fuscicollis* (Callitrichidae, Primates). *Journal of Chemical Ecology* 4:383–94.

1979. Gonadal control of male scent in the tamarin *Saguinus fuscicollis* (Callitrichidae, Primates). *Chemical Senses and Flavour* 4:15–20.

1980. Relationships between aggression, scent marking and gonadal state in a primate, the tamarin *Saguinus fuscicollis*. In *Chemical signals, vertebrates and aquatic invertebrates*, ed. D. Müller-Schwarze and R. M. Silverstein, pp. 87–105. New York: Plenum.

1981a. Effects of prepubertal castration on the development of the scent glands, scent marking and aggression in the saddle back tamarin (*Saguinus fuscicollis*, Callitrichidae, Primates). *Hormones and Behavior* 15:54–67.

1981b. Effect of pair bonding with adults on the ontogenetic manifestation of aggressive behavior in a primate, *Saguinus fuscicollis*. *Behavioral Ecology and Sociobiology* 8:117–23.

Epple, G., and Cerny, V. A. 1979. Effects of castration and social change on scent marking behavior of *Saguinus fuscicollis* (Callitrichidae). *Folia Primatologica* 32:252–62.

Epple, G., and Katz, Y. 1980. Social influences on first reproductive success and related behaviors in the saddle back tamarin (*Saguinus fuscicollis*, Callitrichidae), *International Journal of Primatology* 1:171–83.

Epple, G., and Lorenz, R. 1967. Vorkommen, Morphologie und Funktion der Sternaldrüse bei den Platyrrhini. *Folia Primatologica* 7:98–126.

Epple, G., Golob, N. F., and Smith, A. B., III. 1979. Odor communication in the tamarin *Saguinus fuscicollis* (Callitrichidae): behavioral and chemical studies. In *Chemical ecology: odour communication in animals*, ed. F. J. Ritter, pp. 117–30. Amsterdam: Elsevier/North-Holland Biomedical Press.

French, J. A., and Snowdon, C. T. 1981. Sexual dimorphism in responses to unfamiliar intruders in the tamarin, *Saguinus oedipus*. *Animal Behaviour* 29:822–9.

Green, R., Whalen, R. E., Rutley, B., and Battie, C. 1972. Dominance hierarchy in squirrel monkeys (*Saimiri sciureus*). *Folia Primatologica* 18:185–95.

Harrington, J. E. 1975. Field observations of social behavior of *Lemur fulvus fulvus* (E. Geoffrey 1812). In *Lemur biology*, ed. I. Tattersall and R. W. Sussman, pp. 259–79. New York: Plenum.

Hart, B. 1974. Gonadal androgens and socio-sexual behavior of male mammals. *Psychological Bulletin* 81:383–400.

Hershkovitz, P. 1977. *Living New World monkeys (Platyrrhini)*, Vol. 1. Chicago: University of Chicago Press.

Izawa, K. 1978. A field study of the ecology and behavior of the black-mantle tamarin (*Saguinus nigricollis*). *Primates* 19:241–74.

Jolly, A. 1967. Breeding synchrony in wild *Lemur catta*. In *Social communication among primates*, ed. S. Altmann, pp. 3–14. Chicago: University of Chicago Press.

Joslyn, W. D. 1973. Androgen-induced social dominance in infant female rhesus monkeys. *Journal of Psychology and Child Psychiatry* 14:137–45.

Kleiman, D. G. 1977. Monogamy in mammals, *Quarterly Review of Biology* 52:39–69.

1978. The development of pair preferences in the lion tamarin (*Leontopithecus rosalia*): male competition or female choice? In *Biology and behaviour of marmosets*, ed. H. Rothe, H. J. Wolters, and J. P. Hearn, pp. 203–7. Göttingen: Eigenverlag Rothe.

Leshner, A. J. 1975. A model of hormones and agonistic behavior. *Physiology and Behavior* 15:225–35.

Mazur, A. 1976. Effects of testosterone on status in primate groups. *Folia Primatologica* 26:214–26.

Mirsky, A. F. 1955. The influence of sex hormones on social behavior of monkeys. *Journal of Comparative and Physiological Psychology* 48:327–35.

Moynihan, M. 1970. Some behavior patterns of platyrrhine monkeys. II. *Saguinus geoffroyi* and some other tamarins. *Smithsonian Contributions to Zoology* No. 28:1–77.

Mykytowycz, R. 1970. The role of skin glands in mammalian communication. *Advances in chemoreception,* Vol. 1: *Communication by chemical signals,* ed. J. W. Johnston, D. G. Moulton, and A. Turk, pp. 327–60. New York: Appleton-Century-Crofts.

Neyman, P. F. 1978. Aspects of the ecology and social organization of free-ranging cotton-top tamarins (*Saguinus oedipus*) and the conservation status of the species. In *The biology and conservation of the Callitrichidae,* ed. D. G. Kleiman, pp. 39–71. Washington, D.C.: Smithsonian Institution Press.

Perkins, E. M. 1966. The skin of the black-collared tamarin (*Tamarinus nigricollis*). *American Journal of Physical Anthropology* 25:41–69.

1968. The skin of the pygmy marmoset – *Callithrix* (= *Cebuella*) *pygmaea*. *American Journal of Physical Anthropology* 29:349–64.

1969a. The skin of the cotton-top pinché *Saguinus* (= *Oedipomidas*) *oedipus*. *American Journal of Physical Anthropology* 30:13–27.

1969b. The skin of Goeldi's marmoset (*Callimico goeldii*). *American Journal of Physical Anthropology* 30:231–49.

1969c. The skin of the silver marmoset – *Callithrix* (= *Mico*) *argentata*. *American Journal of Physical Anthropology* 30:361–87.

Poole, T. B. 1978. A behavioral investigation of "pair bond" maintenance in *Callithrix jacchus jacchus*. In *Biology and behaviour of marmosets,* ed. H. Rothe, H. J. Wolters, and J. P. Hearn, pp. 209–16. Göttingen: Eigenverlag Rothe.

Quadagno, D. M., Briscoe, R., and Quadagno, J. S. 1977. Effect of perinatal gonadal hormones on selected nonsexual behavior patterns: a critical assessment of the nonhuman and human literature. *Psychological Bulletin* 84:62–80.

Ralls, K. 1971. Mammalian scent marking. *Science* 171:443–9.

Rose, R. M., Bernstein, I. S., Gordon, T. P., and Catlin, S. F. 1974. Androgens and aggression: a review and recent findings in primates. In *Primate aggression, territoriality and xenophobia,* ed. R. L. Holloway, pp. 275–304. New York: Academic Press.

Schilling, A. 1974. A study of marking behaviour in *Lemur catta*. In *Prosimian biology,* ed. R. D. Martin, G. A. Doyle, and A. C. Walker, pp. 347–62. Gloucester-Crescent: Duckworth.

Siegel, S. 1956. *Nonparametric statistics for the behavioral sciences.* New York: McGraw-Hill.

Smith, A. B., III, Yarger, R. G., and Epple, G. 1976. The major volatile constituents of the marmoset (*Saguinus fuscicollis*) scent mark. *Tetrahedron Letters,* 983–6.

Starck, D. 1969. Die circumgenitalen Duftdrüsenorgane von *Callithrix* (*Cebuella*) *pygmaea* (Spix 1823). *Der Zoologische Garten* 36:312–26.

Sutcliffe, A. G., and Poole, T. B. 1978. Scent marking and associated behaviour in captive common marmosets (*Callithrix jacchus jacchus*) with a description of the histology of scent glands. *Journal of Zoology* 185:41–56.

Thiessen, D. D. 1976. *The evolution and chemistry of aggression.* Springfield, Ill.: Thomas.

Thorington, R. W., Jr. 1968. Observations of the tamarin *Saguinus midas*. *Folia Primatologica* 9:95–8.

Wilson, A. P., and Vessey, S. H. 1968. Behavior of free-ranging castrated rhesus monkeys. *Folia Primatologica* 9:1–14.

Wislocki, G. B. 1930. A study of scent glands in the marmosets, especially *Oedipomidas geoffroyi*. *Journal of Mammalogy* 11:475–82.

Yarger, R. G., Smith, A. B., III, Preti, G., and Epple, G. 1977. The major volatile constituents of the scent mark of a South American primate *Saguinus fuscicollis*, Callitrichidae. *Journal of Chemical Ecology* 3:45–56.

13 · A cognitive analysis of facial behavior in Old World monkeys, apes, and human beings

SUZANNE CHEVALIER-SKOLNIKOFF

The facial expressions of Old World primates have been described in considerable detail by Hinde and Rowell (1962), Andrew (1963, 1964), Van Hooff (1967), Chevalier-Skolnikoff (1973a, 1974), Redican (1975), and others. Comparisons have failed to reveal salient phylogenetic differences because no measuring tool of appropriate scope has hitherto been devised to evaluate these differences. Piaget's model of cognitive development during the sensorimotor period provides a promising hierarchical framework for classifying and comparing behavior (Piaget 1951, 1952, 1954). This study represents an initial attempt to examine phylogenetic differences in facial behavior using Piaget's model as a framework for comparison.

PIAGET'S MODEL

Piaget's model of sensorimotor development provides a systematic and relatively objective framework for comparison. It is made up of six hierarchical and qualitatively different stages that appear sequentially during the first two years of human development. Each stage in the series can be defined by a set of behavioral parameters specific to that stage. These parameters reflect the development of increasingly complex cognitive functioning, since the six stages are delineated by *increasing voluntary control, increasing elaboration of the number of motor patterns and contextual variables* involved in a specific act, and *increasing variability of motor patterns.* Consequently, each stage in the series is characterized

I thank the San Francisco Zoological Society, Saul Kitchner, and Steve Taylor for providing access to the orangutans as subjects. I thank John Alcaraz and the orangutans' many surrogate mothers for the time and effort they graciously gave to this project. Paul Ekman, Mardi Horowitz, Alan Skolnikoff, and Charles Snowdon read early drafts of this manuscript and provided useful criticism. I wish to thank William Demarest for statistical assistance, Phyllis Cameron and Janice Andersen for typing and editing the manuscript, Jeff Burner (Figs. 13.5a, 13.6, 13.8a,b), Margaret Burks (13.9b), Stephen Longstreth (13.5b), and Edward Nyberg (13.8c,d, 13.9a, 13.12) for taking photographs. All photographs in this chapter are © 1981, Suzanne Chevalier-Skolnikoff. This research was partially supported by the Academic Senate Committee on Research, University of California, San Francisco, and the Endowment Fund of the Hooper Foundation, University of California, San Francisco.

by a new, more advanced level of functioning, as new abilities become incorporated into the behavior.

Piaget's model of sensorimotor development is made up of several parallel series, each consisting of six stages that progress at about the same rates in human infants. Among them are the Object Concept Series (the development of the infant's understanding that out-of-sight objects continue to exist), the Imitation Series (a series of increasingly complex modes of imitation), and the Sensorimotor Intelligence Series (general sensory and motor adaptations to the environment). This chapter will examine the ontogeny of facial behavior mainly in terms of sensorimotor intelligence. I will be examining the development of emotional facial expressions and nonemotional facial behaviors, such as oral object manipulation, but will not focus on feeding behavior.

Central to Piaget's conception of sensorimotor intelligence is environmental adaptation. He perceives the individual's adaptation to his environment as a process that includes two complementary subprocesses: *accommodation* and *assimilation*. Accommodation refers to the individual's tendency to modify his behavior according to his new and evolving perceptions of his environment. Assimilation involves using current understanding and modes of coping to deal with the environment. By means of accommodation and assimilation the individual balances his level of understanding with the requirements of his world. Through repeated sequences of accommodation and assimilation, the maturing individual repeatedly attains new levels of understanding and environmental equilibrium, or new stages of cognitive development, that repeatedly become unbalanced as the youngster's further development enables him to perceive the world in new ways to which he must again accommodate.

The six stages of the Sensorimotor Intelligence Series are: (1) *reflex* — unlearned, stereotyped responses to stimuli; (2) *primary circular reaction* – repetitive self-oriented behavior with acquired adaptations to impinging stimuli; (3) *secondary circular reaction* – reaching out to the environment and repeatedly eliciting reactions from objects; (4) *coordinations* – combining behaviors to achieve goals; (5) *tertiary circular reaction* – experimenting with the relationships between objects, space, and force; (6) *insight* – inventing or solving problems through mental representation (see Table 13.1).

METHODS

Subjects

This study is based on approximately 1,500 hr of observation of 65 adult and infant subjects: 12 captive monkeys (stump-tailed macaques and Hanuman langurs, c. 700 hr); 42 captive apes (chimpanzees, orangutans, and gorillas, c. 800 hr); and 11 human infants (c. 25 hr). See the endpapers figure for the probable phylogenetic relationships of these species (Wilson and Sarich 1969; Sarich 1970). Five monkeys and ten apes were observed longitudinally for all or most of the sensorimotor period; the other subjects were studied cross-sectionally during one or two developmental periods. Approximately half of the nonhuman infants were reared by their mothers in social groups, and half were reared in zoo nurseries.

The data on stump-tailed macaques (*Macaca arctoides*) are derived from Parker's (1977) study on the cognitive development of one infant living with his mother through the first five months of life and on my study on the communicative development of three infants living in a social group of seven adults, including their mothers, through the first six months of life (Chevalier-Skolnikoff 1974). The data on langurs (*Presbytis entellus*) are based on my observations on the cognitive development of two infants nursery-reared by human caretakers. The langurs were observed for the first nine and ten months, respectively.

The data on apes are derived from my current observations of 42 great apes: 20 chimpanzees (*Pan troglodytes*), 10 lowland gorillas (*Gorilla gorilla gorilla*), and 12 orangutans (*Pongo pygmaeus*). Half the ape subjects are adults and half are infants and juveniles. Of the 22 infants and juveniles, 9 are being nursery- or home-reared by human caretakers, and the other 13 are being reared by their mothers, generally in small social groups. Of the infants, 10 are being observed longitudinally: 4 have been observed for 4- to 6-month periods, 4 have been observed throughout most or all of the first 18 months, and 2 have been studied for the first 3 and 4 years, respectively. The other 12 infants and juveniles have been studied during one or more cross-sectional periods.

The data on humans are derived primarily from the literature. In addition, I have studied 11 human infants cross-sectionally, observing one or more subjects during each of the six sensorimotor stages.

Data collection and analysis

Several approaches are used in collecting data. Subjects living with conspecifics are observed as they behave spontaneously in their social groups. The behavior of focus subjects and the contexts in which it occurs are recorded on tape in running narrative form. This narrative describes both general sensorimotor behavior and facial behavior. Objects, such as toys, are introduced to give the subjects a better opportunity to demonstrate their cognitive capacities. The human infants, as well as the nursery-reared nonhuman subjects, are given formal Piagetian tasks derived from Užgiris and Hunt (1966, 1975). These include problem-solving tasks (such as using a tool to obtain an out-of-reach object). The results of these tests are recorded on check sheets. Thirteen hours of film have been collected on the behavior of stump-tailed macaques, orangutans, gorillas, and human neonates. The descriptions of all the macaque facial expressions and some of the contextual information were based on frame-by-frame analysis of the film record (Chevalier-Skolnikoff 1971). Fewer ape expressions were recorded on film, and only some descriptions and contexts were derived from film analysis.

Table 13.1 *The Sensorimotor Intelligence Series as manifested by human infants*

Stage	Age (months)	Description	Major distinguishing behavioral parameters	Example
1. Reflex	0–1	Unlearned, involuntary, invariable responses	Unlearned, involuntary, stereotyped	Roots and sucks
2. Primary circular reaction	1–4	Action centers about infant's *own* body (primary); learns to *repeat* (circular) action to reinstate an event; first acquired adaptations	Self-oriented, recognizes contexts, acquires adaptations to impinging stimuli	Repeats hand–hand clasping
3. Secondary circular reaction	4–8	Repeats (circular) actions to reproduce environmental (secondary) events first discovered by chance	Environment-oriented; single behaviors toward a single object or person; semi-intentional (initial act not intentional but subsequent acts are); establishes object- or person-action relationships through attempts to reproduce interesting environmental events	Swings objects and attends to the swinging spectacle or to the resulting sound; repeats

4. Coordinations	8–12	Coordinates two or more behavioral acts, one serving as instrument to the other	Intentional; goal established from the outset; establishes relationships (coordinations) between two objects; coordinates or relates multiple aspects of single objects; combines behaviors; begins to attribute cause of environmental change to others	Sets aside an obstacle to reach an object behind it
5. Tertiary circular reactions	12–18	Becomes curious about the functions of objects, object–, space–, gravity–force relationships (tertiary); repeats behavior (circular) with variation in exploring the potentials of objects through trial-and-error experimentation	Behavior becomes variable and non-stereotyped as the infant invents new behavior patterns; trial-and-error experimentation in play and to solve problems; coordinates object–, person–, space–, and gravity–force relationships	Experimentally discovers that one object (e.g., stick) can be used to obtain another
6. Invention through mental combinations (insight)	18–24	Arrives at solution mentally, not through experimentation	Symbolically represents objects and events not present; solves problems mentally	Mentally figures out how one object can be used to obtain another

Note: The table was abstracted from Piaget and adapted from Chevalier-Skolnikoff 1976.

A corpus of approximately 1,280 facial acts was analyzed for this chapter: 700 were manifested by stump-tailed macaques, 60 by langurs, and 529 by orangutans. In addition, the ages of the first appearance of the various facial behaviors of the gorilla and chimpanzee subjects, and the contexts in which these behavioral acts occurred, were analyzed.

Data analysis for this chapter consisted of examining both general cognitive development and the development of facial behavior in terms of sensorimotor stages. I analyzed the behaviors and the contexts in which they occurred in terms of their behavioral parameters (presented in Table 13.1) and thereby classified them into one of the six stages of the Sensorimotor Intelligence Series. (See Table 13.2, a transcription of 3 min of the taped narrative. This demonstrates my methods of categorizing data recorded in running narrrative form.)

Facial behavior was classified both by its behavioral parameters, that is, according to the way it functioned, and by the contexts in which it occurred. For example, a human infant smiles at his mother; when she smiles back at him, he smiles at her again. In terms of their behavioral parameters (Table 13.1), such smiles are environmentally oriented, repetitive, single behavioral acts directed toward a single object, and they establish person-action relationships. They function as secondary circular reactions and can be functionally classified as stage 3 facial behaviors. In another example, an infant sees a bell, reaches out and touches it; the bell rings, he touches it again and smiles as it rings again. In this case the smile is environmentally oriented (a stage 3 behavioral parameter, suggesting stage 3 classification), but it does not function as a secondary circular reaction. However, because it occurs in the context of the secondary circular reaction, as the infant repeatedly rings the bell with his hand (repetitive behavior directed toward a single object, establishing an object-action relationship), it is classified as a stage 3 behavior.

The ontogenetic development of general sensorimotor behavior and facial behavior is examined in infants and juveniles, whereas the capacity of the species is evaluated from the behavior of adults.

A subject was judged to have entered a specific stage when it unambiguously manifested at least one behavioral parameter characteristic of that stage in at least two separate bouts during a single day. Using these criteria, it is theoretically possible for an animal judged as being in one stage to occasionally, though not repeatedly, manifest behavior classifiable at a higher stage. Such behavior represents the "slop" in the classification or in the developmental systems. Examining their frequencies gives some indication of how behavior develops – whether it is a gradual process or whether it occurs in sudden steps. Analyzing the frequency also is a way of testing the validity of the stage system of classifying development. If an infant judged to be in one stage occasionally manifests behavior characteristic of several other higher stages, it would put the entire model into question. But if the behavioral parameters of the various stages appear in an orderly sequence, either abruptly or gradually, we can assume the model provides a fair assessment of behavioral development.

Table 13.2. *Three-minute sample of data analysis demonstrating method of categorizing narrative data into sensorimotor stages*

Narrative data[a]	Categorization into sensorimotor stages	Number of acts
10:37–Pogo picks up ball, puts it into water, and rubs it briskly several times with both hands	Stage 4: intentional; establishment of relationships between two objects–ball and water; coordination of two behaviors toward an object–putting ball in water and rubbing it	1
Pogo brings the ball to her mouth and chews on the ball	Stage 3: "linear reaction,"[b] that is, intentional environmental orientation toward single object	1
She lies on her back and places the ball between her feet and rolls it between her feet	Stage 5: coordination of object-force relationship	1
She picks up the ball and rolls it between her hands three times	Stage 5: coordination of object-force relationships	3
Repeats this five more times	Stage 5: coordination of object-force relationships	5
She sits up, bounces the ball, making a play face, and catches it	Stages 5 and 6: coordination of object–force and object–gravity relationships indicate stage 5, but the accurate anticipation of these relationships – she successfully catches the ball – indicates stage 6; the play face is also stage 5/6 since it occurs in the context of other stage 5/6 behavior	1
10:40– She puts the ball in her mouth, looks at the other gorillas and zoo visitors, and chest-beats	Stage 4: intentional; goal established from the outset; coordination of several behaviors – puts ball in mouth, beats chest toward others	1

[a] San Francisco Zoological Gardens: Gorillas, March 26, 1975. Focus subject: Pogo.

[b] Term used by Parker (1977) to designate environmentally oriented (stage 3) behaviors that are not circular and do not reproduce environmental events or spectacles.

Reliability

Three kinds of reliability tests are in progress. I am collecting film records on three species from which to perform an interobserver reliability study. Independent observers and I will record in narrative notes the behavior observed in the film and categorize it in terms of its behavioral parameters into sensorimotor stages. Interobserver reliability will be examined. This study is still in progress.

I am also doing a separate reliability study on categorizing into sensorimotor stages the data collected in narrative notes. A pilot has been run on this study. One 6-min sample on the behavior of an adult female gorilla, consisting of 24 behavioral acts representing stages 3 to 6 of the Sensorimotor Intelligence Series, has been examined (3 min of this sample are presented in Table 13.2). When this sample was categorized by an independent skilled observer (a Piagetian psychologist who received no instruction from me), agreement was 100%. A number of samples of the behavior of animals of different ages and different species will be examined in this manner.

Both independent observers and I are studying several of the same animals. I am observing from a Piagetian perspective, and the other observers, who are unfamiliar with Piaget's theory, are studying the animals from other perspectives. After completely independent data collection, we are comparing the ages of appearance of sensorimotor behaviors. Data have been analyzed for one subject so far, an infant orangutan reared by her mother and observed by myself and an independent observer (Defiebre 1980) for 18 months. Of 25 sensorimotor behaviors for which I have calculated ages of appearance, the independent observer had data on 8 behaviors, representing stages 2 to 5 in the Sensorimotor Intelligence Series. Even though we used different criteria to assess ages of appearance (mine being the first *day* the behavior was observed in two separate acts and hers being the *month* by which the behavior had become well established and a regularly occurring part of the subject's behavior repertoire, Defiebre, personal communication), we both observed the same order of appearance of the 8 behaviors and agreed closely on the ages at which they appeared, having recorded them within an average of less than one month of one another. My ages were consistently less than the other observer's whenever they differed. Agreement between the two observers is very high, as reflected by the Spearman Rank Correlation of .98 ($p < .01$) (Siegel 1956).

RESULTS

The Old World monkeys facially manifested the first five stages of sensorimotor development. The apes and humans facially manifested all six stages.

In all species, the levels of facial ability developed in the same stage-by-stage sequence that Piaget has described for general sensorimotor development in human infants. However, in their facial behavior, the monkeys and apes did not manifest all the behavioral parameters characteristic of every stage of development in human infants. Some stages *were only partially manifested* (see Figure 13.1).

Furthermore, the rates of development differed in the three species. The Old World monkeys completed the sensorimotor period in about 6 months. The apes progressed through the first four stages a little more

Stage	Type of Behavior	Examples		
		Monkey	Ape	Human
1 Reflex	Reflexive expressions	Sucking Scream grimace	Sucking Crying pout face	Sucking Crying face REM Smile
2 Primary Circular Reaction	Responds to impinging familiar stimuli Repetitive self-oriented expressions	Puckers lips to a familiar monkey's approach	Puckers lips to mother	Smiles at mother Repetitive self-smiling
3 Secondary Circular Reaction	Initiates emotional interactions Expressions in circular reactions	Expressions of emotion	Expressions of emotion	Expressions of emotion Smiles in circular reactions with smiling adults
4 Coordinations	Voluntarily masks or uses expressions (consistent with emotional states) to attain goals Attempts to make new expressions	Obtains eye contact before threatening Masks expressions	Masks expressions Extends lower lip and looks down at it (new expression)	Smiles at adult to obtain social contact Sticks out tongue (new expression)
5 Tertiary Circular Reaction	Expressions to attain person-person & -object interactions Makes new expressions	Lipsmacks to enlist support	Holds pen in mouth and draws Trial & error Makes new faces (play)	Reaches toward object, looks at adult & smiles to obtain object Trial & error Makes new faces
6 Invention through mental combination	Invents expressions (inconsistent with emotional states) to achieve goals		Makes friendly expressions to observer-- bites & laughs (Lying, teasing)	Offers toy & smiles, withdraws toy & laughs (Teasing, lying)

Figure 13.1. Sensorimotor analysis of the development of facial behavior in monkeys, apes, and humans. (Shaded area indicates that the stage is manifested; hatched area indicates that the stage is partially manifested.)

rapidly than did human infants, entering the fourth stage at about 6 months, rather than 8 months as human infants do. But the apes were slower than human infants to achieve the last two stages, completing the fifth stage at 3 to 4 years (rather than 18 months) and the sixth stage at about 7 years (rather than 2 years). In addition, the data suggest that some emotional expressions emerge during more advanced sensorimotor stages in apes and humans than in monkeys.

A summary of the data will be presented, stage by stage.

Old World monkeys: Stump-tailed macaques

Stage 1. The first stage, which occurs between birth and 1 month in human infants is the reflex stage. During this stage, infants make *unlearned, involuntary, and stereotyped responses to stimuli.* Typical behaviors of this stage are the sucking and grasping reflexes (see Table 13.1). Assessed by nonfacial criteria (e.g., the grasping reflex), sensorimotor stage 1 occurs from birth through about 2 weeks in stump-tailed monkeys.

During the first two weeks, the three stump-tailed monkeys I observed manifested only reflexive facial behaviors. They showed stereotyped rooting (opening of the mouth followed by sweeping side-to-side movements of the head) and sucking (as in nursing) in response to any nonspecific auditory, tactile, or kinesthetic stimuli, such as loud noises or being touched, grasped, groomed, or hit, or the mothers' shifts in position or locomotion. The neonates also occasionally made pucker-lip expressions (a static portion of the sucking response formed by protruding closed lips, with lip corners brought forward, resulting in a pursed or puckered expression) and sucking movements in response to nonspecific but salient visual stimuli, such as the moving body parts (legs) of other monkeys, the faces of other monkeys, swinging objects (tires), and shiny objects (the chrome water faucet) (Chevalier-Skolnikoff 1974). Characterized by stage 1 behavioral parameters (being unlearned, stereotyped responses to generalized stimuli), these facial behaviors are classified as stage 1 behaviors (see Table 13.3). Although rooting and sucking behaviors occurred in contexts other than the nursing context, they were indistinguishable in form from rooting for, and sucking on, the mother's nipple, and their function appears to be to obtain nourishment.

Whereas the facial behaviors observed in the stump-tailed neonates were generalized responses to relatively nonspecific stimuli, not all the reflexes observed during the postnatal period were as general. Neonates screamed with accompanying grimaces and ear flattening in response to startling or painful stimuli (grimaces unaccompanied by screams were never seen during this period). They also manifested a combined startle and grasp reflex, which is probably homologous with the Moro reflex described in humans in response to loss of support. This reflex was not recorded on film and occurred too quickly for me to describe its facial configuration. (Facially, the human Moro reflex features only partial-to-complete eye closure with no salient movements about the mouth.) It is not known whether these neonatal scream grimaces or startle reflexes are related to adult fear grimaces or startles. Adult fear grimaces are characterized by partial-to-complete eye closure, alternating gaze avoidance and eye-to-eye contact, slight lowering of the brow, ear flattening, closed to slightly open jaws with lips retracted vertically and horizontally, and are sometimes accompanied by screams (Chevalier-Skolnikoff 1971). Andrew (1972) has described the adult startle as consisting of eye closure, contraction of orbicularis oris, brow lowering, ear flattening, and horizontal retraction of the lip corners. The startle response described by Landis and Hunt (1939), based on frame-by-frame film analysis of startles made by two monkeys of different species, was characterized by eye closure and widening of the mouth as though in a grin, the same facial configuration shown in human startles.

Stage 2. The second stage, which occurs between 1 and 4 months in human infants, is that of the primary circular reaction. During this stage, the infant's actions are centered about his *own body* (primary) and he learns to *repeat* them (circular). Typical examples of this stage are repeated hand-to-mouth behaviors and hand-hand clasping. During this stage, behavior is *self-oriented.* Stage 2 infants show *repetitive coordina-*

Table 13.3. *Sensorimotor analysis of the ontogeny of facial behavior in stump-tailed macaques*

Stage	Age	Facial behavior	Behavioral parameters
1. Reflex	0–2 wk	Nursing: roots, sucks; scream grimace	Stereotyped, involuntary, unlearned responses to generalized stimuli
2. Primary linear reaction	2–3 wk	Puckers lips; self-mouthing	Recognition of familiar contexts; adaptations to impinging stimuli; self-oriented
3. Secondary linear reaction	3 wk– 2 mo	Object-mouthing Emotional expressions: lip-smacks (friendly); square mouth (friendly); open mouth, eyelids down (play); grimace (fear); open-mouthed stare (threat); teeth chatter (fear, affection)	Environmentally oriented; initiated by the subject; single behavior toward single animal
4. Coordinations	2–4 mo	Emotional expressions: whispered chirl (subordinate threat); backing bared-teeth stare (subordinate threat); round-mouthed stare (dominant threat)	Establishes relationships between animals
5. Tertiary linear reaction	4 mo–	Lip-smacks to enlist support	Coordinates animal–animal relationships

tions of their own bodies. They also come to *recognize familiar contexts* (such as the nursing position), and they show *acquired adaptations to impinging stimuli* (Table 13.1).

Judged by nonfacial criteria (such as the tactile investigation of the animal's own body – hand, genitals), the transition to sensorimotor stage 2 occurred at about 2 weeks in the four stump-tailed monkeys (see Table 13.3). However, although the stump-tailed monkeys manifested coordination of their own bodies, these coordinations were rarely repetitive and therefore differed from the human stage 2 primary circular reactions. Parker (1977) has called these nonrepetitive body coordinations "primary linear reactions."

By 2 weeks of age, the three infants I observed began to make pucker-lip and lip-smacking movements fairly regularly in a limited number of friendly and presumably familiar social contexts, as when other lip-smacking monkeys approached them. Although similar in form to neonatal sucking, this stage 2 lip-smacking behavior does not include sucking air into the mouth. It also differed from the adult emotional lip-smacking expression, the tempo being less regular and bouts consisting of fewer smacks (Chevalier-Skolnikoff 1974). Stage 2 lip-smacking behavior occurred only in a limited number of contexts, rather than in response to the diverse and generalized stimuli that elicited reflexive neonatal sucking. Furthermore, lip smacking by stage 2 infants occurred in positive, familiar social contexts, as does the adult emotional lip-smacking expression, and never in noxious contexts, which may elicit sucking behavior in neonates. Thus, infants of the second stage began to show facial expressions, as well as nonfacial behaviors, in response to impinging familiar stimuli —behavioral parameters characteristic of stage 2 (see Table 13.3). However, stage 2 stump-tailed monkeys were not observed to show the repetitive self-oriented facial behaviors (i.e., primary circular reactions) characteristic of human infants.

Stage 3. The third stage, which occurs between 4 and 8 months in human infants, is that of the secondary circular reaction. During this stage human infants show *repeated* (circular) attempts to *reproduce environmental* (secondary) *events initially discovered by chance.* A typical behavior of this stage is repeatedly shaking a rattle or ringing a bell. Stage 3 behaviors establish *object–action relationships.* They are *semi-intentional;* that is, the initial acts are unintentional but subsequent repetitions are intentionally initiated by the subject (see Table 13.1).

Assessed by nonfacial criteria (e.g., visually directed reaching for and grasping objects), the transition to sensorimotor stage 3 occurred between 2 and 4 weeks in the four stump-tailed monkeys. These stage 3 behaviors were environmentally oriented and voluntary (being visually guided and sometimes attempted several times before they were successfully achieved). However, as in stage 2, Parker (1977) found that the stage 3 behavior manifested by this monkey species was rarely repetitive and did not clearly represent secondary circular reactions. Parker has called these behaviors "secondary linear reactions."

During the third stage, the stump-tailed infants began to manifest adult affect expressions. By 2 to 6 weeks of age, the lip-smacking behavior of the three stump-tailed infants I observed had achieved its adult form. Smacking occurred in a longer, more rapid series, the muzzle extended forward and up, the ears retracted, and the tongue, which alternately protruded and retracted, produced a smacking noise as it broke contact with the hard palate. Contextually, the infants began to lip-smack as they approached other monkeys, thereby initiating social interactions, and when other monkeys, who sometimes also were lip-smacking, approached them. So during stage 3 lip smacking became environmentally oriented and came to be used to initiate environmentally oriented interactions. Thus by stage 3 the facial and nonfacial behaviors began to incorporate the linear behavioral parameters of the third stage of sensorimotor development (see Table 13.3). Lip smacking in response to other lip-smacking monkeys may represent secondary circular reactions.

Two other facial expressions, the open-mouth, eyelids down (play) expression and the open-mouthed stare (threat) expression, followed a developmental sequence similar to that of lip smacking (Chevalier-Skolnikoff 1974). The earliest developmental manifestation of these expressions, in sensorimotor stage 1, was reflexive rooting for the nipple. During stage 2 infants mouthed and chewed on whatever objects they appeared to accidentally encounter, such as their own body parts, the body parts of their mothers or other monkeys, or inanimate objects in their environment. During stage 3 the infants began to initiate environmentally oriented social interactions – both positive and negative – with open-mouthed approaches. Within a few days, the two different affect expressions, the open-mouth, eyelids down (play) and open-mouthed stare (threat) expressions emerged.

By the end of stage 3 the three stump-tailed macaque infants were using eight emotional expressions in social interactions: pucker lips, lip smacking, and square mouth (affectionate expressions); open mouth, eyelids down (play); open-mouthed stare (threat); stare (mild threat); grimace (fear); and teeth chattering (ambivalent fear, affection) (see Table 13.3). (These emotional interpretations are based on contextual evidence, Chevalier-Skolnikoff 1974.)

Stage 4. The fourth stage, which occurs between 8 and 12 months in human infants, is called the stage of coordinations. During this stage, human infants begin to *combine behaviors, coordinating two or more acts to achieve predetermined goals*. They also become capable of understanding that objects can be coordinated, that is, combined. They also begin to *attribute cause of environmental change to others*. Furthermore, stage 4 behavior is considered to be *intentional*. This stage is characterized by the hand-to-hand transfer of objects, removing an obstacle to obtain an object situated behind it, obtaining an object attached to a string by pulling the string, and pushing another's hand to induce him to repeat an activity (Table 13.1).

According to nonfacial criteria (such as hand-to-hand transfer of objects and bimanually pulling objects apart), the stump-tailed infant observed by Parker (1977) achieved stage 4 at about 2 months of age.

In previous papers (1976, 1977) I reported only the first three stages of facial development in Old World monkeys because it is very difficult to verify *volition*, or *purposefulness* (a major distinguishing behavioral parameter of stage 4), in the behavior of nonverbal subjects. However, I think such interpretations can be made if one has sufficiently detailed contextual information. Some contextual variables are:

1. The subject appears to "study" the situation before acting (suggesting that he may be purposefully informing himself about its different aspects before choosing a goal-oriented strategy).

2. The subject initiates the behavior (indicating that the behavior is not some kind of automatic response, such as a reflex or an "innate releasing mechanism").
3. The subject is looking at the animal, person, or object toward whom the behavior is directed (indicating that the behavior is not accidental or merely coincidental with the presence of the receiver).
4. The behavior is part of a sequence or a simultaneous combination of behaviors (implying the purposeful organization of several behaviors, often to achieve a goal).
5. The behavior is repeated in several different bouts (indicating that it is reproducible and not merely an accident or a coincidence).
6. The behavior is repeated – sometimes with modification – before the presumed "goal" is achieved (implying that the goal was established from the outset and that the behavior is intentional and not accidental).

Recent analysis of the film records indicates that stump-tailed macaques do mask and produce emotional expressions voluntarily. They also indicate that the macaques may combine behaviors to achieve predetermined goals.

Examples of both probable *masking* and *purposeful direction* of expressions and of combining behavior to achieve predetermined goals occurred in the following interaction (this description is based on motion picture film and simultaneous narrative notes):

Stanford, April 16, 1968, 2 p.m.

Mom, the most subordinate adult female in the stumptail group is sitting next to the dominant male, subadult female, Girl, and subadult male, Boy. Mom looks about (nervously) as the dominant male forcefully restrains and grooms her 5-week-old infant. She looks directly at Girl, the most subordinate animal in the vicinity, but Girl is looking the other way and does not receive the look. Mom shoves her nose directly into Girl's face, but Girl ignores her and continues to look the other way with a neutral expression on her face. Mom retrieves her infant and again looks directly at Girl, grasps her chin, and turns her face toward her [Figure 13.2]. As Mom starts to deliver an open-mouthed stare threat toward Girl (redirected aggression directed toward the most subordinate animal present), Girl, with head correctly oriented toward Mom, again averts her gaze, avoids receiving the threat, and maintains her neutral facial expression [Figure 13.2]. Mom nips Girl on the shoulder and Girl draws away. Boy (who often supports Girl in dominance interactions) has been sitting behind Mom and now nips Mom on the head. Girl (now supported by Boy) turns toward Mom with an open-mouthed stare and both animals rush at each other with attack faces [Figure 13.2]. Just as the two animals are about to make contact, Mom leaps back and off the shelf with a shriek. [Chevalier-Skolnikoff 1974:26.]

In this example, Mom (1) studies the situation (looks at Girl before acting); (2) directs her gaze toward the potential receiver; (3) performs a sequence of behaviors (shoves her face into the potential receiver's face, turns the receiver's face toward her, nips the receiver, and then directs an open-mouthed stare at the receiver); and, presumably performing the first three behaviors of the series to achieve the goal of delivering the threat, she (4) modifies the behavior employed to achieve the goal (i.e., she

Figure 13.2. An adult female stump-tailed macaque, Mom (extreme right), in attempting to deliver a threat to a subadult female, Girl (center left), stares at Girl, who is looking away to avoid the stare, and grasps Girl's chin to turn her head (Frames from super 8 film).

shoves her face into the receiver's, turns her head and nips her, all in order to catch her gaze so she can deliver her threat). That Girl repeatedly averts her gaze suggests voluntary gaze aversion. She also appears to mask her angry emotion (which would be manifested by a threat expression) until she has the support of the subadult male.

Voluntary attempts to deliver fear grimaces were also observed in the stump-tailed monkeys. Animals under attack made fear grimaces and often turned their heads to look up at their attackers. They also occasionally reached up to grasp their attackers' faces to orient them to receive their grimaces.

Ontogenetically, it was observed that from 2 months of age, the contexts in which specific affect expressions occurred increased. For example, lip smacking began to accompany grooming, huddling, and embracing, as well as friendly approaches. In addition, the infants began to show that they distinguished subtle differences in dominance status and contextual differences between agonistic situations. They displayed subordinate whispered chirl and backing bared-teeth stare expressions and the highly dominant round-mouthed stare-threat expression (Chevalier-Skolnikoff 1974), as well as the moderately dominant stare and open-mouthed stare-threat expressions displayed during stage 3, (see Table 13.3). These changes in specificity of response probably are due to the infants' new abilities to process multiple contextual variables. During stage 3 the infant is capable of dealing with only single orientations toward a single object or animal. As the infant enters stage 4, he becomes capable of coordinating several behavioral acts and of understanding the relationships between two or more objects. Similarly, the stage 4 infant evidently becomes capable of processing several contextual variables, such as those involved in appraising dominance interactions.

Stage 5. The fifth stage, which occurs between 12 and 18 months in human infants, is called the stage of the tertiary circular reaction. The infant becomes curious about the *functions* of objects and about object-object relationships (i.e., tertiary). He *repeats* his behavior (i.e., circular) with *variation,* as he explores the potentials of objects through *trial-and-error experimentation.* Stage 5 behaviors characteristically involve *new* and *variable* motor patterns and the *coordination of object –, person –, space –,* and *gravity – force relationships.* During this stage infants begin to employ multiple means to achieve desired ends. Examples of stage 5 behaviors are tool use (the use of one object to modify another; an object–object coordination) and ball play (experimental object–, space–, gravity–force relationships; see Table 13.1).

Assessed by nonfacial criteria, such as trial-and-error approach–avoidance play ("keep away"), the infant stump-tailed monkey studied by Parker achieved stage 5 at 4 months of age. However, Parker has emphasized that although stump-tailed macaques do manifest experimental interanimal coordinations involving variable motor patterns, they only manifest tertiary circular reactions socially. They do not perform them with objects. They do not show experimentation with animal–object, object–object, object–space, or object–force relationships. Consequently, they manifest only some of the distinguishing behavioral parameters of stage 5, which is the stage of the tertiary circular reaction.

In their facial behavior, also, stump-tailed macaques appear to incorporate only some of the behavioral parameters characteristic of stage 5. No novel or variable facial behaviors were observed, and stump-tailed macaques do not appear to incorporate novel motor patterns into their facial expressions. Nor were any facially facilitated animal-object coordinations observed (e.g., making an expression to persude another animal to do something with an object). However, facial expressions were used to facilitate animal–animal social interactions, as when one animal solicited the support of others in a dominance context by lip-smacking to uninvolved monkeys to enlist their support. Attributing stage 5 to such animal–animal interactions is a new interpretation of macaque facial behavior.

Stage 6. The sixth.stage, the stage of invention (insight), occurs between 18 and 24 months in human infants. During this stage, infants become capable of inventing new behaviors and of solving problems through *mental representation* and *mental combinations,* that is, *insight.* An example of

Table 13.4. *Sensorimotor development in Old World monkeys (range in parentheses)*

Sensorimotor series	Age at first occurrence	
	Macaque ($N = 4$)	Langur ($N = 2$)
Stage 1	0–2 wk	0–$1\frac{3}{4}$ wk ($1\frac{1}{2}$–2)
Stage 2	2–3 wk (2–4)	$1\frac{3}{4}$ wk–$2\frac{1}{4}$ wk (2–$2\frac{1}{2}$)
Stage 3	3 wk–2 mo[a]	$2\frac{1}{4}$ wk–$2\frac{1}{4}$ mo (2–$2\frac{1}{2}$)
Stage 4	2 mo[a]–4 mo[a]	$2\frac{1}{4}$ mo–4 mo[a]
Stage 5	4 mo[a]–	4 mo[a]–

[a] Precise age of transition for one subject only.

insightful behavior is choosing a stick of appropriate length to use as a tool for obtaining an out-of-reach object on the first try (see Table 13.1).

It is uncertain whether stump-tailed monkeys attain stage 6 in the Sensorimotor Intelligence Series. Perhaps they never reach this cognitive level. It is also possible that they are capable of some kinds of mental representation but that these abilities are difficult to detect in animals that do not manifest object–object coordinations. Parker speculated that the planning of pathways and the detouring she observed in the stump-tailed monkeys might represent stage 6 spatial–mental representation. Similarly, I have speculated (1976) that macaques may mentally plan tactics for approach–avoidance play (such as reversals around a tree in a game of "tag" or "keep away").

However, there is no evidence that stump-tailed monkeys manifest stage 6 facial behavior.

Old World monkeys: Langurs

Sensorimotor Intelligence developed in the same five-stage sequence at approximately the same rates in the two langurs as it did in the stump-tailed macaques (see Table 13.4). However, there was one major difference in sensorimotor development in these two species. Whereas stump-tailed macaques showed only secondary linear reactions during the third stage, the langurs manifested secondary circular reactions.

As far as the data are comparable, the langurs and the macaques seemed to follow a similar sequence of facial development. During stage 1 the langurs manifested only reflexive rooting, sucking, and scream grimaces. During stage 2 the nursing reflexes regressed. However, the stage 2 langurs did not show facial expressions as adaptations to familiar impinging stimuli as had the macaques, perhaps because their repertoire of facial expressions is different (they evidently don't make pucker-lip expressions, Jay 1965) or because of the conditions in which the langurs were reared. Cared for by 21 human caretakers (multiple mothering by surrogates of a different species), they may not have experienced an environ-

Table 13.5. *Sensorimotor development in apes (range in parentheses)*

Sensorimotor series	Age at first occurrence		
	Orangutans (*N* = 3)	Gorillas (*N* = 1)	Chimpanzees (*N* = 1)
Stage 1	0–1 mo	0–1 mo	0–2 mo
Stage 2	1–2¼ mo (2–2¾)	1–3 mo	2–4¾ mo
Stage 3	2¼–5¼ mo (5–5½)	3–less than 7 mo	4¾–6 mo
Stage 4	5¼–10½ mo (9½–11½)	7–14 mo[a] (13½–14½)	6–11½ mo
Stage 5	10½ mo–3½ yr (3¼–4 yr)	14 mo–4 yr	11½ mo–
Stage 6	18 mo[b]–	18 mo–? yr	

[a] Two animals. [b] One animal.

ment appropriate to the development of stage 2 facial expressions. During stage 3 the langurs, like the macaques, began to manifest emotional expressions, especially grimace (fear) and open-mouthed, eyelids-down (play) expressions. However, unlike macaques who showed only secondary linear reactions, the langurs also manifested facial expressions of emotion in the context of the secondary circular reaction. For example, these animals were occasionally observed to make play faces as they repeatedly rang bells. They did not, however, use facial expressions in social secondary circular reactions, as do human infants.

During stage 4 the langur infants, like the macaques, began to differentiate between dominant and subordinate threats in interactions with infant orangutans and human caretakers and visitors. It was also observed that the langurs, like the macaques, never made novel facial expressions. However, it was not possible to ascertain whether these langurs, who were reared without conspecific animals, were able to manifest stage 4 voluntary expressions or stage 5 expressive behaviors designed to facilitate animal–animal interactions.

Apes: Orangutans

Stage 1. This stage, the reflex stage, occurs from birth to 1 month in orangutans (see Table 13.5).

With respect to facial behavior, both human-reared and mother-reared orangutan infants almost exclusively manifested reflexive rooting, sucking, and crying-pout faces (pouted lips extended in a funnel, like a trumpet, with an "O"-shaped opening, which is accompanied by high-pitched cries) during spontaneous activity (Figure 13.3). Rooting and sucking occurred in response to mild tactile or kinesthetic behaviors of the mother and when the infants appeared to make inadvertent oral contact with objects, the substrate, or their own body parts. Crying-pout faces, always

Figure 13.3. An orangutan neonate manifests a crying-pout face (stage 1).

accompanied by high-pitched cries, occurred when infants were startled or hungry or when they became detached from their mothers or surrogate mothers; neonates always showed this response when prevented from clinging. These types of facial behavior appeared to be reflexive, being stereotyped responses to generalized stimuli, although crying-pout faces occurred in response to generalized negative stimuli. Of the facial movements observed during stage 1, 98% were reflexive. The other 2% were pucker-lip expressions in which the lips were slightly extended with the lip corners brought forward and the mouth either closed or slightly open, forming a small "O"-shaped aperture and resulting in a pursed, puckered, or pouting expression. Thus the facial behavior of neonatal orangutans was similar to that of newborn Old World monkeys.

When orangutan infants were positioned face-to-face with human observers they behaved differently from the monkeys. Whereas the monkeys remained unexpressive, the orangutans were observed to make seemingly random facial expressions, especially mouth movements such as pucker lips, open mouths, and tongue protrusions. This facial mobility was very prominent, with movements occurring at an average frequency of 24/hr during face-to-face orientation as contrasted with rooting and sucking, which occurred at a frequency of 11/hr (range 5 to 14).

Stage 2. The stage of repetitive self-oriented behaviors (primary circular reactions) and acquired adaptations to impinging stimuli, sensorimotor

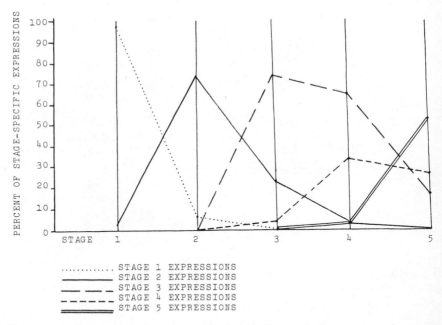

Figure 13.4. Percentage of stage-specific facial behavior during ontogeny in orangutans.

stage 2 was manifested between 1 and about 2 months in the orangutan infants according to nonfacial criteria (such as the regression of the grasp reflex; the appearance of repetitive hand-to-body behaviors, such as foot clasping; and adaptation to the feeding position or recognition of the bottle in the human-reared infants).

During this stage, as reflexive rooting and sucking regressed, the infants began to make pucker-lip and open-mouthed expressions in familiar positive contexts, as, for example, they anticipated the approaching bottle or putting their fingers into their mouths. It appeared that these stage 2 pucker-lip expressions developed from reflexive sucking and that the open-mouthed expressions developed from rooting in the same way that stage 2 pucker lips and mouthing develop in macaques.

During this period (up to 2 months) 76% of the facial expressions recorded in the course of spontaneous activity or during routine care were responses to impinging familiar objects and appeared to represent acquired adaptations, which categorized them as stage 2 behavior (see Figure 13.4). Of the facial behavior, 6% was classifiable as reflexive stage 1 behavior and 18% was cognitively unclassifiable (as when infants opened and closed their mouths in the absence of any known stimuli). Whereas these apes differed from the monkeys by showing prominent manual primary circular reactions during stage 2, they resembled the monkeys in not manifesting repetitive self-oriented facial expressions.

During the second stage also, the orangutan infants showed a great deal of facial mobility (24 expressions per hour) when in a face-to-face orientation with a human observer.

Stage 3. The stage of environmental orientation and secondary circular reactions, sensorimotor stage 3 occurred between about 2 and 5 months in the orangutans as judged by nonfacial criteria (such as visually guided reaching out and grasping objects and repeatedly ringing bells or squeaking squeak-toys). Both secondary linear and secondary circular reactions were observed in the infant orangutans.

During the third stage, the orangutan infants began to make facial expressions as they initiated interactions with objects, people, and other animals, as well as in response to them. During spontaneous activity and routine care, 73% of the expressions observed occurred at the beginning of such bouts of behavior, which were initiated by the infants. Being single (rather than complex) behaviors directed toward single objects in the environment, these expressions were classified as stage 3 behaviors. No stage 1 expressions were observed during this stage, but 22% of the expressions could be classified as stage 2 expressions. A few expressions (2%) appeared to be stage 4 behavior, but these occurred only once and could not be verified. In terms of sensorimotor stages (see Figure 13.4), 3% of the behavior could not be classified.

During this stage the number of expresive forms increased, and expressions began to appear in obviously affective contexts (see Table 13.6). The mother-reared infants began to initiate mouth-to-mouth kisses with their mothers. Play attacks by one infant on her mother and father were initiated by approaches accompanied by open-mouth, eyelids-down (play) expressions. Tickling brought forth open-mouth, bared-teeth (laugh[1]) expressions, sometimes accompanied by vocal laughter. Object play that elicited reactions to objects (secondary circular reactions), such as ringing bells, squeaking squeak-toys and waving branches, was responded to with closed-mouth, horizontally retracted lip (smiles), open-mouth, bared-teeth (laugh), and open-mouth, eyelids-down (play) expressions (see Figure 13.5*a*). The "smiles" occurred in less intense emotional contexts, as at the beginning of such play bouts when the behavior was still tentative, and the "laugh" and "play" expressions occurred when play became more intense. Of the expressions observed during spontaneous

---

[1] Whether to use descriptive terms commonly applied to the facial expressions of nonhuman primates to describe ape expressions or to use the names of presumably homologous human expressions is a difficult question. Although descriptive terms have the advantage of being more precise and more objective, the application of descriptive terms to monkey and ape facial behavior and of common names (e.g., laugh, smile) to human expressions implies that ape facial behavior is more similar to that of monkeys than to that of humans, and this tends to distort the data. Ideally, descriptive terms should be used for all species, but only a few ethologists (e.g., Blurton Jones 1972; McGrew 1972; Young and Gouin Décarie 1977) have applied descriptive terms to human expressions. I have chosen the less-than-ideal solution, using the human names for the presumably homologous ape expressions and using descriptive terms for expressions that do not have obvious human homologues.

Table 13.6. *Sensorimotor and facial development in primates*

Stage	Monkey	Ape	Human
1	Nursing: roots, sucks; crying face	Nursing: roots, sucks; crying-pout face	Nursing: roots, sucks; crying face; REM smile
2	Puckers lips; self-mouthing	Puckers lips; self-mouthing	Smile; self-mouthing
3	Object-mouthing Emotional expressions: lip-smacks (friendly); square mouth (friendly); open mouth, eyelids down (play); grimace (fear); open-mouthed stare (threat); teeth chatter (fear, affection)	Object-mouthing Emotional expressions: kiss; smile; open mouth, eyelids down (play); open-mouthed bared teeth (laughing face)	Object-mouthing Emotional expressions: smile, laughing face
4	Emotional expressions: whispered chirl (subordinate threat); backing bared-teeth stare (subordinate threat); round-mouthed stare (dominant threat)	Emotional expressions: grimace (fear) Novel behaviors: blows bubbles (once); purses out lower lip and looks down at it (once)	Emotional expressions: fear face, angry face, sad face, surprise face Novel behaviors: spitting sound with lips
5	Lip-smacks to enlist support	Novel behaviors: spits, spitting sound with lips, kissing sound with lips, sticks out tongue, sticks tongue in jar to get honey, holds pen in mouth and draws	Novel behaviors: makes bubbles, sticks out tongue
6	None	Facial lying: puckers lips for subversive ends	Facial lying, teasing: offers toy and smiles, withdraws toy and laughs

activity and routine care, 17% were obviously affect expressions (see Van Hooff 1967, 1971, 1972; van Lawick-Goodall 1968a, and Chevalier-Skolnikoff 1974 for descriptions of ape affect expressions; see MacKinnon 1974, Rijksen 1978 and Maple 1980, for descriptions of orangutan affect expressions). Most affect expressions (14% of recorded expressions) were classified as stage 3 behavior, since they were environmentally oriented actions directed toward single objects (i.e., animals or people).

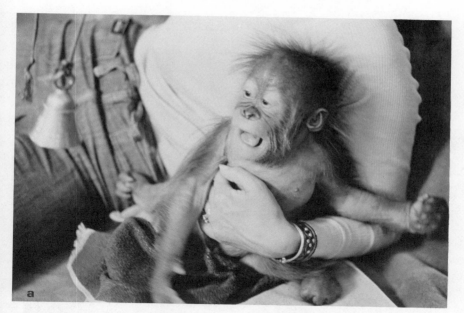

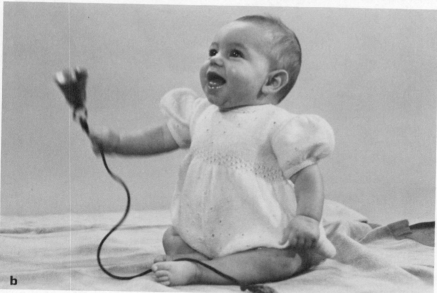

Figure 13.5. (a) The orangutan infant makes an open-mouthed, eyelids-down play face after ringing a bell (stage 3). (b) The human infant shows an open-mouthed smile or laugh face as she rings a bell (stage 3).

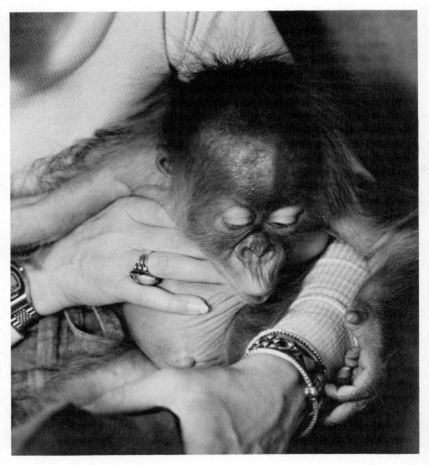

Figure 13.6. The orangutan infant shows a pucker-lip expression as she examines the researcher's bracelet (stage 3).

Of the recorded expressions, 78% were pucker-lip and open-mouth expressions (without bared teeth), which occurred most often in conjunction with approaches to objects (see Figure 13.6), though 22% also accompanied the initiation of primary circular reactions such as hand and foot sucking. The pucker-lip expression, which represented 69% of the recorded expressions, is similar to the kiss expression and may represent positive affect expression in these stage 3 infants, though this could not be deduced with certainty from the contexts in which they occurred. The open-mouth expressions (15% of the expressions) were anticipatory to mouthing, which always followed, and evidently did not represent affect expressions.

During stage 3, facial as well as nonfacial behaviors began to incorpo-

Figure 13.7. An orangutan infant shows a laugh play-face blend as he responds to the open-mouthed expression of his surrogate mother. This may be an example of a facial behavior functioning as a secondary circular reaction.

rate the behavioral parameters of stage 3, showing the initiation of environmentally oriented social and nonsocial interactions with single objects, persons, and animals. However, whereas smiles, laugh faces, and play faces often accompanied manual secondary circular reactions, facial behavior did not appear to actually function as secondary circular reactions. Human infants repeatedly smile at caretakers who smile back at them, and these smiles actually function as secondary circular reactions. This kind of facial interaction was not obvious in the infant orangutans. It is possible that play faces and laugh faces sometimes function in this way, however, for these expressions were occasionally responded to with similar expressions by stage 3 infants (see Figure 13.7).

The advent of kissing and play faces in infant orangutans during the third stage appeared to be further developments of the stage 2 pucker-lip and open-mouth behaviors (as the rhythmic adult lip smacking and play faces of the third stage in macaques appeared to develop from pucker lips and nonrhythmical lip smacking and mouthing).

As in stages 1 and 2, stage 3 infants also manifested a great deal of "random" facial mobility (now as frequent as 132 expressions per hour) when face-to-face with a human observer. Besides showing puckered lips, open mouths, and tongue protrusions, the infants showed bared-teeth grimaces, yawns, closed-mouth lip retractions, upper-lip raising, squared-mouth puckers, closed-mouth, inverted-lip expressions, and "fish faces" (pucker-lip expressions with cheeks sucked in), as well as combinations of expressive elements such as open-mouth puckered lips and unilateral expressions such as unilateral closed-mouth lip retractions and unilateral bared-teeth grimaces. The Old World monkeys never manifested this kind of "random" facial mobility.

Stage 4. According to nonfacial criteria, the orangutans achieved stage 4, the stage of coordinations, at about 5 months of age. At this age they began putting objects such as branches on their heads, rubbing objects on the substrate, pulling strings to obtain objects attached to them, and manifesting hand-to-hand and hand-to-foot transfers.

As the orangutans began to manifest coordinations and the behavioral parameters of stage 4 in their general behavior, they also began demonstrating them in their facial behavior. The differences between the stage 3 and stage 4 expressions were that affect expressions began occurring in new contexts, and novel types of facial behavior began to occur in contexts that suggested they were voluntary behaviors.

One of the most obvious changes was that the infants began to demonstrate an understanding of relationships between objects or between an animal and an object. Infants began to make play faces while they grabbed at objects held by others, and they made play faces or laughed as they watched others do things with objects, like waving branches or hitting the ground with sticks.

They also demonstrated that they attributed cause of environmental change to others. After being tickled by his human surrogate mother, a human-reared infant took his mother's hand and placed it on his body again and laughed, indicating that he wanted her to tickle him again. Similarly, a stage 4 group-reared infant watched her father as he waved a branch and then pushed his hand and laughed, indicating that she wanted him to wave the branch again. Another time, this same infant escaped from her mother and climbed to the top of the cage where she clung tenaciously. Her mother tried to retrieve her, but after several attempts at detaching her, the mother gave up. As the mother turned to leave, her infant looked back at her and laughed.

The first fear grimaces (closed mouth with lips retracted horizontally and vertically, exposing the teeth) occurred during the fourth stage. One group-reared infant grimaced and squeaked as her father grabbed at her. Evidently she anticipated that he might handle her roughly or restrain her against her will. Attributing cause of environmental change to others may have been involved in this fear response. On another occasion this infant grimaced as she reached for a stick her father was holding. The context in which this grimace occurred suggests that an understanding of the coordination between the stick and her father and an ability to attribute cause of environmental change to others were involved in making this event emotionally fearful.

During the fourth stage, infants also demonstrated, through their facial behavior, that they could conceptualize the different three-dimensional aspects of animals as well as those of objects. This was exemplified by a group-reared infant when she was observed to laugh as she saw her father hang upside down in front of her, demonstrating that she understood that he was strangely oriented. Similarly, stage 4 infants also demonstrated a unitary body concept – that is, an understanding that all parts of the bodies of others are united physically and behaviorally. For example, a group-reared infant made a play face at her father's face and grabbed at his toes as he wiggled them.

At the end of the fourth stage, variable new motor patterns that evidently are not part of the species-specific repertoire of affect expressions and appeared to be voluntary behavior were first seen, though none occurred regularly during this stage. On one occasion an infant blew bubbles in her milk cup. On another occasion, when she was given a new kind of food, she protruded her lower lip with the morsel of food on it and looked down at the new item before eating it. This latter behavior occurs in both wild orangutans and wild chimpanzees. It appears to be learned during stages 4 and especially 5, when animals begin to manifest, and to gain voluntary control over, variable new motor patterns.

During the fourth stage, 33% of the facial acts manifested incorporated behavioral parameters characteristic of stage 4. Stage 3 facial behavior was still prominent during the fourth stage, comprising 63% of the facial acts recorded. Facial behavior characterized by stage 2 behavioral parameters made up 2%. Irregularly occurring behaviors incorporating stage 5 behavioral parameters made up 2% of the facial acts manifested during this stage (see Figure 13.4).

During the fourth stage, "random" facial movement became less pronounced. At the beginning of the stage such movements were recorded at a frequency of 120/hr, but within a couple of weeks they were no longer observed.

This chapter focuses on the development of sensorimotor intelligence and its relationship to the development of facial behavior, but since studies on humans have examined the relationship between facial development and cognition almost exclusively in terms of the Object Concept Series, it is pertinent to mention a few observations made during this study on facial and object-concept development in apes.

The Object Concept Series is made up of six stages that unfold at approximately the same rate as the six stages of the Sensorimotor Intelligence Series. The stage that has been related to the development of affect expressions in human infants is stage 4. During the third stage a human infant will visually follow the movement of an object, and his gaze will linger at the point at which the object disappears as it moves behind a screen. Furthermore, he will search for an object that escapes him while his hand is touching it, and he will search for an object he loses track of as he is performing an action with it – such as a secondary circular reaction. He also will uncover a partially hidden object. In addition, he will remove an obstacle that is interfering with his perception. Thus, the stage 3 infant is beginning to acquire the earliest rudiments of active search. But once

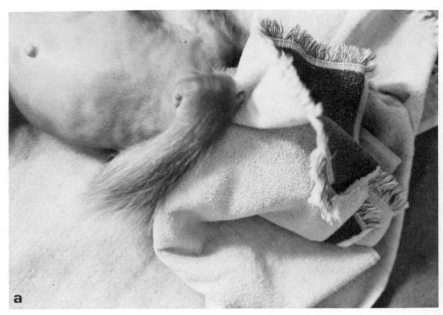

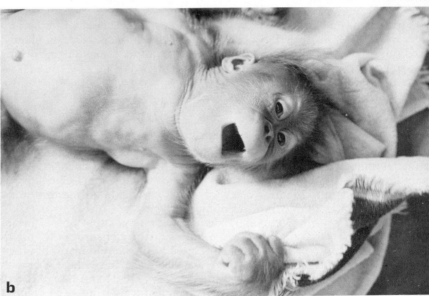

Figure 13.8a,b. The orangutan infant makes a play face as she lifts the cloth in a solitary game of peek-a-boo (stage 3).

Figure 13.8c,d. The baby shows a mild open-mouthed smile as he uncovers the investigator in a social game of peek-a-boo (stage 4).

an object is completely covered, the stage 3 infant acts as though he has never seen it or as though it had never existed – out of sight, out of mind. It is not until the fourth stage that the infant will uncover an object that he has seen being hidden. During stage 4 he begins to understand that out-of-sight objects continue to exist. However, he still conceives of objects as existing only where he has seen them being hidden, and it is not until he reaches stage 5 that he comes to understand that objects can be hidden in different places. When a stage 4 infant watches an object as it is placed first under one cloth and then under another, he will search only under the first cloth. Not until stage 6 will he look under a cloth for an object that has been surreptitiously hidden to determine whether in fact it has been hidden there (Piaget 1954; Gouin Décarie 1965).

During the third stage, the orangutan infants began to play peek-a-boo. As they lay on their backs in their playpens by themselves, they pulled cloths over and then off their faces, making play or laugh faces each time they removed the cloths (see Figure 13.8*a, b*). The peek-a-boo behavior may be affectively arousing for several reasons: It causes the repeated appearance of the patterned stimuli in the environment; it repeatedly removes an obstacle interfering with the infants' perception of the environment; and it functions as a secondary circular reaction (reinstating an interesting spectacle).

The orangutan infants generally played peek-a-boo by themselves. However, the hand-reared infants began to play peek-a-boo with their human surrogate mothers and with the observer occasionally during stage 4 (Figure 13.8*c, d*). After removing the cloths from their faces, they looked at the humans and made laugh or play faces. They also removed the cloths from the humans' faces (stage 4: uncovering of a hidden object), making play and laugh faces as they saw their human playmates. When the infant who was reared by her parents was in the fifth stage, her father played peek-a-boo with her on several occasions, covering and uncovering her face with a gunnysack, which elicited laugh faces from the infant each time he removed the sack and she looked at him.

Stage 5. Assessed by nonfacial behavior, the orangutans attained stage 5, the age of experimentation, at about 10 months of age. At this age they began to experiment with objects and space: collecting several branches and piling them up, scattering the piles, crumpling branches into small bunches and straightening them out, arranging them in circles around their bodies, putting objects into containers, and weaving branches and ropes in and out of the holes in the cage wire. They also began experimenting with objects and their relationships to gravity and force: tossing branches into the air and watching them fall, catching water drops from the faucet, balancing branches on their heads as they walked quadrupedally, and tossing, rolling, spinning, bouncing, and catching balls. At this age they also began to experiment with object–object relationships and with tool use – poking sticks into holes in the cement, hitting objects or other animals with sticks, and drawing on paper with crayons or pens.

As was the case with the other stages, the infants began to manifest facial behaviors characterized by the behavioral parameters of stage 5 at about the same time they began to show them in their general behavior. The two major advances in facial behavior during stage 5 were that novel, nonemotional, facial behaviors became prominent and that affect expressions began to be used in stage 5 contexts and to be functionally incorporated into tertiary circular reactions.

During this stage, 27% of all facial acts manifested were innovative expressions. These accounted for 51% of the stage 5 facial acts recorded. They included spitting (silently); making spitting noises with the lips; spitting objects, such as peas; blowing bubbles in water or milk; sticking the tongue out and sometimes peering down at it (see Figure 13.9a); using the extended tongue to obtain honey from a small-mouthed jar; and using the lips to hold a pen or crayon to draw on a piece of paper (see Figure 13.10). The use of the mouth as a manipulative organ for tool use represented 10% of all facial acts recorded during this stage and 20% of the facial acts classified as stage 5 behavior.

Of all facial acts manifested, 73% were affect expressions, and affect expressions classified as stage 5 expressions made up 25% of all facial acts recorded. These represented 49% of the facial acts classified as stage 5 behavior. Most of these accompanied stage 5 manual behavior and were probably dependent upon affective arousal based on an understanding of stage 5 cognitive concepts. For example, infants manifested laugh and play expressions as they repeatedly rolled sticks, threw balls (see Figure 13.11), or hit a xylophone with a stick. Stage 4 infants occasionally rolled objects or hit one object with another, but they seemed to do so inadvertently, for they did not watch the objects' movements, they did not repeat the behaviors, and they showed no affective responses to these events. Stage 5 infants, on the other hand, were obviously aroused by such events, and their interest probably was due to their new level of understanding object-, space-, and force–gravity relationships. Similarly, stage 5 infants often made laugh expressions as they dangled by their arms, rotating or swinging their bodies back and forth (experimenting with body–force–gravity relationships) as they learned to brachiate. Stage 5 infants also made laugh faces as they played "games" involving new motor patterns with their parents and peers. Sitting face-to-face, they took turns flailing their hands on the substrate or clapped each other's hands rhythmically in pat-a-cake.

During this stage, group-reared infants also began to use emotional expressions in social communications accompanied by gestures "as tools" to attain goals. For example, one infant reached for the food in her mother's mouth with a palm-up, open-hand gesture accompanied by a pout or puckered-lip expression as a begging gesture. I have suggested elsewhere that this use of gestures and expressions is comparable to the use of signs by apes and humans who are taught sign language (Chevalier-Skolnikoff 1979a, 1981).

Of all facial behaviors observed, 3% were of this nature; of the stage 5 facial behaviors, 5% were. During stage 5, 52% of the recorded facial acts incorporated stage 5 behavioral parameters, 27% incorporated stage 4 pa-

Figure 13.9. (a) An orangutan makes a novel expression, sticking out his tongue in a game with the photographer (stage 4–5). (b) The baby makes a novel expression in a social game with a relative (not shown in photo) (stage 4).

Figure 13.10. An orangutan uses her mouth as a manipulative organ to draw with a pen (stage 5).

Figure 13.11. An orangutan shows a laugh play-face blend as she runs and throws a ball (stage 5).

rameters, and 18% incorporated stage 3 parameters. For 3% of the facial behaviors, insufficient contextual data were available to distinguish whether they were stage 3, 4, or 5.

Stage 6. The first activities that could be interpreted as stage 6 behavior— showing mental solutions to problems, or insight – occurred at 18 months in one infant. However, the orangutan infants rarely manifested stage 6 behavior during the study period, although the adults displayed them frequently. Most commonly, adults used sticks as tools to recover objects, such as food, that had fallen outside their cages, choosing implements of correct size and shape and successfully recovering the objects on the first try.

Orangutans over 18 months of age were occasionally observed to manifest stage 6 facial behaviors. One day an adult female picked up an egg that was covered with ants. Before eating it, she carefully blew off all the ants (demonstrating her figuring out the solution to a problem mentally). On another occasion, I made the inexcusable social transgression of bringing a cup of coffee into the observation room, which is separated from the animals by only wire. The adult male looked at my cup and held out his hand in a begging gesture. When I ignored him, he went to his water faucet, got a mouthful of water and spat it (about 1½ m) into my face and the coffee cup. This behavior showed skillful facial motor coordination in which he achieved his goal on the first try and indicated a stage 6 level of functioning.

Apes: Gorillas and chimpanzees

The gorillas and chimpanzees followed essentially the same courses of sensorimotor and facial development as the orangutans (Chevalier-Skolnikoff 1976, 1977, 1979b). Although one cannot make conclusive statements from such a small number of subjects, it appears that gorillas and chimpanzees develop a little more slowly than orangutans (see Table 13.5).

There also appears to be some differences in the repertoires of emotional expressions in the three species. Only the orangutans were observed to make closed-mouth, retracted-lip (smile) expressions in playful emotional contexts. Similarly, only in orangutans were silent open-mouth, bared-teeth (laugh) expressions prominent during play; gorillas and chimpanzees manifested either open-mouth, eyelids-down (play faces or open-mouth, bared-teeth expressions accompanied by vocal laughter in similar contexts.

It also was noted that gorillas manifested a greater amount of deceptive behavior during the sixth stage of sensorimotor development than did orangutans or chimpanzees. Deceptive tactics to achieve goals were observed facially as well as in other modes. Both gorillas observed during this stage were seen to make positive expressions in negative social contexts. For example, on several occasions an 18-month-old infant made puckered-mouth expressions, and even hugged and kissed the observer during psychological tests, as he surreptitiously attempted to take her watch or tape recorder. Similarly, a caged 4-year-old occasionally made pucker-lip expressions and kissing noises with her lips at the observer, but when the observer reached her fingers through the wire to pet her, the gorilla bit the observer or pressed the observer's fingers against the wire of the cage with her nose and then laughed.

Random facial movements seemed to be less prominent in the gorillas and chimpanzees during the first few months of life than they were in the infant orangutans, although no data on the frequencies of these behaviors were collected on the other apes. Nevertheless, novel and variable voluntary behavior patterns were manifested by gorillas and chimpanzees during stage 5, as they were in orangutans. On a number of occasions stage 5 gorillas were observed to purposefully make novel, nonemotional faces to themselves in play. On one occasion a stage 5 infant sat and looked up toward the sky and made a series of facial contortions, sticking out his tongue several times, grimacing unilaterally, repeatedly closing his eyes and opening his mouth, and finally rolling his head all around. As soon as he stopped, an adult female who had been sitting and watching him made a play face, looked skyward, and made a series of facial contortions similar to those just made by the infant. A few days later, I observed this same infant look at his reflection in the drinking pond and again make a series of facial contortions.

Human beings

Stage 1. As assessed by nonfacial criteria, stage 1, the reflex stage, occurs between birth and around 1 month in human infants (Piaget 1952, 1954).

Rooting and sucking. The most salient facial behaviors in the human neonate are rooting for the nipple and sucking. Rooting and sucking are unconditioned (unlearned) reflexes in the Pavlovian sense. Touching the cheeks or lips is the unconditioned stimulus that elicits rooting, and touching the oral cavity, especially the tongue, elicits sucking (Prechtl 1958; Blauvelt and McKenna 1961; Peiper 1963). In the neonate, neither response appears to be affected by learning of any kind (Sameroff 1972). Both reflexes are mediated by reflex arcs that pass through the hindbrain and neither involves the cortex (Prechtl 1958; Peiper 1963). Both kinds of behavior function in nursing: rooting in the location of the nipple and sucking in expressing the milk. Both behaviors are unlearned, involuntary, and stereotyped and can be classified as stage 1 reflexive behaviors.

Expressions of emotion. Several early studies on facial expression in infants suggest that most emotional expressions are present at birth (see review by Charlesworth and Kreutzer 1973). Crying, as described by Darwin (1877), was characterized by a frown, firmly closed eyes, retracted lips, and raised tissue on upper cheeks, and was reported to occur at birth. Smiling, elicited by tickling, shaking, or patting, was reported immediately after birth by Watson (1925), and Dennis and Dennis (1937) report smiling from week one. Laughter was reported from week three by Dennis and Dennis (1937) and from 23 days by Preyer (1892). Anger, as described by Watson (1919), was characterized by stiffening of the body, holding the breath, flushed face, and slashing movements of the hands and arms, and was reported at birth. Fear was reported during birth by Stirnimann (1940), and Valentine (1930) reported fear in response to loud noises at 10 days. Watson (1919) reported fear during the first weeks in response to sudden loss of support, a push, or loud noises, and described it as a catching of breath, closing of the eyelids, and puckering of the lips. Darwin (1877) also reported fear during the first weeks and described it as a startle followed by crying. However, with the exception of Darwin's description of the crying face and Watson's description of the fear expression, these reports do not describe the facial expressions upon which the affect interpretations are based, and the criteria used for assessing emotional state appear to have been intuitive or impressionistic. Of the two expressions described, the description of fear does not fit the fear faces that have been elicited cross-culturally in experimental situations or cross-culturally judged as fear in adults (Ekman 1972). The infantile cry face evidently is the same as the adult "distress" expression, an intense manifestation of "sadness" (Ekman and Friesen 1975; P. Ekman, personal communication).

Bridges (1932) reported that adult expressions of emotion cannot be observed in the neonate. She proposed that young infants are capable of experiencing and manifesting only the global states of undifferentiated distress and undifferentiated nondistress and that specific emotions develop out of these global reactions. However, Bridges did not describe the process of differentiation.

Although a number of recent studies, based on detailed descriptions of facial movements and the contexts in which they occur, support Bridges' general proposal that the emotions differentiate gradually, they are beginning to give us a more detailed picture of the ontogeny of human emotional expressions (Peiper 1963; Wolff 1963, 1966, 1969; Bronson 1968b, 1974; Ricciuti 1968; Sroufe and Wunsch 1972; Emde, Gaensbauer, and Harmon 1976; Sroufe and Waters 1976; Cicchetti and Sroufe 1978; Emde, Kligman, Reich, and Wade 1978; Sroufe 1979). The new picture is that of a gradual stage-by-stage unfolding of the affect expressions, most of which appear to attain forms and contexts recognizably similar to those of adults at periods corresponding to stages 3 and 4 of Piaget's Sensorimotor Intelligence Series. After their appearance, the contexts in which the expressions occur continue to change, as higher-stage cognitive mechanisms apparently become involved in their elicitation.

Despite the recent flurry of studies on infant emotional development much is still unknown. We still do not know precisely how or when each of the emotional expressions that have now been verified cross-culturally as emotional responses in adults – sadness, happiness, disgust, surprise fear, and anger (Ekman 1972) – develop and emerge. Detailed descriptive studies have focused only on the development of smiling, between birth and 3 months (Peiper 1963; Wolff 1963, 1966; Emde, Gaensbauer, and Harmon 1976) and on the expressive repertoire of 9 to 12-month-olds (Young and Gouin Décarie 1977). Contextual studies have focused on the development of the smile from birth to 3 months (Wolff 1963, 1966; Sroufe and Wunsch 1972; Emde, Gaensbauer, and Harmon 1976; Sroufe and Waters 1976), on laughter from 4 months to 1 year (Wolff 1969; Sroufe and Wunsch 1972; Rothbart 1973; Sroufe and Waters 1976; Cicchetti and Sroufe 1978), and on "fear of strangers" from 5 months to 1 year (Bronson 1968a, b, 1974; Scarr and Salapatek 1970; Kagan 1974; Ricciuti 1974; Sroufe, Waters, and Matas 1974; Emde, Gaensbauer, and Harmon 1976; Campos, Hiatt, Ramsay, Henderson, and Svejda 1978; Cicchetti and Sroufe 1978).

Thus the ontogeny of the affect expressions for anger, disgust, sadness, and surprise has hardly been studied, either descriptively or contextually, and the ontogeny of the expressions of fear and laughter have not been studied descriptively. Likewise, there are virtually no studies on the ontogeny of nonemotional facial behavior (such as feeding or the use of the mouth as a manipulatory organ), other than studies on the nursing complex.

A number of other issues relevant to the nature and ontogeny of emotion have also been addressed but remain to be solved. It still is not understood precisely what emotions are. Nor is it known how they are experienced at different developmental periods (do neonates have emotional experiences?); nor are the conscious feelings and psychological meanings of the emotions, and their motivational, social, and communicative significance, at different stages of development understood.

Recently a number of researchers have reported on the striking facial mobility of the newborn (Wolff 1966; Korner 1969; Oster 1978; Oster and Ekman 1978). Oster and Ekman note that "prolonged [facial] quiescence is rare in young infants; and there is often much low-level transitory, indeterminate activity, especially around the mouth region (e.g., chewing and munching movements, compression of the inside of the cheeks against the gums, tongue movements inside and outside the lips, etc)" (1978:254).

Neonatal facial mobility is especially striking during rapid-eye-movement (REM) sleep (Wolff 1966; Korner 1969; personal observation). During this state, Korner (1969), Peiper (1963), and Wolff (1966) have noted that neonates make facial movements that resemble smiles, frowns, cry faces, and sucking. In this study, on examining 20 min of film on the facial behavior of one neonate during REM sleep, such expressions were also observed, as well as expressions resembling startles, "sneers," "pouts," and fear and disgust expressions. Some expressions involved simultaneous (organized?) movements of several muscle groups and appeared to be the same as corresponding affect expressions described by Ekman (1972) for adults. For example, expressions resembling "disgust," observed in REM sleep between 6 and 10 days after birth, involved tightly closed eyes, vertical creases between the brows, horizontal wrinkles over the bridge of the nose, and raising of the upper lip with raising of the cheeks and deepening of the nasolabial folds. Broad "smiles," observed in REM sleep between 11 and 18 days of age, involved broad lateral stretching of the lips with simultaneous bulging of the cheeks below the eyes (action of zygomaticus major) and lateral lowering of the eye corners (action of orbicularis oculi).

Using the Facial Action Coding System developed by Ekman and Friesen (1976), a system designed to measure the movement of individual muscles, Oster and Ekman found that practically all discrete facial muscle actions visible in adults can be identified in newborns (Oster 1978; Oster and Ekman 1978; Ekman and Oster 1979). Similarly, using the methods of scoring facial movements developed on nursery-school children by Blurton Jones (1971), Haviland (1976) found that the young infant shows nearly the complete repertoire shown by older children. However, no one has yet investigated systematically whether organized patterns of facial movement comparable to adult affect expressions occur in human infants before 1 month of age.

Peiper (1963) and Steiner (1979) have described specific neonatal facial configurations in response to specific tastes and odors. These expressions can be elicited during the first hours of life, before the first feeding, and are unlearned behaviors. Among tastes, Steiner has found that sweets were responded to by retracted mouth corners (smiles?), sucking, and licking. Sours elicited lip pursing, nose wrinkling, and blinking. Bitter tastes elicited depressed mouth corners and elevated upper lip, visible flat tongue, and platysmal contraction, retching, and spitting (disgust?). Among odors, banana and vanilla odors elicited expressions similar to those elicited by sweet tastes, and fishy and rotten-egg odors elicited expressions similar to those elicited by bitter tastes. Though their forms have not been studied in detail, the former expressions appear to be similar to adult "happy" affect expressions, and the latter resemble adult "disgust" expressions as described by Ekman (1972). Steiner has observed these facial responses in anencephalic and hydroanencephalic neonates possessing only intact brainstems, as well as in normal neonates. He concludes that the manifestation of these behaviors does not require cortical functioning, that they are inborn unconditioned reflexes involving specific facial configurations in response to specific gustatory and olfactory stimuli. In addition, he proposes that processing these pleasant versus aversive stimuli occurs subcortically in human neonates.

Cry faces and frowns. Cry faces, with and without accompanying vocal crying, are prominent in neonates. Although there has been little descriptive research on cry faces since Darwin, several studies have investigated the contexts in which these expressions occur. Wolff (1969) has found that neonatal crying is correlated with hunger, cold, absence of soft body contact, pain, and with spontaneous jerks during drowsiness that subsequently awaken the infants. By 3 weeks of age an array of nonnoxious stimuli, such as inanimate sounds, voices, pat-a-cake, or a human face (including mother's), presented when a baby was in a state of wakeful, diffuse activity, caused crying, though these same stimuli often elicited "smiles" if the baby was in a state of alert inactivity. Crying appears to function as a state-dependent reflexive response to generalized stimuli. Contextually, crying is interpretable as diffuse distress.

Emde, Gaensbauer, and Harmon (1976) found that neonatal frowning occurs at specific rates during REM sleep and drowsiness but increases in frequency before awakening, crying, and feeding. Frowning appears to be a spontaneous (reflexive) REM state of behavior as well as an indicator of distress. It is likely that the frowns that occur before crying are less intense manifestations of the same affect as crying and are also contextually interpretable as diffuse distress.

Smiles. The earliest "smiles," reported during the first week of life, are correlated with REM sleep and drowsy REM (Wolff 1963; Emde, Gaensbauer, and Harmon 1976). They occur at a regular frequency of about six per hour throughout the feeding and 24-hour cycles (Emde, Gaensbauer, and Harmon 1976). These first smiles have been described as involving an oblique upward movement of the mouth corners, caused by the contraction of the zygomaticus major (Wolff 1963; Emde, Gaensbauer, and Harmon 1976). Neonate smiles have been called "spontaneous," because they occur in the absence of known stimulation and are not correlated with drive states (e.g., time since feeding) but rather appear to be caused by spontaneous central nervous system activity, especially during REM sleep.

Like the facial expressions in response to tastes reported by Steiner, neonate smiles appear to be independent of the cortex and probably are mediated by the brainstem. They are more frequent in prematures (Emde, McCartney, and Harmon 1971) in which the limbic systems are relatively immature (Rabinowicz 1964), and they have been recorded at normal rates in a microcephalic infant in which only the brainstem was well developed (Harmon and Emde 1972; Emde, Gaensbauer, and Harmon 1976).

The first elicited smiles also occur during irregular sleep and drowsiness. Wolff (1963) reports that smiles can be elicited in the first week during these states by a variety of high-pitched sounds.

The first waking smiles (also involving only mouth corners) are reported to occur irregularly during the second week. They occur spontaneously when babies are satiated and drowsy or in response to generalized stimuli: high-pitched sounds (Wolff 1963) and low-level tactile or kinesthetic stimulation (Emde and Koenig 1969). These smiles involve greater stretching of the mouth than those of the first week, but the eyes appear glassy and unfocused.

These data on facial behavior during the neonatal period indicate that early smiling is unlearned, involuntary, stereotyped behavior that occurs in response to generalized stimuli and can be classified as sensorimotor stage 1 reflexive behavior.

However, not all contemporary researchers hold the same view of emotional development. Although some emphasize a gradual development of individual emotions (e.g., Wolff 1963; Emde, Gaensbauer, and Harmon 1976), others take a more ethological view, emphasizing the innate genetic programming that guides their emergence through maturation (e.g., Izard 1978, 1980) or proposing that the facial affect repertoire of the neonate is complete (Haviland 1976). Furthermore, not all contemporary researchers view the behavior of the neonate as merely reflexive. Demos (in press) and Stechler and his colleagues (Stechler, Bradford, and Levy 1966; Stechler and Carpenter 1967) view the emotional behavior of the neonate as more complex. Citing Wolff's studies on smiling, Demos interprets the wakeful 2-week-old infant's smiling response to the high-pitched human voice as a social response having psychological significance. Stechler views the young infant's capacity for focused attention as an aspect of a complex sensory-affective self-regulating system in which the neonate seeks sensory stimuli, processes this stimulus information, reacts emotionally to it, and regulates his activity accordingly. Stechler and his colleagues and Kagan (1971) view sensory perception, rather than sensorimotor behavior, as the prime avenue of cognitive development in human infants. In addition, there is a controversy as to whether the affect behaviors of the newborn are generalized responses to nonspecific stimuli (Kagan 1971; Emde, Gaensbauer, and Harmon 1976; Sroufe 1979), or whether they are relatively specific responses to specific stimuli (Izard 1979; Demos, in press). The former researchers propose or imply that the behavior of the neonate is reflexive and subcortically mediated, the latter that the neonate's behavior is complex and based on something more than subcortical reflexes.

Neurological studies show that the human brain is immature at birth. Only some of the cranial nerves that carry mode-specific information are myelinated. The optic nerve is only partially myelinated, and the olfactory nerve is not myelinated at all. Consequently, these structures are capable only of slow and diffuse transmission of information. Furthermore, the motor area of the cortex, the sensory association areas, and the limbic system (which mediates emotional behavior) have not yet begun to myelinate. This indicates that the organism is not yet capable of voluntary fine motor coordination, learned associations, or adult emotional responses (Flechsig 1920; Yakovlev and Lecours 1967). Research on the electrical activity of the neonate's brain supports this view, for only from the primary sensory areas of the neonate's cortex can electrical responses be evoked (Ellingson 1964). These neurological reports and the behavioral findings that the nursing behaviors – rooting and sucking, specific facial responses to specific tastes and odors, and REM smiles and frowns— occur in anencephalic infants, all point to subcortical mediation of behavior in the neonate. The data also indicate that these subcortically mediated responses include both diffuse responses to nonspecific stimuli (e.g., rooting and sucking) and more specific responses to more specific stimuli (e.g., the startle, which will be discussed later, and specific facial expressions in response to specific tastes). As Steiner's (1979) studies suggest, however, mediation of a behavior by lower brain centers does not preclude the processing of information. Evidently the neonate is capable of performing an array of subcortically mediated reflexive responses that have been selected through evolution and are (or were) adaptive for the newborn. Some of these are adaptive responses to more general stimulus conditions and others to more specific ones.

Stage 2. The stage of self-oriented behaviors (primary circular reactions) and acquired adaptations to impinging stimuli, sensorimotor stage 2 occurs between about 1 and 4 months in human infants. During the second stage the rooting and sucking reflexes gradually regress, that is, these behaviors can no longer be elicited by generalized stimuli. Rooting, a side-to-side sweeping of the head, drops out of the infant's behavioral repertoire, and nursing and sucking are elicited first by familiar tactile-kinesthetic contexts (the mother's nursing position) and later by the visual stimulus of the breast or bottle (Piaget 1952; personal observation).

Smiles. During the second stage the most prominent facial change is in the development of the smile. At 3 to 4 weeks, during wakefulness, infants begin to manifest alert-looking smiles. These smiles involve crinkling the muscles around bright, focused eyes (contraction of the orbicularis oculi) and lateral upward movement of the mouth corners (Wolff 1963). Despite changes in form and the physiological state in which they occur, these early "alert" smiles are elicited only irregularly by generalized multimodal stimuli – tactile, auditory, kinesthetic, and visual (Emde, Gaensbauer, and Harmon 1976). Wolff (1963) reports that eliciting stimuli include pat-a-cake, voices – especially high-pitched – a nodding head or moving faces, and an object or hand moving across the visual field. Wolff found that whereas the first smiles elicited during irregular sleep and drowsiness during the first week of life had 6- to 8-sec latencies between stimulus and response, the first alert smiles elicited during wakefulness at 3 to 4 weeks had 1- to 5-sec latencies (though REM smiles could still be elicited at this age, and they still had 6- to 8-sec latencies). These data indicate that early REM smiles and alert smiles at 3 to 4 weeks are mediated by different neural mechanisms, as the Piagetian model also implies.

During the second and third months, the smile becomes more reliably elicited (Emde, Gaensbauer, and Harmon 1976). And it is most readily elicited by patterned, three-dimensional visual stimuli (Wolff 1963; Emde, Gaensbauer, and Harmon 1976). However, abstract, nonhuman, two-dimensional stimuli also elicit smiles at this age (Salzen 1963). At about the same time that infants begin to smile differentially to patterned stimuli, they also begin to coo and gurgle, often in association with smiling (Wolff 1963). Oster (1978) found that between 1 and 3 months infants' smiles are preceded by a knitting of the brows, either alone or in conjunction with raised eyebrows. (One example of brow knitting was observed as early as the second week.[2] Oster, personal communication.) She speculates that these knitted and raised brows reflect an active and discriminating appraisal of the stimuli that elicit the smiling. As patterned visual stimuli be-

---

[2] This early example of brow knitting may represent an instance of early cortical participation in mediating behavior.

come the most reliable elicitors, the stationary human face becomes the best stimulus for eliciting smiles (Ambrose 1961; Wolff 1963; Gewirtz 1965; Spitz, Emde, and Metcalf 1970; Emde, Gaensbauer, and Harmon 1976). Wolff (1963) reports that by the beginning of the second month he was able to elicit 23 smiles in succession by repeatedly presenting a human face. He found that varying the stimulus (as with masks) would again elicit more smiles when smiling began to wane (or habituate). Similarly, Schultz and Zigler (1970) found that the frequent changes in stimulus caused by the motion of a moving clown were most effective in eliciting smiling during the second month. However, they found that a stationary stimulus (e.g., a stationary clown) become more effective at 3 months. This suggests that by 3 months, Piaget's (1952) concept of recognition and assimilation are operating, and it also suggests that active appraisal becomes more important than passive stimulation. Gewirtz (1965) found that institutional infants, compared with home-reared babies, show a delay of several weeks in responding to the immobile human face. These infants presumably have less experience with familiar patterned visual stimuli (i.e., constant familiar caretakers). Gewirtz's data therefore support the hypothesis that recognition of patterned stimuli is involved in differential smiling during this period.

Whereas normal 2- and 3-month-old infants have been reported to smile most readily at patterned visual stimuli, blind infants smile at those who hold, pet, and speak to them (Fraiberg and Freedman 1964; Freedman 1965; Fraiberg, Siegel, and Gibson 1966; Fraiberg 1968, 1971), indicating that the smile is used to respond to nonvisual stimuli as well.

Emde and his co-workers report that infants first begin to recognize their mothers between 1 and 4 months, during the second stage of sensorimotor development. Mothers recounted that their infants began to be more easily soothed and fed by them than by other relatives or strangers after 1 month of age. Presumably such recognition is based on tactile, kinesthetic, and possibly auditory stimuli. Of 14 mothers, 4 reported differential soothing ability at 1 month, 8 at 2 months, and 10 by 3 months (Emde, Gaensbauer, and Harmon 1976). Later, at around 3 and 4 months, infants begin to show differential smiling in response to their mothers' faces, which are patterned visual stimuli (Ainsworth 1967; Bowlby 1969; Emde, Gaensbauer, and Harmon 1976). Emde and his co-workers found that 5 of 14 mothers reported differential smiling at 2 months, 11 at 3 months, and 13 at 4 months.

During the period from 1 to 4 months, the smile becomes an acquired adaptation to impinging stimuli, demonstrating the infant's new ability to recognize familiar contexts. During this period, infants also begin to smile in the context of the primary circular reaction (Figure 13.12; Wolff 1963; personal observation). Thus smiling comes to be characterized by the behavioral parameters of stage 2.

Neurologically, the onset of the second stage probably represents a shift from primarily subcortical mediation to participation of the cortex. This shift evidently is gradual, because occasional instances of cortical participation apparently occur as early as the second or third week in

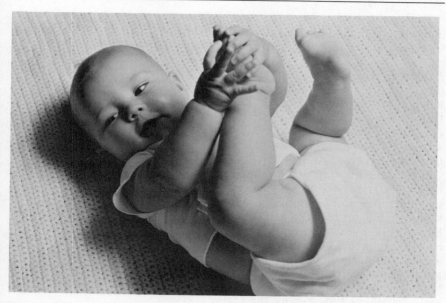

Figure 13.12. A baby smiles to herself as she performs primary circular reactions (stage 2).

some infants and become increasingly frequent around the end of the first month. The infant's motor cortex gradually begins to myelinate around the end of the first month. As the motor cortices mature, they apparently repress or reorganize the activity of the corresponding subcortical brain centers that mediate the reflexes (Yakovlev and Lecours 1967; Vlach 1971; Wolff 1974). At the same time, electrical response latencies, which are relatively long in the neonate, show a rapid decrease (Ellingson 1964). During this period the infant's sensory association areas begin to myelinate and become electrically responsive (Ellingson 1964; Yakovlev and Lecours 1967), maturational advances that are probably the basis for early stimulus recognition and association learning in the form of conditioned responses.

The facial developments of the second stage reflect these neurological advances. The sucking reflexes regress. As facial expressions such as frowning and smiling begin to occur in non-REM states, their latencies decrease. Furthermore, facial expressions (smiles) begin to function as acquired adaptations, reflecting recognition of familiar contexts and stimuli; for example, infants begin to show differential smiling toward their mothers.

Stage 3. Sensorimotor stage 3, the stage of environmental orientation and secondary circular reaction (repeated behaviors employed to make interesting spectacles last), occurs between about 4 and 8 months in human infants.

During the third stage, three possibly interrelated developments occur: an understanding of contingency relationships; the manifestation of anticipation; and the appearance of the emotional expression laughter.

Laughter. As mentioned earlier, Preyer (1892) and Dennis and Dennis (1937) reported early laughter at 3 weeks, and Wolff (1963, 1969) reports laughter in some 4- and 5-week-olds in response to pat-a-cake and tickling (gentle, firm, and rapid scratching movements of the blunt fingertips applied to the armpits). Other researchers do not report laughter before 3 or 4 months (Darwin 1877; Washburn 1929; Sroufe and Wunsch 1972; Sroufe and Waters 1976; Sroufe 1979).

Early laughter expressions have not been described, but Young and Gouin Décarie (1977) have described "play faces," often accompanied by laughter, in 9- to 12-month-old infants. Eyes sparkle, often with a wrinkling of the outer corners, and the eyelids may be narrowed, with pouches beneath the eyes; the mouth is open wide with the teeth not visible or barely visible and the lips slightly rolled into the mouth; the cheeks bulge and form a nasolabial fold. These authors also found laughter to accompany open-mouthed smiles, which resemble the play face but are characterized by upward and lateral retraction of the lips, which are not rolled in and partially or fully expose the teeth.

Contextual studies on laughter indicate that the eliciting stimuli change during ontogeny. The first manifestations of laughter, recorded through the fourth month, are in response to direct tactile/kinesthetic stimulation, such as pat-a-cake, tickling, and kissing or blowing on the stomach (Wolff 1963, 1969; Sroufe and Wunsch 1972; Sroufe and Waters 1976). This early laughter appears to function as a response to impinging stimuli and is characteristic of stage 2 behaviors. By 4 months, infants also begin to laugh in response to a mask with human likeness, suggesting laughter (like smiling) in response to quasi-familiar three-dimensional patterned stimuli.

During the fourth month, laughter also begins to occur in the context of the secondary circular reaction. Piaget (1952) describes laughter in the 4-month-old in response to making interesting spectacles last. Presumably, the 4-month-old's new understanding of the contingency relationships involved in the secondary circular reaction gives these events affective value. And the infant evidently develops expectations and anticipation through exercising these contingencies.

During the fourth month infants also begin to respond with laughter in contexts in which they evidently anticipate the recurrence of events caused by others. In this case their expectations and anticipation give these events their affective value. For example, Sroufe and Wunsch (1972) found that 4-month-olds began to respond to their mothers' saying "I'm gonna get you," followed by grabbing their stomachs, and by walking their fingers toward their babies and poking them in the ribs. The babies laugh when the mothers repeat the sequences but not during the first trials (my interpretation of these data). During the third stage it appears that the anticipation of repetitions and the social use of laughter to make interesting spectacles last or repeat become increasingly involved in laughter, as the 5-month-old baby begins to laugh at additional repetitive tactile/kinesthetic stimuli, such as mothers' bouncing their babies on their knees or jiggling them, and repetitive auditory stimuli, such as mothers' lip popping repetitively and saying "boom, boom, boom" (also my interpretation of Sroufe and Wunsch's data). By the fifth month, infants are no longer laughing at the human mask, although direct tactile stimuli, such as mothers' kissing their stomachs or tickling them, are still effective elicitors (Sroufe and Wunsch 1972).

Though the appearance of laughter coincides with the advent of stage 3, the contexts that elicit it continue to change during development. Some changes in stimulus effectiveness appear to be due to the development of increasingly complex cognitive structures. We will examine some of these changes in the contexts that elicit laughter as we investigate stages 4 and 5.

Stage 4. Both sensorimotor stage 4, the stage of coordinations, and stage 4 of the Object Concept Series, the stage in which infants first begin to understand that out-of-sight objects continue to exist, occur between about 8 and 12 months in human infants.

The available data on human facial development indicate that most of the affect expressions emerge during the fourth stage in human infants. The "anger face," the "sad face," the "surprise face," and the "fear face" are first described in infants of this age. During the fourth stage human infants also begin to manifest novel, nonemotional expressions. And laughter begins to occur in new contexts.

Laughter. By 7 and 8 months, presumably the beginning of stage 4, laughter, which first appeared in stage 3, begins to occur in new contexts (Sroufe and Wunsch 1972). One can interpret these new contexts as responses to incongruities and to unexpected coordinations. At this age laughter is elicited when mother speaks in a squeaky voice, makes a sound like a horse sneezing, or sticks out her tongue (incongruities). Unexpected coordinations, such as mother putting a cloth into her mouth or sucking on a baby's bottle, also elicit laughter. Babies of this age sometimes removed the cloth or baby bottle from their mothers' mouths, demonstrating their understanding of these coordinations, and followed this behavior with additional laughter.

Angry and sad faces. The first convincing, though incomplete, description of an infant's anger is by Darwin (1877). He describes his son screaming, showing reddening of the face and scalp, and beating objects at 11 months.[3] Young and Gouin Décarie (1977) also describe "negative expressions" that may be anger expressions in 9- to 12-month-old infants. These expressions are the clenched-teeth face (eyelids narrowed, brows

---

[3] Darwin also attributes anger to his son at 4 and 7 months, but because neither the expression nor the contexts are described in detail, it is difficult to interpret and categorize the behaviors to which he refers.

drawn inward and down, mouth opened with upper lip squared, and teeth clenched, accompanied by facial reddening) and the tight-lip face (same as clenched-teeth face but with lips closed and tightly pressed together). These descriptions are virtually identical to those described for adults and judged cross-culturally as anger in studies by Ekman (1972).

The sad face has also been recorded and described by Young and Gouin Décarie (1977) in 9- to 12-month-old infants. This expression is characterized by eyes lacking luster and sometimes filmy with tears, brows drawn together with inner corners raised and outer corners appearing lowered, mouth closed or open with outer corners pulled down, and lower lip sometimes drawn down or slightly everted. This description also is similar to the descriptions of adult expressions judged as sad in studies by Ekman (1972).

Although these expressions are descriptively similar or identical to the adult affect expressions, it is not known whether they correspond to the same feeling states or whether they occur in comparable contexts as the adult expressions. Young and Gouin Décarie (1977) were unable to distinguish the precise affective meanings of 12-month-old infants' emotional reactions (happiness, anger, disgust, sadness, and fear) on the basis of facial expression alone and classified the facial expressions they observed according to hedonic tone, that is, positive, undifferentiated, or negative.

Surprise. The classical surprise face (brows raised, jaw lowered) is first reported in 8-month-old infants (Sroufe 1979). It also has been described by Young and Gouin Décarie (1977) in 9- to 12-month-old infants, and the description corresponds to the description of adult expressions judged as surprise (Ekman 1972). This affect, like the smile, appears to develop gradually. In a study of 4- to 12-month-old infants that examined their reactions to the fate of objects that disappeared from view, Charlesworth (1966) found that some elements of the surprise expression first appeared at 7 or 8 months and unexpected appearances generally were reacted to with increased attention. Sroufe (1979) has hypothesized that surprise, which in accordance with Piagetian theory is presumably elicited by "misexpected" events, could not appear until infants were cognitively capable of forming expectancies. His hypothesis was confirmed in games of peek-a-boo in which he found that only infants who demonstrated "person permanence" showed elements of the surprise face. Those infants who had not shown the ability to understand that their out-of-sight mothers still existed and could be found in their hiding places seemed to forget their mothers as soon as they hid and did not show surprise faces when they reappeared.

Fear. Despite numerous studies on the development of "fear of strangers," little is known about the development of the fear affect expression. Emde, Gaensbauer, and Harmon (1976) report in a developmental study starting at birth that mothers and observers first identified "fear expressions" (not described), accompanied by other manifestations of distress (frowning, averting gaze, or crying), in response to strangers at 8 months.

Campos et al. (1978) and Hiatt, Campos, and Emde (1977) report that infants placed on the "visual cliff"[4] first show fear expressions (open mouth with tense, drawn-back lips) at 10 to 12 months, whereas 6- and 7-month-olds do not manifest this expression. In generally negative contexts, as when a stranger approaches or mother leaves, infants display a fear expression described by Young and Gouin Décarie (1977) as eyes wide open, mouth open with lateral lip retraction, and a squaring of the upper lip with a slight downward pull of the mouth corners, sometimes accompanied by wailing vocalizations, an expression that is similar, or identical, to the adult fear expressions described by Ekman (1972). Human infants evidently express the fear affect expression by 8 to 12 months of age in negative contexts; however, it is not clear whether these contexts are specifically fearful or whether they are more generally distressful.

Although the preceding data suggest that the fear expression appears between 8 and 9 months and presumably during the fourth stage of sensorimotor development, these data are based on tests in situations (such as stranger approach and reactions to the visual cliff) that presumably require relatively complex cognitive abilities, such as the ability to have specific expectations, to perceive incongruities, to understand three-dimensional spatial relationships, and a basic (stage 4) understanding of object permanence. It is possible there are two or more levels of fear manifestation. Charlesworth and Kreutzer (1973) have suggested an early display (stage 3 or earlier) in response to sudden, intense, impinging, or aversive stimuli, such as loud noises or painful tactile stimuli, and a later (stage 4) manifestation based on the infant's perception of incongruity between his expectations and the characteristics of the stimuli.

In the Landis and Hunt (1939) study on responses to unexpected, sudden, and intense stimuli, such as a gunshot, a spray of ice water, bright lights, or a pinprick, in which the reactions of infants, children, and adults were recorded on film, the earliest response to these stimuli was the Moro reflex. It was readily elicited in neonates during the first month, then gradually dropped out by the end of the fourth month. It was characterized by partial to complete eye closure (but no other salient facial configurations; see photos in Prechtl 1963, 1965) and abduction and extension followed by flexion of the limbs and trunk. "Startle" responses were first observed during the second month and became the normal response to these stimuli by the end of the fourth month. Startles were characterized by eye closure followed by lip parting and retraction of the mouth corners, as in a "grin," accompanied by limb and trunk flexion. This facial response is similar in form to the fear expression, except that it is very brief, less than one-half second in duration. From the second month there were no ontogenetic changes in the startle reaction. Neurologically, both the Moro reflex and the startle are mediated subcortically and can be elicited in anen-

[4] The visual cliff is an apparatus consisting of a glass surface, one portion of which rests directly on an opaque surface and the other portion of which lies some distance above the opaque surface. An infant placed on the glass has the illusion that he will fall if he is placed on the portion that lies some distance above the opaque surface.

cephalic individuals. Whether the Moro reflex, the startle, and the fear response are phylogenetically, ontogenetically, or neurologically related to one another is not known (see the discussion in Landis and Hunt 1939).

The literature on the development of "stranger distress," usually called "fear of strangers" even though it may or may not be synonymous with the development of fear (the accompanying expressions are rarely described), indicates that this phenomenon, like the smile, laughter, and surprise, develops gradually.

As noted previously, as early as the second stage of sensorimotor development (1 to 4 months), infants begin to react differentially toward strangers, showing more distress and less soothability in response to their tactile, kinesthetic, and auditory stimulation than to that of mother (Kagan 1974; Emde, Gaensbauer, and Harmon 1976). Presumably these infants are distressed by the unfamiliarity of these impinging stimuli or by their inability to exercise their newly developing adaptations to familiar impinging stimuli. Kagan (1974) suggests that infants of this age are alerted by unexpected changes and may cry, since these events are discrepant with their earlier experiences and they are unable to relate them to previous events. This interpretation is similar to a Piagetian interpretation of an infant's inability to accommodate to his newly evolving perceptions of his environment.

Between 4 and 6 months, the beginning of the third stage of sensorimotor development, infants begin to show differential responses to the visual stimulus of the stranger, as smiling to strangers wanes and sobering appears (Bridges 1932; Bronson 1968a, 1974; Emde, Gaensbauer, and Harmon 1976). At this age infants also begin to look back and forth between their mothers and a stranger as though comparing the two (Emde, Gaensbauer, and Harmon 1976). Several workers have reported some actual distress beginning at 3 to 5 months (Bronson 1974; Ricciuti 1974; Campos, Emde, Gaensbauer, and Henderson 1975; Campos et al. 1978). Bronson (1974) reports that this early distress is always preceded by a period of intense visual examination. He suggests that at this age some infants have difficulty fitting the unfamiliar faces into existing facial schema. By 6½ months, Bronson found that the long periods of initial staring had vanished and that the infants immediately turned away from the stranger and did not cry if the stranger did not insist that the baby look at him. At this age the infants no longer had difficulty identifying the stranger as different. One might interpret their immediate turning away and showing no distress as characteristic stage 3 object-concept behavior. Once the stranger was out of view, he was out of mind; the baby could make him "go away" by merely turning his head.

By 9 months, presumably early in stage 4, Bronson (1974) reports that infants turned away from the stranger immediately but often began to cry as well. He interprets this behavior as the baby's anticipating that he would not like the encounter, the baby already having assimilated the advent of the stranger into a schema derived from previous unhappy encounters with strangers. The 9-month-old's behavior toward strangers could also be related to new, stage 4 developments of his object-concept.

He now knows that out-of-sight objects continue to exist, so he cannot merely put the stranger out-of-sight and out-of-mind by turning away.

Most other researchers have reported onset of stranger distress, characterized by frowning, fussing, or crying, at 8 to 12 months, the average age usually being 8½ months (Spitz 1950; Freedman 1961; Ainsworth 1967; Scarr and Salapatek 1970; Schaffer 1974; Emde, Gaensbauer, and Harmon 1976). Distress (fear?) in response to other stimuli appears during this same period. Scarr and Salapatek (1970) found that distress in response to strangers, the visual cliff, a jack-in-the-box, a mechanical dog, masks, and loud noises all appeared after 7 months. Campos et al. (1978) and Schwartz, Campos, and Baisel (1973) found distress in response to the visual cliff, and Cicchetti and Sroufe (1978) and Hruska and Yonas (1972) report distress in response to the visual "loom"[5] at this age. Thus a new level of distress appears at about the same age that infants are entering the fourth stage of sensorimotor and object-concept development.

A number of researchers have postulated that stranger distress is founded on a new level of cognitive functioning. For example, Kagan (1976a, b) speculates that it is based either on a new ability to form hypotheses or a new level of memory ability. Campos et al. (1978) suggest that the fear system is similar, and possibly related, to the distress system of early infancy, but that by around 9 months fear becomes possible as infants become cognitively capable of reacting to threat, as well as to direct unpleasant or noxious stimuli. Schaffer (1974) proposes that stranger distress is based on a new ability to consider several events simultaneously and to detect discrepant events. In support of this hypothesis, he notes that studies on adults show associated cardiac acceleration with mental work. Studies on infants show that before 9 months discrepant events produce cardiac deceleration, whereas after that age the same stimuli produce cardiac acceleration (Hruska and Yonas 1972; Schwartz, Campos, and Baisel 1973; Emde, Gaensbauer, and Harmon 1976). This suggests that the 9-month-old is able to apply a new level of cognitive functioning to processing discrepant events.

Another line of evidence supporting a relationship between onset of distress and fearfulness and cognitive development comes from studies on Down's syndrome infants. Cicchetti and Sroufe (1978) found that infants with Down's syndrome, who had low baby intelligence quotients, on an average showed distress to the looming stimulus and the visual cliff later than normal infants. Furthermore, the Down's syndrome infants showing the earliest distress had higher baby intelligence quotients than the others.

However, two studies suggest that the onset of stranger distress is not due to a new level of cognitive functioning and is not related specifically to attaining the fourth stage of the Object Concept Series. Scarr and Salapatek (1970) did not find relationships between distress (fear?) in response to multiple stressful stimuli and the attainment of stage 4 of the Object Concept Series. Similarly, Emde, Gaensbauer, and Harmon (1976) report

---

[5] The visual loom is a shadow cast on a rear projection screen that appears to be rapidly approaching the subject on a collision course.

that 8 of 13 infants in a longitudinal study manifested stranger anxiety before they attained stage 4 abilities in the Object Concept Series, and 6 infants showed it 2 months before attaining the fourth stage. In view of these data, both groups of researchers propose that the maturation of the emotional system that mediates fear determines the onset of stranger distress and fearfulness. In support of this hypothesis, Emde and his coworkers emphasize that stranger distress emerges in the same sequence and at the same age in blind infants (Fraiberg 1971) and that identical twins show higher concordance for age of onset than fraternal twins (Freedman 1961, 1971). Fraiberg (1969) reports that stranger distress develops at the same age in blind and normal infants, whereas the object-concept is greatly retarded.

Some researchers (Paradise and Curcio 1974; Vaughn and Sroufe 1976; Sroufe 1979) suggest that stranger distress is based on "person permanence" rather than object permanence, a conceptual ability that might develop earlier than object permanence. Bell (1970) found this to be the case. Fraiberg (1969) suggests that object permanence, which implies stage 6 of the object-concept, and the ability to evoke the memory of the mother through mental representation, which develops at about 18 months, are not involved in the development of fear and stranger distress. Rather, she proposes that a more primitive "recognition memory," in which a stimulus has characteristics or signs that revive memory traces from previous experiences, may be involved. Recognition memory probably is not related to the development of the object-concept, however, since blind infants, who show great lags in object-concept development, manifest stranger distress at the same age as normal infants.

It is also possible that stranger distress is based on stage 4 abilities in the Sensorimotor Intelligence Series, rather than the Object Concept Series, a hypothesis that has not been tested. Schaffer's (1974) proposal that stranger distress is based on a new ability to consider several events simultaneously suggests that stage 4 sensorimotor abilities to coordinate several aspects of a situation may be involved. Clearly, further work is necessary before we will understand the determinants of the emergence of stranger distress and fear.

Novel facial behaviors. During the fourth stage, the infant I observed began to use novel, nonemotional facial expressions in social interactions with adults. After initiating a back-and-forth laughing game, he began responding to the adult's laughter by making spitting noises, through vibrating lips, with the tongue stuck out (see Figure 13.9b). Thus new motor patterns, a behavioral parameter first seen in stage 4, became incorporated into the secondary circular reactions manifested by this baby during the fourth stage.

Stage 5. Sensorimotor stage 5, the stage of experimentation and the tertiary circular reaction, occurs between about 12 and 18 months in human infants.

During this stage the contexts that elicit emotional expressions continue to change and begin to include the tertiary circular reaction. For example, in their tests to elicit laughter, Sroufe and Wunsch (1972) found that after the age of 12 months a rag in a mother's mouth might elicit some laughter but that babies of this age often removed the rag and then stuffed it back into their mothers' mouths (a tertiary circular reaction exercising person- and object-space relationships), which elicited still more laughter. Similarly, both stage 5 infants I observed laughed in the context of the tertiary circular reaction as they played with balls (exercising object-space-, gravity-force relationships.)

In experiments on babies' reactions to the visual cliff, Campos et al. (1978) found that after 12 months some infants do not merely cry, as do stage 4 infants; instead, they began to use stage 5 solutions to the fearful situation, as they detoured around the deep side of the cliff, employing multiple means to achieve the desired end, that is, to reach their mothers on the other side of the cliff.

During this stage, both babies I observed showed novel facial behaviors – making spitting sounds with their lips, making bubbles, and sticking out their tongues in social play. They usually laughed afterward, indicating that they were aware of what they were doing and were amused by, or pleased at, making these novel behaviors.

Stage 6. As human infants attain stage 6, the stage of mental representation and insight, they evidently become capable of imagining situations different from what they actually are. Thus, they become capable of imagining how they might obtain an out-of-reach object by finding or making a stick of the correct length to use as a tool for raking in the object (Piaget 1952). Verbally or by the use of sign language, the mental representation of nonexisting situations is a logical prerequisite to conceptualizing displacements in time (past and future) and place (Piaget 1952), to joking, lying, arguing, negating, correcting (Chevalier-Skolnikoff 1981), and to asking questions. One would expect affect expressions to be incorporated into the infants' insightful activities during the sixth stage. Only a few hours were spent observing a human child in the sixth stage during this study and only one example of stage 6 facial behavior was observed. The child showed facial joking or teasing as he repeatedly smiled and held out a toy to his mother and then withdrew his hand and laughed each time she reached for the toy.

DISCUSSION

The stage concept

A discussion of developmental stages elicits a number of questions, some of which cannot be solved with data from human development alone. Among these questions are: Do the stages emerge in the same orderly progression in different primate species? Can more than one stage ever appear simultaneously? Can a stage be skipped? Is each stage manifested in the same manner in different species? Is there change within a stage? Is there intergradation from one stage to another? Do the behaviors of one stage drop out when a subsequent stage is attained? And does a species attain the same level of functioning in all dimensions?

This study, which has examined the cognitive development of several species in terms of the behavioral parameters that characterize the different stages, sheds some light on these issues.

Cross-specifically, I find that cognitive ability unfolds ontogenetically in the same general sequence in the different primate species. The stages emerge in the same order. I have never found that a stage was skipped, nor have two or more stages ever appeared (i.e., begun to be manifested) simultaneously. As we have seen in this chapter, however, different species may attain different levels of ability. Furthermore, not all the behavioral parameters manifested in each stage by humans are necessarily manifested in the corresponding stages by other species. For example, during stage 3 stump-tailed macaques manifested environmentally oriented behavior but did not show secondary circular reactions.

I find changes within the stages in all the species I have observed. During the fifth stage, for example, the kinds of tertiary circular reactions that both apes and humans perform with balls change. At the beginning of the stage they throw balls, later they bounce them, and not until the fifth stage is well under way do they attempt to catch them (catching requires more highly developed motor coordination than throwing and bouncing).

This study also demonstrates an intergradation between stages. In other words, because new levels of ability are attained gradually, the boundaries between stages are "fuzzy." Infants in one given stage begin to irregularly manifest behavioral parameters characteristic of the next stage. An average of 1.5% of the facial behaviors manifested by infant orangutans during each stage incorporated behavioral parameters characteristic of the next stage (Figure 13.4). No subject was observed to manifest behaviors incorporating behavioral parameters characteristic of stages beyond the next stage. Thus, an infant in stage 3 might occasionally manifest behaviors incorporating stage 4 behavioral parameters but would never manifest behaviors incorporating stage 5 or 6 parameters.

The behaviors of a given stage do not disappear immediately from the repertoire once a subsequent stage is attained. In fact, stage 3, 4, 5, and 6 behaviors persist into adulthood, although stage 1 and 2 behaviors eventually drop out of the repertoires of older animals.

This chapter has focused on facial behavior, relating its development to that of "general" sensorimotor functions. This actually has been a simplification of the data, since not all species attain the same level of ability in all dimensions. Some species achieve different stages of cognitive ability in different sensory modalities or in different body parts (Chevalier-Skolnikoff 1976). For example, stump-tailed macaques achieve stage 5 and possibly stage 6 abilities in the gross body tactile/kinesthetic modality but evidently do not possess advanced manual/gestural or vocal abilities. Apes achieve advanced (stages 5 and 6) body tactile/kinesthetic and manual abilities but do not attain advanced vocal abilities. Among the primates, only humans achieve advanced abilities in all these dimensions.

Table 13.7. *Development of facial expressions of emotion in macaques: Ranges of first appearance*

Facial expression	Age at first appearance	
	Macaca arctoides[a]	*Macaca mulatta*[b]
Slow lip smacking	2 day– 2 wk	2 day– 2 wk
Adult lip smacking (affection)	2– 6 wk	– 6 wk
Grimace (fear)	2– 6 wk	3– 11 wk
Open mouth, eyelids down (play)	3– 6 wk	$2\frac{1}{2}$– 12 wk
Stares (threats)	3 wk– 4 mo	$8\frac{1}{2}$ wk– 5 mo

[a] Chevalier-Skolnikoff 1974.
[b] Based on studies by Lashley and Watson 1913; Hines 1942; Harlow and Zimmermann 1959; Bernstein and Mason 1962; Hansen 1962; Rowell 1963; Hinde, Rowell, and Spencer-Booth 1964; Harlow, Dodsworth, and Harlow 1965; and Sackett 1970.

Comparison with other studies

There are no other studies on the relationship between cognitive and facial, or communicative, development in nonhuman primates. Furthermore, only a few studies have yielded data relevant to this field of investigation.

Several studies reviewed by Bronson (1968b), Rosenblum and Alpert (1974), Redican (1975), and Emde, Gaensbauer, and Harmon (1976) have presented data on the ontogeny of facial communication in rhesus macaques, *Macaca mulatta*. The data from these studies are compared with the data on stump-tailed macaques from this study and Chevalier-Skolnikoff (1974) and presented in Table 13.7. They suggest that the emotional facial expressions of both these macaque species show similar developmental sequences, though the development of stump-tailed macaques may be slightly more rapid.

Only two primatologists have proposed ontogenetic models of facial expression. Based on observations of baboons (*Papio cynocephalus*) Anthoney (1968) has proposed a model of lip smacking from infantile sucking similar to the model suggested here and in Chevalier-Skolnikoff (1973b, 1974) for the development of lip-smacking and pucker-lip expressions in macaques and orangutans.

Redican (1975) proposed a model of facial affect development beginning with cohesive emotional expressions (lip smacking and play faces), followed by fear expressions (grimaces) and aggresive threat expressions (stares and yawns). Although cohesive lip smacking appears early and threats are reported to emerge last in all studies, it is unclear whether fear and play expressions emerge at the same time or whether one precedes the other (see Table 13.7).

In comparisons with other studies on facial expressions in orangutans, this study has recorded only some of the emotional expressions described by MacKinnon (1974) and Rijksen (1978) for wild orangutans. The pucker-lip face (or silent pout face, Rijksen), pouting cry face (pout moan face, Rijksen), open-mouth, eyelids down play face (relaxed open-mouth face, Rijksen), open-mouth, bared-teeth laugh face, as well as the fear grimace (horizontal bared teeth, Rijksen) and kissing (MacKinnon) were recorded. The tense-mouth face (anger), and the frown (anger), stare (anger), and bulging lips (during copulation) reported by Rijksen; the kiss face[6] (anger-fear?) reported by both Rijksen and MacKinnon; and the eye flash reported by MacKinnon were not recorded. Kiss faces and bulging lips occur rarely in orangutans. It is likely they never were manifested by the subjects of this study, as they probably emerge late in ontogeny. It is possible that my subjects occasionally made threats or eye flashes that were not recorded; it is also possible that these behaviors appear relatively late. One emotional facial expression recorded during this study, smiling, has not been recorded in wild orangutans. Reviewing the literature on emotional expressions in orangutans, Maple (1980) notes that several of the earlier observers do report smiles (Darwin 1872; Yerkes and Yerkes 1929).

Other than Parker's (1977) study on sensorimotor development in an infant stump-tailed macaque (results described earlier), there are no systematic studies on sensorimotor development in macaques.

In a study of the behavior of a pet pig-tailed macaque, *Macaca nemestrina,* Bertrand (1976) describes a level of intelligent behavior in this animal not previously reported for macaques. Among these behaviors were: giving an object to a recipient to perform some action on it (e.g., giving a nut to be cracked); giving on command; trading (initiated by the monkey or by humans); "cheating," in which the monkey substituted less desirable objects for the requested objects (e.g., giving a banana skin when a banana was clearly requested), and nonverbal "lying" or giving false cues. The monkey had been successfully toilet trained and generally signaled her need to eliminate by making a "flehmen" face (also called protruded-lip face) or by putting both hands on the ground behind her. Sometimes, however, when she evidently wanted to go out of a room, she signaled impending urination but did not urinate when taken to the appropriate place. Giving, including giving an object to a recipient to perform some action on it, can be classified as stage 4 sensorimotor behaviors (giving implies the understanding of a coordination between object and recipient, and giving an object to a recipient to perform an action on the object implies the ability to attribute cause of environmental change to others). Bruner (1977) has found that giving begins at 10 months, the middle of stage 4 in human infants.

[6] This behavior is accompanied by loud kiss-squeak sounds and is part of agonistic encounters in wild orangutans (personal observation). It is different from the kissing behaviors observed between adults and infants in this study.

As we have seen, macaques appear capable of performing some stage 4 behaviors, although coordinations between persons or animals and objects have not been previously recorded in this genus. Trading, cheating, and lying imply even higher abilities (stage 6: figuring out new object-object or person-object relationships to achieve a goal through mental combinations or insight). It is possible that pig-tailed macaques possess higher cognitive abilities than the other macaques. In support of this hypothesis, pig-tailed macaques are the only nonhuman primates that are regularly employed as coconut harvesters in Southeast Asia. It is also possible that this macaque was trained to perform these seemingly advanced behaviors through less advanced cognitive mechanisms (see Chevalier-Skolnikoff 1981 on a Piagetian analysis of methods of training animals). However, Bertrand (1976) has proposed that many monkey species actually possess higher cognitive potentials than they normally manifest either in their natural environments or in most captive situations.

Redshaw (1978) has performed one study on the cognitive development of apes from a Piagetian perspective. She examined the cognitive development of four gorillas from birth to 18 months. Using the Operational Causality and the Development of Means for Achieving Environmental Events (Means-Ends) Scales of Užgiris and Hunt (1975), she found that the gorillas progressed through essentially the same developmental levels in the same order as human infants. The ages at which the gorillas in Redshaw's study passed the different developmental milestones are similar to those of this study (see Table 13.8). However, Redshaw found her gorillas deficient in constructive play and in the use of other individuals as agents of action. Although it is likely that apes are deficient in the frequencies with which they manifest these kinds of behaviors compared with humans, I found that the ape subjects of my study did manifest these kinds of behaviors. However, they appeared late in ontogeny compared with humans (who begin to manifest them at around a year), being rarely observed until the age of 3 or more. Redshaw's study focused on the first 18 months, which explains why she did not observe these behaviors.

Facial mobility in neonates

An unexpected finding of this study was the observation of relatively high frequencies of apparently random facial mobility in ape and human neonates as compared with neonatal monkeys. Other researchers also have noted the lack of facial mobility in monkey neonates. For example, Rowell (1963) comments that infant rhesus macaques are curiously expressionless, appearing to have blank expressions during the first 2 months of life. As noted earlier, observers of human neonates (Oster and Ekman 1978) have commented on their striking facial mobility.

It may be more than coincidence that the species manifesting a great deal of neonatal facial mobility (apes and humans) are the same species that purposefully manifest novel facial behaviors later in ontogeny, during sensorimotor stages 4, 5, and 6 and as adults. It is difficult to integrate these findings into the Piagetian model, which does not address novel be-

Table 13.8. *Comparison of cognitive development in gorillas in this study with other studies*

Sensorimotor series	Operational causality scale[a]		Means–Ends Scale[a]	
	Development	Age	Development	Age
Stage 2: 1 mo	Hand watching	c. $1\frac{1}{2}$ mo (6 wk)	Hand watching	c. $1\frac{1}{2}$ mo (6 wk)
Stage 3: 3 mo	Uses specific action to reinstate spectacle	c. 3 mo (10–14 wk)	Grasps toy when both toy and hand in view	c. $2\frac{1}{2}$ mo (10 wk)
Stage 4: 7 mo	Touches examiner's hand or object after demonstrating spectacle	c. 7 mo (26–36 wk)	Lets go of one object to grasp another	c. 4 mo (14–20 wk)
Stage 5: 14 mo[b]	Attempts to wind mechanical toy following demonstration	c. 15 mo (56–62 wk)	Uses stick to obtain out-of-reach toy	26 mo (one subject only)

[a] Redshaw 1978.
[b] Attempted to use tools to obtain out-of-reach objects at the beginning of stage 5 but were not successful until 3 to 4 years of age.

havior before stage 4. Using ethological and evolutionary models, however, one could speculate that neonatal facial mobility might be a prerequisite for making novel voluntary facial movements later in ontogeny, as Condon and Sander (1974) have proposed that the human neonate's ability to move in rhythm with human speech is a necessary prerequisite for the subsequent development of vocal speech. If this were the case, the next question would logically be why was it adaptive for apes and humans to voluntarily make novel facial movements. I suspect that the ability to make novel mouth movements[7] voluntarily was primarily an adaptation for arboreal feeding in apes. In a study on pongid lip mobility and feeding, Walker (1979) found that captive great apes, and especially the more arboreal orangutans and chimpanzees, often used their lips to prepare foods while their hands were employed in locomotor or postural activities. The lips were used to pick up foods, to hold pieces of food while other pieces were being processed in the molar region, to store partially masticated boluses of food, to compress boluses of fibrous food to extract the juices, and to remove the skins from fruits (e.g., oranges) and the shells from nuts. Similarly, we have seen in this study that orangutans frequently use their mouths to effect tool use and other tertiary circular reactions. It is unlikely that the primary adaptive function for voluntary novel facial mobility was expressive, since the facial manifestation of emotion occurs without voluntary novel facial mobility.

The ability to voluntarily make novel facial movements was probably a preadaptation for the elaborate facial expressiveness and the evolution of vocal language in humans. The studies of Ekman and others suggest that human adults make more nonemotional expressions than do apes and use them more in social contexts (Ekman 1972, 1978). This difference is probably due to the elaborate cultural and social development of humans.

Although the evolution of vocal language probably involved neurological changes, especially in the brain, and possibly changes in the oral cavity and auditory system as well, the ability to voluntarily make novel mouth movements was probably a prerequisite for its evolution. During ontogeny, infants first attempt to make novel vocalizations and to combine sounds as they begin to babble during the fourth stage of sensorimotor development. This is the same stage during which they begin to manifest voluntary novel facial expressions.

Phylogenetic differences in facial behavior

Whereas previous studies have tended to treat the facial expressions of monkeys and apes as similar phenomena, emphasizing their similarities and implying that they contrast with those of man, this cognitive analysis of facial behavior proposes a new view. First, it shows that the facial expressions of apes are intermediate between those of monkey and man. Second, it suggests that the facial behavior of apes is more similar to man's than to monkeys'.

Ape facial behavior resembles man's and contrasts with that of mon-

---

[7] The adaptive significance of voluntary movements of the brows still remains obscure (Ekman 1979).

keys in a number of ways. The affect expressions of ape and man develop at similar rates and evidently emerge during the same sensorimotor stages. They are used in cognitively similar contexts. Both ape and man manifest novel, nonemotional facial expressions. And both manifest facial expressions that are inconsistent with their emotional states, that is, they are able to facially deceive, lie, joke, and tease.

This new perspective on the evolution of facial behavior, emphasizing similarities between ape and man, is hardly surprising in view of other recent advances in primatology (Gardner and Gardner 1969; Gallup 1970; King and Wilson 1975; Gallup, Boren, Gagliardi, and Wallnau 1977; Zihlman, Cronin, Cramer, and Sarich 1978; and Patterson 1979). Until the mid-1960s, anthropologists, psychologists, and other primatologists emphasized discontinuities between nonhuman primates and man. Man was seen as unique in his culture – he was distinguished by his ability to make tools, his language, his means of subsistence (which included hunting), his self-concept and consciousness, and his intelligence.

Since the mid-1960s, however, new data have invalidated these presumably unique qualities of man. Wild chimpanzees have been observed to make and use tools (van Lawick-Goodall 1968b); chimpanzees, gorillas, and orangutans have been taught sign language (Gardner and Gardner 1969; Patterson 1979; Shapiro, personal communication); wild baboons, chimpanzees, and other species of nonhuman primates have been observed to hunt and eat meat (DeVore and Washburn 1963; van Lawick-Goodall 1968b; Jones 1972); chimpanzees have been shown to possess a self-concept (Gallup 1970); and chimpanzees, gorillas, and orangutans have been found to possess all six levels of sensorimotor intelligence (Chevalier-Skolnikoff 1976, 1977, 1979b). Biochemically, chimpanzees and human beings have been found to possess an identical order of amino acid molecules (Wilson and Sarich 1969), and biochemically and genetically, they are at least as similar as sibling species of other animals, such as dogs and wolves (King and Wilson 1975). Both biochemical and archeological studies suggest a recent five-million-year separation between man, chimpanzee, and gorilla; a ten-million-year separation between these species and orangutans; and a twenty-million-year separation between these species and Old World monkeys (Wilson and Sarich 1969; Sarich 1970; Zihlman et al. 1978). Since the average time period for a species to form is eight million years (Simpson 1945), these data all indicate very close ties between the great apes and man. Consequently, researchers have been forced to reevaluate the relationships between man and his nonhuman-primate relatives. The emerging picture emphasizes the behavioral, biological, and phylogenetic closeness of the great apes and man (Gallup et al. 1977).

REFERENCES

Ainsworth, M. 1967. *Infancy in Uganda: infant care and the growth of love*. Baltimore: Johns Hopkins University Press.
Ambrose, J. A. 1961. The development of the smiling response in early infancy. In

Determinants of infant behaviour, Vol. 1, ed. B. M. Foss, pp. 179–96. New York: Wiley.

Andrew, R. J. 1963. The origin and evolution of the calls and facial expressions of the primates. *Behaviour* 20:1–109.

1964. The displays of the primates. In *Evolutionary and genetic biology of primates,* Vol. 2, ed. J. Buettner-Janusch, pp. 227–309. New York: Academic Press.

1972. The information potentially available in mammal displays. In *Non-verbal communication,* ed. R. A. Hinde, pp. 179–206. Cambridge University Press.

Anthoney, T. R. 1968. The ontogeny of greeting, grooming, and sexual motor patterns in captive baboons (superspecies *Papio cynocephalus*). *Behaviour* 31:358–72.

Bell, S. M. 1970. The development of the concept of object as related to infant–mother attachment. *Child Development* 41:291–311.

Bernstein, S., and Mason, W. A. 1962. The effects of age and stimulus conditions on the emotional responses of rhesus monkeys: responses to complex stimuli. *Journal of Genetic Psychology* 101:279–98.

Bertrand, M. 1976. Acquisition by a pigtail macaque of behavior patterns beyond the natural repertoire of the species. *Zeitschrift für Tierpsychologie* 42: 139–69.

Blauvelt, H., and Mckenna, J. 1961. Mother–neonate interaction: capacity of the human newborn for orientation. In *Determinants of infant behaviour,* Vol. 1, ed. B. M. Foss, pp. 3–29. New York: Wiley.

Blurton Jones, N. G. 1971. Criteria for use in describing facial expressions of children. *Human Biology* 43:365–413.

1972. Categories of child–child interaction. In *Ethological studies of child behaviour,* ed. N. Blurton Jones, pp. 97–127. Cambridge University Press.

Bowlby, J. 1969. *Attachment* Vol. 1. New York: Basic Books.

Bridges, K. M. B. 1932. Emotional development in early infancy. *Child Development* 3:324–41.

Bronson, G. W. 1968a. The fear of novelty. *Psychological Bulletin* 69:350–8.

1968b. The development of fear in man and other animals. *Child Development* 39:409–31.

1974. General issues in the study of fear, section II. In *The origins of fear,* Vol. 2, ed. M. Lewis and L. A. Rosenblum, pp. 254–8. New York: Wiley.

Bruner, J. S. 1977. Early social interaction and language acquisition. In *Studies in mother–infant interaction,* ed. H. R. Schaffer, pp. 271–89. New York: Academic Press.

Campos, J. J., Emde, R. N., Gaensbauer, T., and Henderson, C. 1975. Cardiac and behavioral interrelationships in the reactions of infants to strangers. *Developmental Psychology* 11:589–601.

Campos, J. J., Hiatt, S., Ramsay, D., Henderson, C., and Svejda, M. 1978. The emergence of fear on the visual cliff. In *The development of affect,* Vol. 1, ed. M. Lewis and L. A. Rosenblum, pp. 149–82. New York: Plenum.

Charlesworth, W. R. 1966. Persistence of orienting and attending behavior in infants as a function of stimulus–locus uncertainty. *Child Development* 37: 473–91.

Charlesworth, W. R., and Kreutzer, M. A. 1973. Facial expressions of infants and children. In *Darwin and facial expression: a century of research in review,* ed. P. Ekman, pp. 91–168. New York: Academic Press.

Chevalier-Skolnikoff, S. 1971. The ontogeny of communication in *Macaca speciosa.* Ph.D. thesis, University of California, Berkeley.

1973a. Facial expression of emotion in nonhuman primates. In *Darwin and fa-*

cial expression: a century of research in review, ed. P. Ekman, pp. 11–90. New York: Academic Press.

1973b. Visual and tactile communication in *Macaca arctoides* and its ontogenetic development. *American Journal of Physical Anthropology* 38:515–18.

1974. The ontogeny of communication in the stumptail macaque (*Macaca arctoides*). *Contributions to Primatology*, Vol. 2. Basel: Karger.

1976. The ontogeny of primate ingelligence and its implications for communicative potential: a preliminary report. In *Origins and evolution of language and speech*, ed. S. R. Harnad, H. D. Steklis, and J. Lancaster. *Annals of the New York Academy of Sciences* 280:173–211.

1977. A Piagetian model for describing and comparing socialization in monkey, ape, and human infants. In *Primate bio-social development: biological, social, and ecological determinants*, ed. S. Chevalier-Skolnikoff and F. E. Poirier, pp. 159–87. New York: Garland.

1979a. Intelligence: A basis for language in apes and humans. Paper presented at the Origins of Language conference, L. S. B. Leakey Foundation, Eureka, Calif.

1979b. Kids: zoo research reveals remarkable similarities in the development of human and orangutan babies . . . and one very special difference. *Animal Kingdom* 82:11–18.

1981. The Clever Hans phenomenon, cuing, and ape signing: a Piagetian analysis of methods for instructing animals. In *The Clever Hans phenomenon: communication with horses, whales, apes and people*, ed. T. A. Sebeok and R. Rosenthal. *Annals of the New York Academy of Sciences* 364:60–94.

Cicchetti, D., and Sroufe, L. A. 1978. An organizational view of affect: illustration from the study of Down's syndrome infants. In *The development of affect*, Vol. 1, ed. M. Lewis and L. A. Rosenblum, pp. 309–50. New York: Plenum.

Condon, W. S., and Sander, L. W. 1974. Neonate movement is synchronized with adult speech: interactional participation and language acquisition. *Science* 183:99–101.

Darwin, C. 1872. *The expression of the emotions in man and animals*. London: Murray.

1877. A biographical sketch of an infant. *Mind* 2:285–94.

Defiebre, L. 1980. Development of the adult male–female offspring relationship in a captive orangutan family. Master's thesis, California State University, San Jose.

Demos, E. V. In press. Affect in early infancy: physiology or psychology? *Psychoanalytic Quarterly*.

Dennis, W., and Dennis, M. G. 1937. Behavioral development in the first year as shown by forty biographies. *Psychological Record* 1:349–61.

DeVore, I., and Washburn, S. L. 1963. Baboon ecology and human evolution. In *African ecology and human evolution*, ed. F. C. Howell and F. Bourliere, pp. 335–67. Chicago: Aldine.

Ekman, P. 1972. Universals and cultural differences in facial expressions of emotion. In *Nebraska symposium on motivation, 1971*, ed. J. Cole, pp. 207–83. Lincoln: University of Nebraska Press.

1978. Facial signs: facts, fantasies, and possibilities. In *Sight, sound and sense*, ed. T. Sebeok, pp. 124–56. Bloomington: Indiana University Press.

1979. About brows: emotional and conversational signals. In *Human ethology*, ed. M. von Cranach, K. Foppa, W. Lepenies, and D. Ploog, pp. 169–249. Cambridge University Press.

Ekman, P., and Friesen, W. V. 1975. *Unmasking the face: a guide to recognizing emotions from facial expressions*. Englewood Cliffs, N.J.: Prentice-Hall.

1976. Measuring facial movement. *Journal of Environmental Psychology and Nonverbal Behavior,* 1:56–75.

Ekman, P., and Oster, H. 1979. Facial expressions of emotion. *Annual Review of Psychology* 30:527–54.

Ellingson, R. J. 1964. Cerebral electrical responses to auditory and visual stimuli in the infant (human and subhuman studies). In *Neurological and electroencephalographic correlative studies in infancy,* ed. P. Kellaway and I. Petersen, pp. 78–114. New York: Grune & Stratton.

Emde, R. N., and Koenig, K. L. 1969. Neonatal smiling and rapid eye movement states. *Journal of American Academy of Child Psychiatry* 8:57–67.

Emde, R. N., Gaensbauer, T. J., and Harmon, R. J. 1976. *Emotional expression in infancy: a biobehavioral study. Psychological Issues Monograph* 37. New York: International Universities Press.

Emde, R. N., Kligman, D. H., Reich, J. H., and Wade, T. D. 1978. Emotional expression in infancy. I: Initial studies of social signaling and an emergent model. In *The development of affect,* Vol. 1, ed. M. Lewis and L. A. Rosenblum, pp. 125–48. New York: Plenum.

Emde, R. N., McCartney, R. D., and Harmon, R. J. 1971. Neonatal smiling in REM states. IV: Premature study. *Child Development* 42:1657–61.

Flechsig, P. E. 1920. *Anatomie des menschlichen Gehirns und Rückenmarks auf myelogenetischer Grundlage.* Leipzig: Thieme.

Fraiberg, S. 1968. Parallel and divergent patterns in blind and sighted infants. *Psychoanalytic Study of the Child* 23:264–300.

1969. Libidinal object constancy and mental representation. *Psychoanalytic Study of the Child* 24:9–47.

1971. Smiling and stranger reaction in blind infants. In *Exceptional infant.* Vol 2: *Studies in abnormalities,* ed. J. Hellmuth, pp. 110–27. New York: Brunner/Mazel.

Fraiberg, S., and Freedman, D. A. 1964. Studies in the ego development of the congenitally blind child. *Psychoanalytic Study of the Child* 19:113–69.

Fraiberg, S., Siegel, B. L., and Gibson, R. 1966. The role of sound in the search behavior of a blind infant. *Psychoanalytic Study of the Child* 21:327–57.

Freedman, D. G. 1961. The infant's fear of strangers and the flight response. *Journal of Child Psychology and Psychiatry and Allied Disciplines* 2:242–8.

1965. Hereditary control of early social behavior. In *Determinants of infant behavior,* Vol. 3, ed. B. M. Foss, pp. 149–56. London: Methuen.

1971. Genetic influences on development of behavior. In *Normal and abnormal development of brain and behaviour,* ed. G. B. A. Stoelinga and J. J. van der Werff ten Bosch, pp. 208–29. Leiden: Leiden University Press.

Gallup, G. G., Jr. 1970. Chimpanzees: self-recognition. *Science* 167:86–7.

Gallup, G. G., Jr., Boren, J. L., Gagliardi, G. J., and Wallnau, L. B. 1977. A mirror for the mind of man, or will the chimpanzee create an identity crisis for *Homo sapiens? Journal of Human Evolution* 6:303–13.

Gardner, R. A., and Gardner, B. T. 1969. Teaching sign language to a chimpanzee. *Science* 165:664–72.

Gewirtz, J. L. 1965. The course of infant smiling in four child-rearing environments in Israel. In *Determinants of infant behaviour,* Vol. 3, ed. B. M. Foss, pp. 161–84. London: Methuen.

Gouin Décarie, T. 1965. *Intelligence and affectivity in early childhood,* Trans. E. P. Brandt and L. W. Brandt. New York: International Universities Press.

Hansen, E. W. 1962. The development of maternal and infant behavior in the rhesus monkey. Ph.D. thesis, University of Wisconsin, Madison.

Harlow, H. F., and Zimmerman, R. R. 1959. Affectional responses in the infant monkey. *Science* 130:421–32.
Harlow, H. F., Dodsworth, R. O., and Harlow, M. K. 1965. Total social isolation in monkeys. *Proceedings of the National Academy of Sciences* 54:90–7.
Harmon, R. J., and Emde, R. N. 1972. Spontaneous REM behaviors in a microcephalic infant. *Perceptual and Motor Skills* 34:827–33.
Haviland, J. 1976. Looking smart: the relationship between affect and intelligence in infancy. In *Origins of intelligence: infancy and early childhood,* ed. M. Lewis, pp. 353–77. New York: Plenum.
Hiatt, S., Campos, J., and Emde, R. 1977. Fear, surprise, and happiness: the patterning of facial expression in infancy. Paper read at the Society for Research in Child Development Meeting, New Orleans, March 1977 (cited in Campos et al. 1978).
Hinde, R. A., and Rowell, T. E. 1962. Communication by postures and facial expressions in the rhesus monkey (*Macaca mulatta*). *Proceedings of the Zoological Society of London* 138:1–21.
Hinde, R. A., Rowell, T. E., and Spencer-Booth, Y. 1964. Behaviour of socially living rhesus monkeys in their first six months. *Proceedings of the Zoological Society of London* 143:609–49.
Hines, M. 1942. The development and regression of reflexes, postures and progression in the young macaque. *Contributions to Embryology* 30:153–209. Washington, D.C.: Carnegie Institution.
Hruska, K., and Yonas, A. 1972. Developmental changes in cardiac responses to the optical stimulus of impending collision. Abstract 24. *Psychophysiology* 9:272.
Izard, C. E. 1978. On the development of emotions and emotion-cognition relationships in infancy. In *The development of affect,* Vol. 1, ed. M. Lewis and L. A. Rosenblum, pp. 389–413. New York: Plenum.
1978. Emotions as motivations: an evolutionary-developmental perspective. In *Nebraska symposium on motivation,* Vol. 26, ed. R. H. Dienstbier, pp. 163–200. Lincoln: University of Nebraska Press.
1980. The emergence of emotions and the development of consciousness in infancy. In *The psychobiology of consciousness,* ed. J. M. Davidson and R. J. Davidson, pp. 193–216. New York: Plenum.
Jay, P. 1965. The common langur of North India. In *Primate behavior: field studies of monkeys and apes,* ed. I. DeVore, pp. 197–249. New York: Holt, Rinehart and Winston.
Jones, C. 1972. Natural diets of wild primates. In *Pathology of simian primates.* Part I: General pathology, ed. R. N. T-W. Riennes, pp. 58–77. Basel: Karger.
Kagan, J. 1971. *Change and continuity in infancy.* New York: Wiley.
1974. Discrepancy, temperament, and infant distress. In *The origins of fear,* Vol. 2, ed. M. Lewis and L. A. Rosenblum, pp. 229–48. New York: Wiley.
1976a. Emergent themes in human development. *American Scientist* 64:186–96.
1976b. Three themes in developmental psychobiology. In *Developmental psychobiology: the significance of infancy,* ed. L. P. Lipsitt, pp. 129–37. Hillsdale, N.J.: Erlbaum/New York: Halsted.
King, M.-C., and Wilson, A. C. 1975. Evolution at two levels in humans and chimpanzees. *Science* 188:107–16.
Korner, A. F. 1969. Neonatal startles, smiles, erections, and reflex sucks as related to state, sex, and individuality. *Child Development* 40:1039–53.
Landis, C., and Hunt, W. A. 1939. *The startle pattern.* New York: Farrar & Rinehart.

Lashley, K. S., and Watson, J. B. 1913. Notes on the development of a young monkey. *Journal of Animal Behavior* 3:114–39.

McGrew, W. C. 1972. *An ethological study of children's behavior*. New York: Academic.

MacKinnon, J. 1974. The behaviour and ecology of wild orang-utans (*Pongo pygmaeus*). *Animal Behaviour* 22:3–74.

Maple, T. L. 1980. *Orang-utan behavior*. New York: Van Nostrand Reinhold.

Oster, H. 1978. Facial expression and affect development. In *The development of affect*, Vol. 1, ed. M. Lewis and L. A. Rosenblum, pp. 43–75. New York: Plenum.

Oster, H., and Ekman, P. 1978. Facial behavior in child development. In *Minnesota symposia on child psychology*, Vol. 11, ed. W. A. Collins, pp. 231–76. Hillsdale, N.J.: Erlbaum.

Paradise, E. B., and Curcio, F. 1974. Relationship of cognitive and affective behaviors to fear of strangers in male infants. *Developmental Psychology* 10: 476–83.

Parker, S. T. 1977. Piaget's sensorimotor series in an infant macaque: a model for comparing unstereotyped behavior and intelligence in human and nonhuman primates. In *Primate bio-social development: biological, social, and ecological determinants*, ed. S. Chevalier-Skolnikoff and F. E. Poirier, pp. 43–112. New York: Garland.

Patterson, F. G. 1979. Linguistic capabilities of a lowland gorilla. Ph.D. thesis, Stanford University, Stanford, Calif.

Peiper, A. 1963. *Cerebral function in infancy and childhood*, trans. B. Nagler and H. Nagler, ed. J. Wortis. New York: Consultants Bureau.

Piaget, J. 1951. *Play, dreams and imitation in childhood*, trans. C. Gattegno and F. M. Hodgson. New York: Norton.

  1952. *The origins of intelligence in children*, trans. M. Cook. New York: International Universities Press.

  1954. *The construction of reality in the child*, trans. M. Cook. New York: Ballantine.

Prechtl, H. F. R. 1958. The directed head turning response and allied movements of the human baby. *Behaviour* 13:212–42.

  1963. The mother–child interaction in babies with minimal brain damage. In *Determinants of infant behaviour*, Vol. 2, ed. B. M. Foss, pp. 53–66. London: Methuen.

  1965. Problems of behavioral studies in the newborn infant. In *Advances in the study of behavior*, Vol. I, ed. D. S. Lehrman, R. A. Hinde, and E. Shaw, pp. 75–98. New York: Academic Press.

Preyer, W. 1892. *The mind of the child*. New York: Appleton.

Rabinowicz, T. 1964. The cerebral cortex of the premature infant in the 8th month. In *Progress in brain research*, Vol. 4: *Growth and maturation of the brain*, ed. D. Purpura, and J. Shadé, pp. 39–86. New York: Elsevier.

Redican, W. K. 1975. Facial expressions in nonhuman primates. In *Primate behavior*, vol. 4: *Developments in field and laboratory research*, ed. L. A. Rosenblum, pp. 103–94. New York: Academic Press.

Redshaw, M. 1978. Cognitive development in human and gorilla infants. *Journal of Human Evolution* 7:133–41.

Ricciuti, H. N. 1968. Social and emotional behavior in infancy: some developmental issues and problems. *Merrill-Palmer Quarterly of Behavior and Development* 14:82–100.

  1974. Fear and the development of social attachments in the first year of life. In *The origins of fear*, Vol. 2, ed. M. Lewis and L. A. Rosenblum, pp. 73–106. New York: Wiley.

Rijksen, H. D. 1978. *A field study on Sumatran orang-utans* (Pongo pygmaeus abellii *Lesson, 1827*). Wageningen: H. Veenman & Zonen.

Rosenblum, L. A., and Alpert, S. 1974. Fear of strangers and specificity of attachment in monkeys. In *The origins of fear*, Vol. 2, ed. M. Lewis and L. A. Rosenblum, pp. 165–93. New York: Wiley.

Rothbart, M. K. 1973. Laughter in young children. *Psychological Bulletin* 80: 247–56.

Rowell, T. E. 1963. The social development of some rhesus monkeys. In *Determinants of infant behaviour*, Vol. 2, ed. B. M. Foss, pp. 35–45. New York: Wiley.

Sackett, G. P. 1970. Unlearned responses, differential rearing experiences, and the development of social attachments by rhesus monkeys. In *Primate behavior: Developments in field and laboratory research*, Vol. 1, ed. L. A. Rosenblum, pp. 112–40. New York: Academic Press.

Salzen, E. A. 1963. Visual stimuli eliciting the smiling response in the human infant. *Journal of Genetic Psychology* 102:51–4.

Sameroff, A. J. 1972. Learning and adaptation in infancy: a comparison of models. In *Advances in child development and behavior*, Vol. 7, ed. H. W. Reese, pp. 169–214. New York: Academic.

Sarich, V. M. 1970. Primate systematics with special reference to Old World monkeys: a protein perspective. In *Old World monkeys*, ed. J. R. Napier and P. H. Napier, pp. 175–226. New York: Academic Press.

Scarr, S., and Salapatek, P. 1970. Patterns of fear development during infancy. *Merrill-Palmer Quarterly of Behavior and Development* 16(1):53–90.

Schaffer, H. R. 1974. Cognitive components of the infant's response to strangeness. In *The origins of fear*, Vol. 2, ed. M. Lewis and L. A. Rosenblum, pp. 11–24, New York: Wiley.

Schultz, T. R., and Zigler, E. 1970. Emotional concomitants of visual mastery in infants: the effects of stimulus movement on smiling and vocalizing *Journal of Experimental Child Psychology* 10:390–402.

Schwartz, A. N., Campos, J. J., and Baisel, E. J., Jr. 1973. The visual cliff: cardiac and behavioral responses on the deep and shallow sides at five and nine months of age. *Journal of Experimental Child Psychology* 15:86–99.

Siegel, S. 1956. *Nonparametric statistics for the behavioral sciences*. New York: McGraw-Hill.

Simpson, G. G. 1945. The principles of classification and a classification of mammals. *Bulletin of the American Museum of Natural History* Monograph 85: 1–350.

Spitz, R. A., 1950. Anxiety in infancy: a study of its manifestations in the first year of life. *International Journal of Psycho-Analysis* 31:138–43.

Spitz, R. A., Emde, R. N., and Metcalf, D. R. 1970. Further prototypes of ego formation: a working paper from a research project on early development. *Psychoanalytic Study of the Child* 25:417–41.

Sroufe, A. L. 1979. Socioemotional development. In *Handbook of infant development*, ed. J. D. Osofsky, pp. 462–516. New York: Wiley.

Sroufe, L. A., and Waters, E. 1976. The ontogenesis of smiling and laughter: a perspective on the organization of development in infancy. *Psychological Review* 83:173–89.

Sroufe, L. A., and Wunsch, J. P. 1972. The development of laughter in the first year of life. *Child Development* 43:1326–44.

Sroufe, L. A., Waters, E., and Matas, L. 1974. Contextual determinants of infant affective response. In *The origins of fear*, Vol. 2, ed. M. Lewis and L. A. Rosenblum, pp. 49–72. New York: Wiley.

Stechler, G., and Carpenter, G. 1967. A viewpoint on early affective development.

In *Exceptional infant. I: the normal infant,* ed. J. Hellmuth, pp. 163–205. New York: Brunner/Mazel.

Stechler, G., Bradford, S., and Levy, H. 1966. Attention in the newborn: effect on motility and skin potential. *Science* 151:1246–8.

Steiner, J. E. 1979. Human facial expressions in response to taste and smell stimulation. *Advances in Child Development and Behavior* 13:257–95.

Stirnimann, R. 1940, *Psychologie des neugeborenen Kindes.* Munich: Kindeer (cited by Charlesworth and Kreutzer 1973).

Užgiris, I. Č., and Hunt, J. McV. 1966. An instrument for assessing infant psychological development. Unpublished.

1975. *Assessment in infancy: ordinal scales of psychological development.* Urbana: University of Illinois Press.

Valentine, C. W. 1930. The innate basis of fear. *Journal of Genetic Psychology* 37:394–420.

Van Hooff, J. A. R. A. M. 1967. The facial displays of the catarrhine monkeys and apes. In *Primate ethology,* ed. D. Morris, pp. 9–88. Chicago: Aldine.

1971. *Aspects of the social behaviour and communication in human and higher non-human primates.* Rotterdam: Privately published.

1972. A comparative approach to the phylogeny of laughter and smiling. In *Nonverbal communication,* ed. R. A. Hinde, pp. 209–41. Cambridge University Press.

van Lawick-Goodall, J. 1968a. A preliminary report on expressive movements and communication in the Gombe Stream chimpanzees. In *Primates: studies in adaptation and variability,* ed. P. C. Jay, pp. 313–74. New York: Holt, Rinehart and Winston.

1968b. The behaviour of free-living chimpanzees in the Gombe Stream Reserve. *Animal Behaviour Monographs* 1:161–311.

Vaughn, B., and Sroufe, L. 1976. The face of surprise in infants. Paper presented at the Animal Behavior Society Meeting, Boulder, Colo.

Vlach, V. 1971. Evolution of the skin reflexes in infants. In *Second Prague international symposium of child neurology,* ed. I. Lesný and M. Lehovský, pp. 261–3. Prague: Universita Karlova Praha.

Walker, P. L. 1979. The adaptive significance of pongid lip mobility. *OSSA: International Journal of Skeletal Research* 6:277–84.

Washburn, R. W. 1929. A study of smiling and laughing of infants in the first year of life. *Genetic Psychology Monographs* 6:397–537.

Watson, J. B. 1919. *Psychology from the standpoint of a behaviorist.* Philadelphia: Lippincott.

1925. Experimental studies on the growth of the emotions. *Pedagogical Seminary* 32:326–48.

Wilson, A. C., and Sarich, V. M. 1969. A molecular time scale for human evolution. *Proceedings of the National Academy of Sciences* 63:1088–93.

Wolff, P. H. 1963. Observations on the early development of smiling. In *Determinants of infant behaviour,* Vol. 2, ed. B. M. Foss, pp. 113–34. London: Methuen.

1966. *The causes, controls, and organization of behavior in the neonate. Psychological Issues Monograph* 17:1–99. New York: International Universities Press.

1969. The natural history of crying and other vocalizations in early infancy. In *Determinants of infant behaviour,* Vol. 4, ed. B. M. Foss, pp. 81–109. London: Methuen.

1974. Autonomous systems in human behavior and development. *Human Development* 17:281–91.

Yakovlev, P. I., and Lecours, A. R. 1967. The myelogenetic cycles of regional maturation of the brain. In *Regional development of the brain in early life,* ed. A. Minkowski, pp. 3–70. Philadelphia: Davis

Yerkes, R. M., and Yerkes, A. W. 1929. *The great apes.* New Haven: Yale University Press.

Young, G., and Gouin Décarie, T. 1977. An ethology-based catalogue of facial/vocal behaviour in infancy. *Animal Behaviour* 25:95–107.

Zihlman, A. L., Cronin, J. E., Cramer, D. L., and Sarich, V. M. 1978. Pygmy chimpanzee as a possible prototype for the common ancestor of humans, chimpanzees and gorillas. *Nature* 275:744–6.

PART V

Single- versus multiple-channel communication: Communication of reproductive state

How does a female communicate to a male that she is in reproductive condition? Over the years studies of sexual behavior have established three aspects of the communication and coordination of reproductive activities between mates:

1. Receptivity. This term refers to behavioral patterns of the female that lead to the consummation of a mating sequence. Receptivity refers to the signals and behaviors by which a female facilitates the male's copulation with her.
2. Proceptivity. This term refers to those signals and behavioral activities of the female that have an appetitive function. Proceptivity describes the activities that arouse and maintain the male's interest in copulation; receptivity represents a consummatory act; proceptivity an appetitive act.
3. Attractivity. This concept refers to the female's stimulus value to the male. A female may show proceptive behavior toward a male, but he may not be aroused if she is not attractive to him.

In theory each concept is separable from the other. One can examine the sequence of courtship and mating to determine which activities have an arousing effect on the male, which allow him to copulate with the female, and which features allow females to choose among males and vice versa.

The authors in Part V present three similar approaches to the determination of attractive, proceptive, and receptive cues yet with somewhat different outcomes. The basic question raised is: Can any set of discrete signals or discrete signal modalities be identified by which female receptivity and proceptivity are communicated? Or, alternatively, is the communication of receptivity and proceptivity a gestalt of behavior, that is, a multidimensional system with both specialized signals and behavioral patterns across several modalities?

The research directed at this question used three genera of monkeys, geographically diverse but closely related: baboons, talapoins, and macaques. Bielert argues that the visual cue of sex-skin (perineal) swelling in female chacma baboons has an arousal function on males who observe this swelling. His model experimental approach demonstrates how to identify and substantiate the effectiveness of a single cue. After discovering a straight-forward, easily measured index of sexual arousal, the presence or absence of seminal emissions beneath a male's cage, he used this simple index, through a series of studies, to show that the swelling of the sexual skin of the female affects male arousal.

To substantiate the validity of his response measure, Bielert first established that seminal emissions were, in fact, correlated with sexual arousal. His first demonstration proves that seminal emissions are more frequent after a male has had direct sexual contact with a female. Next Bielert demonstrates that the mere sight of a female with a swollen sex skin also elicits an increased rate of seminal emissions. Thus direct sexual contact is not necessary for this sort of sexual arousal. When Bielert presented the males with the sight of a female over the course of her ovarian cycle, he found increased rates of seminal emissions only on those days when the female's sex skin was fully turgid. At other times in the cycle seminal emissions were infrequent. Sex tests with females showed that ejaculation frequency was also correlated with the size of the sex-skin swelling. Therefore, both frequency of ejaculations with a female and seminal emissions from visual exposure alone correlated with the size of the female's swelling.

Finally, Bielert shows that visual isolation of the female, though with olfactory and auditory modalities still open, reduced the rate of seminal emissions, even though the female was at her maximal sex-skin swelling. This experiment demonstrates the critical importance of visual cues over olfactory and auditory cues in the arousal of males, but it does not conclusively prove that the sex-skin swelling is the only visual stimulus that is important. Many subtle behavioral actions, as described in Goldfoot's chapter, could be necessary, along with sex-skin swelling, to arouse males. To demonstrate the importance of sex-skin swelling alone would require a playback study using a sort of "Playboon" centerfold. If the male experienced an increased rate of seminal emission after seeing a static photograph of a female with swollen sex skin, the unique importance of the visual signal would be clear. Bielert argues, however, that sex-skin swelling is simply the most prominent of a variety of visual cues that arouse males. Although Bielert concludes that olfaction is not important in arousing males, he leaves open the possibility that olfactory cues may be important in the consummation of sexual activity. Thus, visual cues may be proceptive and operate distally, whereas olfactory cues may be receptive and operate proximally in the sequence of mating of chacma baboons.

On the other hand, Keverne argues that changes in the vaginal chemistry of rhesus macaques during the ovarian cycle lead to olfactory cues that arouse a male at the appropriate time for copulation. In rhesus monkeys there are some changes in sex-skin coloration coincident with ovulation, but the massive sex-skin swelling described by Bielert for the chacma baboon does not appear. However, rhesus males are not affected by manipulations of sex-skin coloration in females. The considerable olfactory investigation of females by males prompted Keverne and his colleagues to investigate the importance of olfactory cues, particulary those from the vagina, in the stimulation of male sexual behavior.

Keverne describes the logical steps he followed in establishing the likelihood of an olfactory arousal cue. Males performed the task of pressing levers to gain access to an estrous female when their olfactory system was intact, but when they were made anosmic, they failed to respond. An analysis of vaginal secretions showed that odorus vaginal components experienced cyclic changes that related to the ovarian cycle. When an artificial mixture of these odorants placed on an ovariectomized female was observed to stimulate male sexual activity, considerable speculation about vaginal pheromones resulted. Other investigators even attempted to demonstrate similar vaginal odorant changes in human females with a corresponding behavioral impact.

Keverne maintains a conservative stance in the interpretation of his results. He does not believe that vaginal odors have an effect analogous to insect sex pheromones. Rather, he argues that olfactory cues are a part of a complex of behavioral acts and specialized signals that lead to sexual arousal and ultimately to copulation. To substantiate his argument, he presents a study on another species of monkey, the talapoin, in which a group of males were made temporarily anosmic. In this social context the results were considerably more complex than those in the pair-testing situation of the rhesus monkeys. First, the dominant male continued to mate as before, although he was much less likely to initiate mating. Instead, the females showed an increase in initiations. Second, when subordinate males were placed alone with females, their sexual performance after anosmia varied according to their previous sexual experience. The second-ranked male showed no change, the third-ranked male increased sexual activity, and the lowest-ranked male decreased activity. Thus, when animals are members of complex social groups, it appears likely that the importance of olfactory cues in stimulating sexual activity may be diminished by the importance of social variables. In complex social groupings, the signals and behavior leading to mating are multiple and complex.

Seriously questioning the early statements concerning the role of vaginal odorants in proceptivity, Goldfoot carried out a partial replication of the original studies along with additional tests. He interprets his results as counter to the vaginal-odor hypothesis. First, observing that some males mounted females outside the periovulatory period and some pairs would not mate during the periovulatory period, he concluded that a characteristic of female attractiveness independent of the ovarian state was active. In testing the stimulus value of vaginal odors, he found that vaginal secretions mixed with semen from another male were somewhat more powerful than vaginal odors alone. Furthermore, the swabbing of a female with a novel odor (smelling like green peppers) also elicited increased sexual activity. From these results he concludes that the arousal of sexual behavior does not depend on the specificity of the vaginal odors. In addition, he reports that females exposed to testosterone during gestation and subsequently lacking vaginas were slightly more effective than normal females in eliciting male sexual behavior. Significantly, he also observed that anosmic males continue to mate with females. Finally, in charting the temporal distribution of vaginal olfactants, he found that they peaked in the luteal phase when any mating is unlikely to produce conception, a factor that casts even further doubt on the biological utility of the putative vaginal sex attractant.

Rather than a model of sexual stimulation based on vaginal olfactory cues, Goldfoot proposes a sequential multichannel model in which several subtle behavioral actions are combined with specialized signals to lead to mating. He describes sequences of mating behavior in which a gestalt of activities and signals is necessary to the successful culmination of mating. Males are most attracted by females in the proper hormonal condition who actively solicit them, and at any point along the mating sequence, the female can terminate by refusing to go further.

In addition to a sequential multiple-channel model, Goldfoot also argues for the role of conditioning and experience. A male can learn which females accepted or rejected him in the past and direct his attention to only the most promising females. To demonstrate that learning influences behavior, Goldfoot shows that sexually experienced males respond more to olfactory cues than do inexperienced males. The combination of multiple channels and learning make a highly complex set of determinants of mating in rhesus monkeys.

Both Keverne and Goldfoot reach the same conclusions: that odors do play a role, that other forms of communication such as subtle behavioral acts occur, and that social status and context affect mating. Continuing arguments on the question of vaginal sex pheromones may not be productive. The notion of a simple determinant of sexual arousal or of a single element of female attractiveness is not acceptable, especially in the socially complex groups common to primates. The multiple-channel model seems to be the only possible model to account for spontaneous sexual behavior.

14 · Experimental examinations of baboon (*Papio ursinus*) sex stimuli

CRAIG BIELERT

OVARIAN HORMONE INFLUENCES ON VISUAL CUES AND THEIR STIMULUS VALUE

Cyclically pronounced swellings of the perineum, or "sex skin," are found in a number of primates (Napier and Napier 1967). They occur in a few prosimians but to a minor degree and usually involve only the vulva (for review see van Horn and Eaton 1979). Visually prominent swellings are widespread in certain genera of catarrhine monkeys including *Cercocebus*, *Macaca*, *Miopithecus*, and *Papio*, whereas in *Colobus* they occur in only a proportion of species, and in *Cercopithecus*, *Erythrocebus*, *Presbytis*, and *Theropithecus* they are absent altogether. Among the apes, swellings are minor or absent in gibbons, gorillas, orangs, and siamangs, and prominent in chimpanzees and pygmy chimpanzees. There is evidence that in most species the swellings are most pronounced during the mid-cycle periovulatory period (Wickler 1967; Rowell 1972).

These swellings have been the source of considerable speculation from the time of Darwin (1871) to the present day (Rowell 1972; Clutton-Brock and Harvey 1976; Bercovitch 1978; Collins 1978; Alexander 1979; Lancaster 1979; Short 1979). Two major hypotheses exist in regard to their occurrence and species distribution. One suggests that the swellings may help a female near ovulation to attract the attention of at least one conspecific male. The second explanation suggests that the swelling may enable a female to attract a number of males and thereby allow either active or passive female choice among them. Whichever alternative is correct, concrete observations are required to determine that the swellings do indeed affect male behavior and, as such, can be held to have "stimulus value."

There are a number of observations of sexual swellings affecting male

This work was supported by funds from the University Council of the University of the Witwatersrand to the Primate Behaviour Research Group. I thank the staff of the Central Animal Unit of the University of the Witwatersrand and, in particular, Mr. M. E. Howard-Tripp, Mr. Elias Malebye, and Mr. Johannes Moeketsi for their cooperation and assistance. I also thank Professor J. A. M. Myburgh for permission to collect data from a number of his adult male baboons, Parke–Davis Laboratories, Ltd., for their kind donation of Ketalar, Mrs. B. Hurwitz for her conscientious typing, and Dr. C. Anderson for her critical review of this chapter.

behavior. As an example, when ovariectomized talapoin monkeys (*Mio-pithecus talapoin*) are injected with estradiol benzoate and begin to show swellings, socially dominant males begin to copulate with them. As long as the dominant males are present, subordinate males make no overt advances, but there is a marked increase in the frequency with which subordinates visually fixate the treated females (Dixson, Everitt, Herbert, Rugman, and Scruton 1973). The frequency with which pig-tailed macaques (*Macaca nemestrina*) visually inspect a female's genital area is higher when she is nearing the peak of the follicular phase and is maximally swollen (Eaton 1973). This is also the case for the Celebes black ape (*Macaca nigra*) (Dixson 1977). The adult male hamadryas baboon (*Papio hamadryas*) mates exclusively with the females that belong to his "harem." Only when a female comes into estrus and shows swelling of the sex skin does the male conduct frequent and prolonged genital inspections and eventually copulate with her (Kummer 1971).

Though the evidence available makes it clear that males of a number of primate species give visual attention to the female's perineal swelling, evidence of sexual arousal is scanty. Experiments in arousal phenomena have focused upon the rhesus monkey (*Macaca mulatta*). In this species only the adolescents show a pronounced cyclic sex-skin swelling (Hisaw and Hisaw 1961). There are, however, cyclic sex-skin color changes that are correlated with ovarian events (Czaja, Eisele, and Goy 1975; Czaja, Robinson, Eisle, Scheffler, and Goy 1977). Vandenbergh (1969) demonstrated that adult males removed from a free-ranging colony during the nonmating season could be returned to breeding condition if they were exposed for 16 days to estradiol-benzoate-treated ovariectomized females. Using rhesus monkeys housed in large outdoor enclosures, workers have demonstrated significant rises in the plasma-testosterone levels of males introduced into a group of sexually receptive females (Rose, Gordon, and Bernstein 1972). In 1974 Vandenbergh and Drickamer found that recrudescence of male reproductive functions could be induced in free-ranging males during the nonmating season by the presence of females implanted with pellets of estradiol benzoate, which were effective in stimulating an increase in the hue and intensity of the color of the perineal areas.

Though vision may have been important in all the previously described studies, the importance of this specific sensory channel was observed directly in a study by Gordon and Bernstein (1973). In their large outdoor compounds at the Field Station of the Yerkes Primate Research Center, their monkeys bred with a distinct seasonal rhythm; when an all-male group of monkeys was housed adjacent to the heterosexual group, the males showed a significant increase in mounting behavior coincident with the reddening of sex skin and the increasing sexual behavior in the heterosexual group. Another all-male group housed nearby that was unable to see the heterosexual group failed to display a seasonal increase in sexual behavior or hormone levels.

Additional supportive evidence of the role of visual cues in the arousal phenomenon is available for the chimpanzee (*Pan troglodytes*). In free-living chimpanzees, males display prominent erections with exposure to estrous (swollen) females (van Lawick-Goodall 1968). Captive males show frequent erections when allowed to view fully swollen females but seldom exhibit erections when viewing ovariectomized females or cycling females in a deflated condition, for example, during the luteal phase of the menstrual cycle (Young and Orbison 1944).

THE BABOON PERINEUM AND ITS STIMULUS VALUE

The sexual swelling of the chacma baboon (*Papio ursinus*) has been described as "an enormous swelling, far larger than that of other species" (Rowell 1972:74). A fully inflated, or turgid, swelling is shown in Figure 14.1. The anatomic labels are the same as those used by Hendricks (1971) in his description of the external anatomy of the closely related olive baboon (*P. cynocephalus*). The role of cues from the perineum is the focus of the research work to be discussed in this chapter. The perineum is a triangular-shaped region at the caudal-most part of the trunk that is bounded dorsally by the root of the tail, ventrolaterally by the ischial callosities, and ventrally by the junction of the lower part of the abdominal wall with the front of the thighs below. The anus and the external genitalia are located within this region, and the anal canal and vaginal vestibule communicate with the exterior. The perineal body is located between the vestibule and the anus. For the most part, the skin of the perineum is hairless, very distendable, and easily lacerated. At maximal turgescence, despite its florid appearance, the sex skin is relatively avascular and contains large quantities of water. These changes are related to active fibroblasts, which synthesize hyaluronic acid, a glycosaminoglycan. This molecule can bind large amounts of water and is generally assumed to give the skin its firm appearance (Rienits 1960; Bentley, Nakagawa, and Davies 1971; Carlisle and Montagna 1979).

Bolwig (1959) described the red swollen genitalia of a female in estrus as usually attracting the attention of other females and juveniles, who examined the genitalia with fingers, nose, and tongue. The role of the perineal swelling of this species in sexually arousing the male was indirectly examined in the field by Saayman (1973), who believed that swollen sex skin was not the only factor responsible for stimulating the sexual activity of the males he studied. He concluded that the stimulation of receptive and appetitive behavior in females is necessary for the effective transmission of olfactory information and for the consequent stimulation of male sexual interest under normal conditions. However, the possibility that the swollen sexual skin constitutes an additional stimulus to the sexual interest of male baboons was not completely excluded. Saayman based his interpretation that sex-skin swelling was relatively unimportant on mounting rate and male initiating behavior. The situation in which he worked did not allow for independent assessment of sexual stimuli because females were giving off visual and olfactory cues simultaneously.

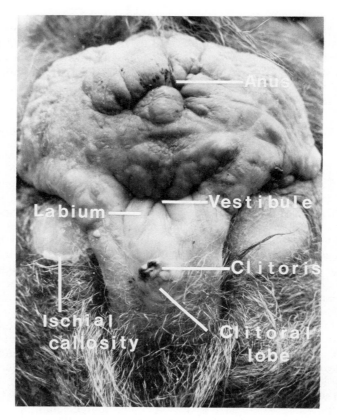

Figure 14.1. The maximally turgid genitalia of an intact adult female chacma baboon (*Papio ursinus*). (From Bielert and van der Walt, 1982. Reprinted with permission.)

In contrast to the baboon, the rhesus monkey, as described earlier, lacks pronounced perineal swelling changes. Instead, it experiences sex-skin color changes. In the laboratory, Herbert (1966) induced perineal color changes like those of a female nearing ovulation by topically applying estrogen cream to the sex skin. However, in spite of these color changes there was no consistent effect on male sexual interest or behavior. Keverne (chap. 15, this volume) has recently emphasized that this work in no way excludes the possibility that sex-skin color changes may act as signals over a distance.

THE EXPERIMENTAL LABORATORY APPROACH

It is well known that in nonhuman primates heterosexual mating activity is not the only form of sexual expression. Males masturbate in free-rang-

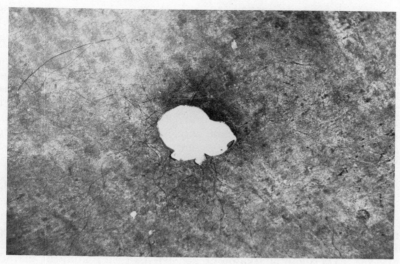

Figure 14.2. An example of the chacma baboon seminal emission termed a "spot." (From Bielert, Howard-Tripp, and van der Walt 1980. Reprinted with permission.)

ing situations with females available (Carpenter 1942; Saayman 1970; Hanby, Robertson, and Phoenix 1971; Hausfater 1975) and in more restricted environments (Phoenix and Jensen 1973; Maple 1977). Coagulated seminal emissions have been used as a dependent measure of sexual arousal in two studies on rhesus monkey sexuality (Phoenix and Jensen 1973; Slimp, Hart, and Goy 1978).

Chacma baboon semen coagulates immediately after mating, forming what is known as an ejaculatory plug in the female's vagina. Fragments of these plugs often accompany penile withdrawal and are frequently picked off and eaten during genital grooming. Semen is also caught and eaten when the ejaculation is the result of masturbation. Following masturbation in the laboratory situation, noticeable amounts of uncaught semen coagulate below the cage's mesh bottom where they cannot be reached by the male. Such emissions are easily recognizable and have been described as "spots." Figure 14.2 illustrates such an emission. In addition, a string-like segment of coagulated semen is usually voided during the first postejaculatory urination and can be found below the cage bottom. This emission can be described as a "squiggle."

Using these coagulated emissions as a dependent variable in studies of sexually arousing stimuli, the laboratory situation allows a manipulation that is impossible in the field. A series of experimental manipulations were carried out to assess how visually mediated female sex stimuli affect male sexual arousal in baboons.

THE EXPERIMENTS

Evidence of arousal as a result of female contact

The initial experiment addressed the question of sexual-arousal increase as assessed by the dependent variable as a result of exposure to estrogen-treated females. For this experiment a group of eight adult males constituted the subject pool. These males were trained to enter a small wheeled transport cage, in which they were transferred from their home cages to the outside housing runs of the females used in the study. A group of eight females was divided into two treatment groups for purposes of experimentation. All the members of a group were treated identically in terms of experimental manipulation. For an experimental run each of the eight males was tested a total of three times in a systematic fashion with each of the four females in a particular group. This testing took 12 days. Following this period, the females received hormone injections for 12 days and then the males were retested with the females for an additional 12-day period, during which time the females continued to receive hormone treatment.' The experimental treatment consisted of 50 μg/day of estradiol benzoate in a volume of 0.2 cc arachis oil. The sexual behavior between the pair stimulated by this dosage resembles that shown by intact females undergoing normal menstrual cycles around the time of ovulation (Bielert, unpublished). The hormonal manipulation is also effective in stimulating full turgidity in the sex skin and results in plasma levels of estradiol similar to those of an intact female near ovulation (Bielert, unpublished; see Figure 14.3). On a test day each male received a single 12-minute pair test during which he had sexual access to one of the four stimulus females in a particular group. Three periods of pretreatment and treatment testing for each of the two groups were included for analysis.

Each morning the floor underneath each male's cage was examined for evidence of spotting. Table 14.1 presents a summary of the results of this experiment. Though previous workers (Phoenix and Jensen 1973) failed to document any increases in seminal-emission frequency as a consequence of concurrent sexual-behavior testing for the rhesus monkey, it is clear that the frequencies of seminal emission associated with masturbation by the male chacma showed a very marked influence of pair testing with estradiol-treated females. It would be difficult to say what specific elements of the female might be serving to arouse the male, since he had complete physical access to the female and could have responded to her behavior, appearance, feel, smell, or some combination of these qualities. It is, however, possible to draw two conclusions from this study. The first is that male sexual arousal is positively affected by estrogen-sensitive female stimuli; the second is that it is only physical contact with a female receiving estrogen that is arousing. This is supported by the little effect on the male's seminal emissions that the females had when they were not receiving treatment. An analysis carried out on how ejaculation during a sex test affected the incidence of spots showed that an ejaculation during a pair test had little to do with the masturbatory activities of the males in

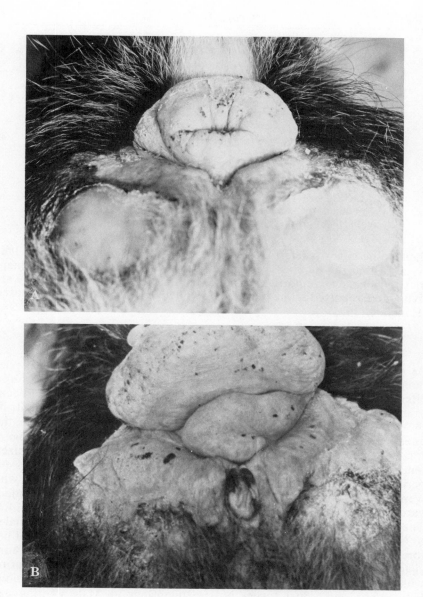

Figure 14.3. The perineal areas of an adult ovariectomized chacma baboon, female B2: (A) in an anhormonal state; (B) after 12 days of treatment with estradiol benzoate (50 μg/day).

Table 14.1. *Median occurrence values of*
seminal emissions recorded as spots for
two 12-day testing periods[a]

	Median spot occurrence	
	Pretreatment	Estradiol treatment[b]
Group I females	0	8
Group II females	1.5	8.5

[a] After Bielert, Howard-Tripp, and van der Walt 1981
[b] Friedman two-way analysis of variance by ranks revealed significance: group I: $\chi_r^2(1) = 6.125$, $p < .02$; group II: $\chi_r^2(1) = 8$, $p < .01$.

their home cages (Bielert, Howard-Tripp, and van der Walt 1981). It is possible that the number of estrogen-primed females in this experiment may have been an important variable. Increases in male masturbatory activity were reported as the number of swollen females in a troop increased (Hausfater 1975).

Female noncontact arousal influences

As previously mentioned, most experimental work on the arousal phenomenon has been done on the rhesus monkey. The short latency arousal effects that noncontact stimuli may have on male primates have not been, to my knowledge, experimentally examined. After the positive results of the initial experiment, another experiment was carried out to examine the question of arousal effects on male baboons from exposure to estrogen-treated females. Eleven adult males were exposed to two ovariectomized females who were then placed on estradiol-benzoate therapy. In this experiment the males were denied physical contact with the females, who were totally unfamiliar to them. Spotting was again used as the dependent variable. The layout of the experimental room is shown in Figure 14.4. The experiment began with the introduction of the two stimulus females to the colony room. A 2-week period for familiarization was allowed before daily morning data collection on spotting was begun. The study consisted of a 2-week pretreatment period, a 3-week treatment period during which time the females received daily estradiol-benzoate injections (50 μg/day), and a final 2-week posttreatment period. Male spotting was not collected during the first week of the treatment period to provide an initial 1-week priming period. The results are summarized in Table 14.2. Treatment of the females resulted in a dramatic increase in spotting by the males.

Though social contact with an estrogen-primed female stimulates increases in the male arousal state, as shown previously, this may be a consequence of increases in the female's proceptive or receptive behavior or her olfactory signals. These results make it clear that actual physical con-

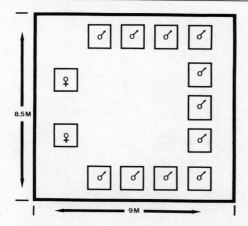

Figure 14.4. The room layout for an experiment in which male chacma baboons were exposed to adult ovariectomized females receiving estradiol-benzoate therapy (50 μg/day).

Table 14.2. *Median occurrence values of seminal emissions recorded as spots for three two-week experimental treatment periods[a]*

Emission category	Stimulus females' condition		
	Pretreatment (anhormonal)	Treatment (50 μg/day estradiol benzoate)[b]	Posttreatment decline period
Spot	0	6	1

[a] After Bielert, Howard-Tripp, and van der Walt 1981.
[b] Wilcoxon matched-pairs-rank tests revealed significant treatment effects ($N = 11$, $t = 0$, $p < .005$).

tact with an estrogen-stimulated female is not necessary for increases in male arousal to occur and suggests that visual stimuli have an important effect on the male's arousal level. Though olfactory cues may play a role in chacma (Bolwig 1959; Hall 1962; Saayman 1971; Bielert and Crewe, in preparation; Bielert, unpublished) and olive baboons (*Papio anubis*) (Michael, Zumpe, Keverne, and Bonsall 1972), these cues probably do not exert a positive effect at the distance involved in this study (3 to 5 m). Evidence for the relatively short-distance nature of olfactory cues comes from the pair-test situation in which the male has physical contact with the female. With the first presentation[1] given at close range to the male by

[1] A posture in which the hindquarters are raised and oriented toward another animal and the front legs are generally rigid or slightly flexed.

Figure 14.5. An adult female chacma baboon sexually "presenting" to her male partner. Note the female's swollen sex skin and the male's erect penis.

the female, the male usually approaches until his nose is only a few centimeters from the female's perineum.

Recently it was suggested that vocal cues, that is, copulatory vocalizations, may serve an arousal function (Hamilton and Arrowood 1978). These calls are rarely given except during coitus and, as such, would have played a minor role in the present experiment. Recent work (Arrowood and Bielert, in preparation) examining the arousal potential of copulatory vocalizations on male baboons failed to demonstrate any effects of such vocalizations on seminal-emission frequencies

I would like to suggest that female visual cues may be much more important as male sexual stimuli than has been suggested previously (e.g., Saayman 1973). It is important to stress that the stimulus females were free to interact with the males from a distance in any way they chose and consequently should be regarded as sources of complex visual stimuli. The female, through her proceptive behavior, increases the probability that the changes in her sex skin will be noticed by the male. Female baboons will sexually present to males whether bilaterally ovariectomized or intact and cyclic (see Figure 14.5). However, it remains to be determined how important the female's behavior is regarding her attractivity, though recent work on the rhesus (Goldfoot, chap. 16, this volume) has made it clear that the female's behavior is quite important.

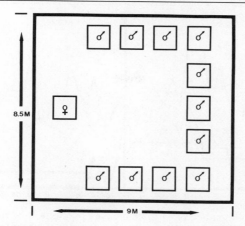

Figure 14.6. The room layout for an experiment in which male chacma baboons were exposed to an intact adult female for one complete menstrual cycle.

Arousal fluctuations coordinated with the stimulus female's menstrual cycle

Subsequent research focused on the stimulus value of the cyclic changes in the swellings of intact females. The cage arrangement was similar to that employed in the previous experiment (see Figure 14.6). A colony female (No. 44) who had shown regular cycles was selected and placed into the central cage. Daily checks were made of the female and her perineal-swelling state was rated on a three-level rating scale, a score of three being equivalent to a fully turgid state (Howard-Tripp and Bielert 1978). Eleven adult males comprised the subject group. The cage areas under each male were checked twice daily for evidence of seminal emission. The animal room floor was hosed down after each check. The morning check occurred at approximately 1000 hours and the afternoon check occurred at about 1530 hours.

The stimulus female's cycle lasted 37 days, beginning with 4 days of menstrual bleeding. The turgescent phase lasted 18 days and the deturgescent phase 19 days, both of which are typical for the chacma (Gillman and Gilbert 1946). She was fully swollen for a total of 9 days. As previously mentioned, the size fluctuations of the female baboon perineum are systematically related to ovarian events. The sex skin increases in size as peripheral estradiol levels rise, and eventually deflates, probably under the influence of falling estradiol levels and rising progestin levels (Hagino 1974). The rapid deflation is what I term sex-skin "breakdown."

Figure 14.7 shows that the pattern of male masturbation matched that of the female's perineal swelling state quite closely. Spotting showed a rise in the percentage of males responding only after the female had been rated as a "2" for several days. Spotting rapidly declined in frequency the day after the occurrence of sex-skin detumescence. As the pattern of re-

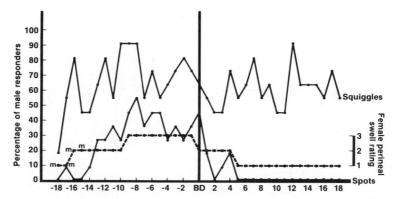

Figure 14.7. The percentage of 11 male chacma baboons exhibiting specific types of seminal emission during the 37-day menstrual cycle of intact female No. 44. The menstrual cycle is aligned to the female's sex-skin breakdown (BD), and days of menstruation are indicated by (m). The female's perineal swell rating is also presented.

sponse for squiggles showed relatively little relation to the swelling state of the female, there was no decline during detumescence.

In chacma baboons the male's masturbatory behavior is closely related to size fluctuations of the female sex skin, with the highest levels of masturbation occurring when the stimulus female's perineum is fully turgid. These observations clearly demonstrate that distance stimuli from an intact cycling female are capable of stimulating male baboon sexual arousal. Furthermore, the data illustrate that the visual attractiveness of the female fluctuates across the menstrual cycle.

Male ejaculation occurrence as related to the female partner's menstrual cycle

How do fluctuations in female attractiveness relate to the sexual interactions of a pair across the menstrual cycle? Data from eight intact females across four successive menstrual cycles are presented in Figure 14.8. The tests were 24 minutes in duration, a time span within which virtually all males will have ejaculated (Bielert, unpublished data). The data for Figure 14.8 are aligned with the day of sex-skin breakdown for the various cycles. These data clearly show that the highest frequencies of ejaculations occur from about five to three days prior to sex-skin breakdown. Laparotomies in this species and the olive baboon during this time period have revealed ovulation (Gillman and Gilbert 1946; Eddy, Turner, Kraemer, and Pauerstein 1976). Others working with olive baboons have had success from single matings during this stage (Hendricks and Kraemer 1969). It is important to remember that matings must *precede* ovulation, often by a period of from 24 to 36 hours, for maximal probability of fertilization. Consequently, peaks in the mating curve must precede

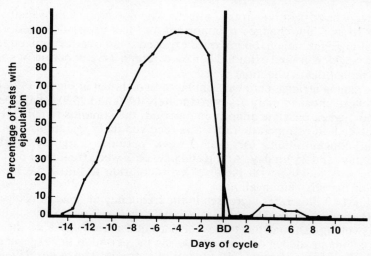

Figure 14.8. The percentage as a 2-day running mean of pair tests with a male ejaculation for eight different male–female pairs of chacma baboons. The menstrual cycles ($N = 32$) are aligned to the females' sex-skin breakdown (BD).

ovulation as revealed by such techniques as laparotomy and laparoscopy. In the laboratory situation the percentge of tests with ejaculation starts to decline from a peak about 3 to 2 days prior to sex-skin breakdown and is at a very low level on the day of sex-skin breakdown, or detumescence. Rowell (1967), working with captive *P. anubis,* reported that males reacted to fully swollen females on the last day of the turgescent phase of the menstrual cycle very much as they reacted to deflating females in the luteal phase. The timing of the decline in the frequency of spotting differed from that of ejaculation. Spotting continued at a fairly high level on the day of breakdown and showed a decline only on the day after detumescence.

In a pair-testing situation in which the male has physical contact with the female, his sexual interest in her begins to decline before her sex skin detumesces. However, when males are denied physical access to naturally cycling females, their sexual arousal declines only with detumescence, suggesting that visual cues can maintain male sexual interest in a situation in which inputs from other sensory channels have been interfered with or eliminated. It is likely that the differences in decline time are related to a time lag between the decline in male arousal level and the decline in spot frequencies.

Threshold limits for male sexual arousal

It is possible that the visual changes accompanying female detumescence are sufficient to negate the female's arousal potential. The results also suggest that the swelling increases occurring postmenstrually at the start of a cycle have positive arousal effects. The question of the sensitivity of males to swelling changes is important and has been addressed experimentally. Eight male baboons were exposed to an ovariectomized female who was given graded estradiol benzoate after a 2-week period of baseline anhormonal data collection.

The caging arrangement was similar to that shown in Figure 14.6. Emissions were checked daily at approximately 1000 and 1530 hours. Following the 2-week baseline anhormonal period, the stimulus female (JP1456) was placed on estrogen therapy. She received daily injections of the following concentrations for four 2-week periods: 1 μg/day, 5 μg/day, 10 μg/day, and 25 μg/day. At the end of each 2-week treatment period the female was sedated with Ketalar[2] hydrochloride to allow the measurement of her sex-skin swelling.

Table 14.3 illustrates increases in the frequency of emissions (spots) as a consequence of visual access to a female showing even very slight perineal swelling. During the two-week baseline period, 50% of the males showed one incidence of spotting. During the period of the experiment, in which the female received 10 μg/day of estradiol benzoate, 75% of the males showed spotting, with a median frequency of 4.5 days per male, although the female's sex-skin swelling increase was only 13% over baseline. This experiment illustrates also that there are swelling increases to which the males will not respond. The results agree with those obtained when an intact female was used as a stimulus. It appears that there is some minimum level of sex-skin swelling that must be attained before a female becomes arousing to males. It is not sex-skin swelling per se that is arousing but rather a certain degree of sex-skin swelling. It is difficult to say if this minimum is related to the overall size of the individual female or not, but it seems reasonable that this would be the case. It should be pointed out that it is possible that the male's response failure in the case of minimum swelling was related to an absence of proceptive behavior by the female. Work on the behavioral effects of graded estradiol treatment is currently planned and should allow the testing of this possibility.

The critical nature of the visual channel in the arousal phenomenon

The effectiveness of the female as a visual stimulus for male sexual arousal is clear-cut, but the contribution of other factors, such as odors or vocalizations, has been difficult to assess. Even though in these experiments the males were exposed to females at a distance, it is possible that they might have been stimulated by odor cues or the vocalization of the females. To demonstrate that visual cues are the primary component in the arousal phenomenon, it was necessary to carry out an experiment in which the visual cues of estrogen-treated females were eliminated.

The cage arrangement is shown in Figure 14.9, which differs from the arrangement in Figure 14.6 only in the addition of partitions, which were placed to eliminate visual contact between the stimulus females and the

---

[2] Ketalar, Parke–Davis Laboratories, Ltd., 4 mg/kg. i.m.

Table 14.3. *Adult male chacma baboon seminal-emission occurrences recorded as spots per 2-week observation period in which the stimulus female received graded estradiol-benzoate therapy*

Stimulus female's condition	Seminal-emission occurrences									Female perineal measurement (cm)	Sex-skin size[a]
	Male A	Male B	Male C	Male DOM	Male E	Male F	Male G	Male H	Median		
Anhormonal (no treatment)	0	1	0	0	1	0	1	1	.5	15	0.0
Estradiol Rx											
1 μg/day	0	0	0	1	0	0	0	1	0	15	0.0
5 μg/day	0	1	0	0	0	0	0	1	0	16	7.0
10 μg/day[b]	0	5	7	4	4	0	8	7	4.5	17	13.0
25 μg/day[b]	1	10	11	11	2	0	6	10	8	22	47.0

[a] Increase divided by baseline as percentage.
[b] Wilcoxon matched-pairs–rank tests revealed a significant increase from the 0-, 1-, 5-μg response levels.

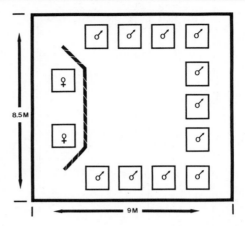

Figure 14.9. The room layout for an experiment in which male chacma baboons were denied visual contact with adult ovariectomized females receiving estradiol-benzoate therapy (50 μg/day).

subject males. The partitions were only slightly higher than the cages in which the females were housed, and they did not fully extend to the floor. Female vocalizations could still be heard by the males, and the opportunity for odor cues to reach the males was still present. The males were exposed to two estrogen-treated females (VP-4, JP-1456). After the placement of the partitions, a 2-week period for familiarization was allowed before daily data collection on male seminal emissions was begun. The experiment consisted of a 2-week pretreatment period, a 3-week treatment period during which the females received daily estradiol-benzoate injections (50 μg/day), and a final 2-week post-treatment period. The two-week estradiol-benzoate period followed a 1-week priming period. The results are presented in Table 14.4. A Friedman analysis of variance (Siegel 1956) failed to substantiate any real changes in the occurrence of spotting across the treatment periods.

The results of this study make it clear that changes in female attractiveness that stimulated the males' emissions were visually, not olfactorily or vocally, mediated. This by no means rules out the significance of cues transmitted during physical contact. It does suggest, however, that visual communication is very important in male baboon sexual arousal. Again, it is important to stress that the female's perineal changes were accompanied by behavioral changes as well. These changes may well have been very important in the effectiveness of the perineum as an arousal stimulus. This interpretation receives support from attempts to determine the effectiveness of various forms of stimuli in producing arousal in human males. The data consistently indicate that erotic motion pictures are more arousing than photographic prints (Kolarsky and Madlafousek 1973; Freund, Langevin, and Zajac 1974; McConaghy 1974), photographic slides (Sanford 1974; Abel, Barlow, Blanchard, and Mavissakalian 1975), and audio tapes (Abel et al. 1975).

Table 14.4. *Adult male chacma baboon seminal-emission occurrences recorded as spots per 2-week observation period in which visual contact with the stimulus females was eliminated*

Stimulus females' condition[a]	Seminal-emission occurrences										Median	% Responders
	Male 2	Male 5	Male 8	Male 11	Male 17	Male 18	Male 30	Male 31	Male 653	Male 661		
Pretreatment (anhormonal)	0	0	0	3	0	0	0	0	0	2	0	20
Treatment (50 µg/day estradiol benzoate)	0	2	0	2	1	0	0	0	0	2	0	40
Post-treatment decline period	0	0	0	0	0	0	0	0	0	0	0	0

[a] Friedman analysis-of-variance tests showed no significant treatment effects ($\chi_r^2(2) = 1.5$, $.5 > p > .3$).

DISCUSSION

In considering the actions of an individual that may be relevant to its mating behavior, it should be remembered that gonadal hormones can:

1. Directly affect the central nervous system of an individual and increase the probability of the exhibition of specific behavior patterns in appropriate settings;
2. Directly affect the sensory systems of an individual and thus allow the perception of relevant stimuli that result in an increase in the probability of the exhibition of specific patterns in appropriate settings;
3. Directly affect the anatomy of an individual in such a way that other individuals initiate behaviors that result in the exhibition of specific behavior patterns in appropriate settings.

Changes in female attractiveness are produced by hormonal actions on the central nervous system as well as on anatomic structures (Keverne, chap. 15, and Goldfoot, chap. 16, this volume). The results discussed in this chapter allow a consideration of those changes in female sex stimuli that might be important in determining the male's sexual responsiveness. It is equally appropriate to consider the fashion in which various communication channels may be coordinated. This approach, first used long ago (Beach 1942a,b), allows an understanding of how the sexual interactions of a species are affected by peripheral hormones.

The data clearly demonstrate that male baboons are aroused sexually by females whose sex skin is only partially swollen. These laboratory observations agree well with field observations that consortships begin in chacmas well in advance of the time of ovulation (Hall 1962; Saayman 1970; Seyfarth 1978). In a natural troop situation males have more than visual access to females. As a result of freedom of interaction, the actual timing of the consortship may be related to the male's perception of olfactory, not visual, changes. The rapid acceptance of estrogen-implanted ovariectomized females into an unfamilar baboon troop is of some relevance here (Saayman 1968, 1970). The initial reactions of the troop males may well have been a result of visually mediated cues. Eventually the males could receive and react to olfactory cues, but their initial approach to, and acceptance of, the strange females may have resulted from their interest in the perineal swellings. The partition experiment demonstrates that the vocal and odor cues of an estrogen-treated female in the absence of actual visual interaction are not sufficient to arouse the male baboon to a point of regular masturbation.

That the stimulus value of the naturally cycling female is probably not based on visual cues alone, however, is made evident by the persistence of male arousal until after sex-skin detumescence. I suggest that the sex skin, in the process of swelling and when fully turgid, serves the important arousal function of increasing the female baboon's attractiveness. The sex skin, however, does not appear to be sufficient in itself to stimulate copulation. It is possible that the males are responding to subtle cues such as color changes, but if that is the case, it is hard to explain why in a pair-test situation they respond to these cues, whereas in a noncontact situation they do not. For the baboon, as for the rhesus (Goldfoot, chap. 16, this volume), an integrated multichannel sexual communication system may be operative. Visual cues probably serve an important attractive function, which is either augmented or negated by olfactory signals. Future experimentation should examine the importance of vaginal odor cues in the chacma and the specific components of the visual stimulus array, such as color, hue, and relative size. Even more important, however, are examinations of those behavioral components of the estrogen-treated female that may be critical to male arousal.

Though those components that determine male masturbation may not be those that determine copulation, the masturbation experiments allow one to generate hypotheses of relevance to the mating behavior in this species. If we seek to understand the sexuality of nonhuman primates with the ultimate goal of understanding our own sexuality, many more experiments aimed at the communicative aspects of reproduction will be necessary.

Wilson (1975) has suggested that different communication channels have different qualities and consequent advantages and disadvantages. The visual channel allows a rapid, fairly long distance transmission. The tonic transmission nature of the sex-skin swelling allows an active or passive selection by the female of a suitable partner from those of either her own or nearby troops as a result of the stimulation of intermale competition. In addition, active communicative events, such as proceptive behaviors, given by a female are probably much more likely to result in male sexual behavior and, possibly, fertilization. The subtle olfactory changes that may occur after the female has formed a consortship might serve to maximize probability of further male arousal and copulation, helping to ensure fertilization. Thus, changes in female attractiveness when coordinated to appropriate changes in proceptivity and receptivity are important aspects of those communicative events between the sexes that are necessary for insemination and fertilization.

The stimuli that determine the relative attractiveness of a female probably vary across species. The baboon, however, may be typical of those species in which perineal elaboration reflects cycle stage. Work on the interaction of the visual communication channel with others that may be involved in mating behavior may help to understand the human situation as much as does the recent emphasis on concealed, or disguised, ovulation (Alexander and Noonan 1979; Benshoof and Thornhill 1979; Burley 1979). The baboon is an appropriate experimental model because of the relatively defined nature of its sexual interactions. The virtual shutdown of reproductively meaningful male behavior (i.e., ejaculation) during the luteal phase of the female's cycle allows the exact determination in the laboratory of those factors that may be critical stimuli to the male. As a consequence, a complete picture of the interactional nature of the sexuality of this species is possible.

REFERENCES

Abel, G. C., Barlow, D. H., Blanchard, E. B., and Mavissakalian, M. 1975. Measurement of sexual arousal in male homosexuals: effects of instructions and stimulus modality. *Archives of Sexual Behavior* 4:623–7.

Alexander, R. D. 1979. Sexuality and sociality in humans and other primates. In *Human sexuality: A comparative and developmental perspective*, ed. H. A. Katchadourian, pp. 81–97. Berkeley: University of California Press.

Alexander, R. D., and Noonan, K. M. 1979. Concealment of ovulation, parental care and human social evolution. In *Evolutionary biology and human social behavior*, ed. N. A. Chagnon and W. Irons, pp. 436–53. North Scituate, Mass.: Duxbury Press.

Arrowood, P. C., and Bielert, C. In preparation. Copulatory vocalizations in primates with specific reference to the chacma baboon (*Papio ursinus*).

Beach, F. A. 1942a. Analysis of the stimuli adequate to elicit mating behavior in the sexually inexperienced male rat. *Journal of Comparative Psychology* 33:163–207.

1942b. Analysis of the factors involved in the arousal, maintenance and manifestation of sexual excitement in male animals. *Psychosomatic Medicine* 4:173–98.

Benshoof, L., and Thornhill, R. 1979. The evolution of monogamy and concealed ovulation in humans. *Journal of Social and Biological Structures* 2:95–106.

Bentley, J. P., Nakagawa, H., and Davies, G. H. 1971. The biosynthesis of mucopolysaccharides and collagen by the sex skin of estrogen-treated animals. *Biochemica and Biophysica Acta* 244:35–46.

Bercovitch, F. B. 1978. Mate selection and sexual selection in nonhuman primates. *Journal of California Anthropology* 8:9–12.

Bielert, C., and Crewe, R. In preparation. Male ejaculation and menstrual-cycle effects on short-chain aliphatic acids in vaginal lavages of the chacma baboon (*Papio ursinus*).

Bielert, C., Howard-Tripp, M. E., and van der Walt, L. A. 1981. Environmental and social factors influencing seminal emission in chacma baboons (*Papio ursinus*). *Psychoneuroendocrinology* 5:287–303.

Bielert, C., and van der Walt, L. A. In press. Male chacma baboon (*Papio ursinus*) sexual arousal: Mediation by visual cues from female conspecifics. *Psychoneuroendocrinology*.

Bolwig, N. 1959. A study of the behaviour of the chacma baboon, *Papio ursinus*. *Behaviour* 14:136–63.

Burley, N. 1979. The evolution of concealed ovulation. *American Naturalist* 114:835–58.

Carlisle, K. S., and Montagna, W. 1979. Aging model for unexposed human dermis. *Journal of Investigative Dermatology* 73:54–8.

Carpenter, C. R. 1942. Sexual behavior of free-ranging rhesus monkeys (*Macaca mulatta*), II. Periodicity of estrus, homosexuality and non-conformist behavior. *Journal of Comparative Psychology* 33:143–62.

Clutton-Brock, T., and Harvey, P. H. 1976. Evolutionary rules and primate societies. In *Growing points in ethology*, ed. P. P. G. Bateson and R. A. Hinde, pp. 195–237. Cambridge University Press.

Collins, T. 1978. Why do some baboons have red bottoms? *New Scientist* 78:12–14.

Czaja, J. A., Eisele, S. G., and Goy, R. W. 1975. Cyclical changes in the sexual skin of female rhesus: relationships to mating behavior and successful artificial insemination. *Proceedings, Federation of American Societies for Experimental Biology* 34:1680–4.

Czaja, J. A., Robinson, J. A., Eisele, S. G., Scheffler, G., and Goy, R. W. 1977. Relationship between sexual skin color of female rhesus monkeys and mid-cycle plasma levels of oestradiol and progesterone. *Journal of Reproduction and Fertility* (U.K.) 49:147–50.

Darwin, C. R. 1871. *The descent of man and selection in relation to sex.* London: Murray; reprinted, New York: Modern Library, 1936.

Dixson, A. F. 1977. Observations on the displays, menstrual cycles and sexual behaviour of the "black ape" of the Celebes (*Macaca nigra*). *Journal of Zoology* 182:63–84.

Dixson, A. F., Everitt, B. J., Herbert, J., Rugman, S. M., and Scruton, D. M. 1973. Hormonal and other determinants of sexual attractiveness and receptivity in rhesus and talapoin monkeys. *Proceedings, IVth international congress of primatology,* Vol. 2, *Primate reproductive behavior,* ed. C. H. Phoenix, pp. 36–63. Basel: Karger.

Eaton, G. G. 1973. Social and endocrine determinants of sexual behavior in simian and prosimian primates. *Proceedings, IVth international congress of primatology,* vol. 2, *Primate reproductive behavior,* ed. C. H. Phoenix, pp. 20–35. Basel: Karger.

Eddy, C. A., Turner, T. T., Kraemer, D. C., and Pauerstein, C. J. 1976. Pattern and duration of ovum transport in the baboon (*Papio anubis*). *Obstetrics and Gynecology* 47:658–64.

Freund, K., Langevin, R., and Zajac, Y. 1974. A note on erotic arousal value of moving and stationary human forms. *Behaviour Research and Therapy* 12:117–19.

Gillman, J., and Gilbert, C. 1946. The reproductive cycles of the chacma baboon with special reference to the problems of menstrual irregularities as assessed by the behaviour of the sex skin. *South African Journal of Medical Science, Biology Supplement* 11:1–54.

Gordon, T. P., and Bernstein, I. S. 1973. Seasonal variation in sexual behavior in all-male rhesus troops. *American Journal of Physical Anthropology* 38:221–7.

Hagino, N. 1974. Follicular maturation, ovulation, luteinization and menstruation in the baboon. In *Physiology and genetics of reproduction,* pt. A, ed. E. M. Coutinho and F. Fuchs, pp. 323–42. New York: Plenum.

Hall, K. R. L. 1962. The sexual, agonistic and derived social behaviour patterns of the wild chacma baboon, *Papio ursinus. Proceedings of the Zoological Society of London* 139:283–327.

Hamilton, W. J., III, and Arrowood, P. C. 1978. Copulatory vocalizations of chacma baboons (*Papio ursinus*), gibbons (*Hylobates hoolock*), and humans. *Science* 200:1405–9.

Hanby, J. P., Robertson, L. T., and Phoenix, C. H. 1971. The sexual behavior of a confined troop of Japanese macaques. *Folia Primatologica* 16:123–43.

Hausfater, G. 1975. *Dominance and reproduction in baboons* (*Papio cynocephalus*): *a quantitative analysis.* Basel: Karger.

Hendricks, A. G. 1971. *Embryology of the baboon.* Chicago: University of Chicago Press.

Hendricks, A. G., and Kraemer, D. C. 1969. Observations of the menstrual cycle, optimal mating time, and pre-implantation embryos of the baboon. *Journal of Reproduction and Fertility* (*U.K.*), *Supplement* 6:119–28.

Herbert, J. 1966. The effects of oestrogen applied directly to the genitalia upon the sexual attractiveness of the female rhesus monkey. *Excerpta Medica International Congress,* ser. 3:212.

Hisaw, F. L., and Hisaw, F. L., Jr. 1961. Action of estrogen and progesterone on the reproductive tract of lower primates. In *Sex and internal secretions,* ed. W. C. Young, pp. 556–89. Baltimore: Williams and Wilkins.

Howard-Tripp, M. E., and Bielert, C. 1978. Social contact influences on the menstrual cycle of the female chacma baboon (*Papio ursinus*). *Journal of the South African Veterinary Association* 49:191–2.

Kolarsky, A., and Madlafousek, J. 1973. Female behavior and sexual arousal in heterosexual male deviant offenders. *Journal of Nervous and Mental Disease* 155:110–18.

Kummer, H. 1971. *Primate societies: Group techniques of ecological adaptation.* Chicago: Aldine.

Lancaster, J. B. 1979. Sex and gender in evolutionary perspective. In *Human sexuality: A comparative and developmental perspective,* ed. H. A. Katchadourian, pp. 51–80. Berkeley: University of California Press.

Maple, T. 1977. Unusual sexual behavior of nonhuman primates. In *Handbook of sexology,* ed. J. Money and H. Musaph, pp. 1167–86. Amsterdam: Excerpta Medica.

McConaghy, N. 1974. Penile volume responses to moving and still pictures of male and female nudes. *Archives of Sexual Behavior* 3:565–70.

Michael, R. P., Zumpe, D., Keverne, E. B., and Bonsall, R. W. 1972. Neuroendocrine factors in control of primate behaviors. *Recent Progress in Hormone Research* 28:665–706.

Napier, J. R., and Napier, P. H. 1967. *A handbook of living primates.* London: Academic Press.

Phoenix, C. H., and Jensen, J. J. 1973. Ejaculation by male rhesus in the absence of female partners. *Hormones and Behavior* 4:231–8.

Rienits, K. G. 1960. The acid mucopolysaccharides of the sexual skin of apes and monkeys. *Biochemical Journal* 74:27–38.

Rose, R. M., Gordon, T. P., and Bernstein, I. S. 1972. Plasma testosterone levels in the male rhesus: Influences of sexual and social stimuli. *Science* 178:643–5.

Rowell, T. E. 1967. Female reproductive cycles and the behavior of baboons and rhesus macaques. In *Social communication among primates,* ed. S. A. Altmann, pp. 15–32. Chicago: University of Chicago Press.

1972. Female reproduction cycles and social behavior in primates. *Advances in the Study of Behavior* Vol. 4, ed. D. S. Lehrman, R. A. Hinde, and E. Shaw, pp. 69–105. New York: Academic Press.

Saayman, G. S. 1968. Oestrogen, behaviour and permeability of a troop of chacma baboons. *Nature* 220:1339–40.

1970. The menstrual cycle and sexual behaviour in a troop of free-ranging chacma baboons (*Papio ursinus*). *Folia Primatologica* 12:81–110.

1971. Behaviour of the adult males in a troop of free-ranging chacma baboons (*Papio ursinus*). *Folia Primatologica* 15:36–57.

1973. Effects of ovarian hormones on the sexual skin and behavior of ovariectomized baboons (*Papio ursinus*) under free-ranging conditions. *Proceedings, IVth international congress of primatology* Vol. 2, *Primate reproductive behavior,* ed. C. H. Phoenix, pp. 64–98. Basel: Karger.

Sanford, D. A. 1974. Patterns of sexual arousal in heterosexual males. *Journal of Sex Research* 10:150–5.

Seyfarth, R. M. 1978. Social relationships among adult male and female baboons: I. Behavior during sexual consortship. *Behaviour* 64:204–26.

Short, R. V. 1979. Sexual selection and its component parts, somatic and genital selection, as illustrated by man and the great apes. *Advances in the Study of Behavior* 9:131–58.

Siegel, S. 1956. *Non-parametric statistics for the behavioral sciences.* London: McGraw-Hill.

Slimp, J. C., Hart, B. L., and Goy, R. W. 1978. Heterosexual, autosexual and social behavior of adult male rhesus monkeys with medial preoptic-anterior hypothalamic lesions. *Brain Research* 142:105–22.

Vandenbergh, J. G. 1969. Endocrine coordination in monkeys: male sexual responses to the female. *Physiology and Behavior* 4:261–4.

van Horn, R. N., and Eaton, G. G. 1979. Reproductive physiology and behavior in prosimians. In *The study of prosimian behavior,* ed. G. A. Doyle and R. D. Martin, pp. 79–122. New York: Academic Press.

van Lawick-Goodall, J. 1968. The behaviour of free-living chimpanzees in the Gombe Stream Reserve. *Animal Behaviour Monographs* 1:161–311.

Wickler, W. 1967. Socio-sexual signals and their intraspecific initiation among primates. In *Primate ethology,* ed. D. Morris, pp. 69–147. London: Weidenfeld & Nicolson.

Wilson, E. O. 1975. *Sociobiology.* Cambridge, Mass.: Harvard University, Belknap Press.

Young, W. C., and Orbison, W. D. 1944. Change in selected features of behavior in pairs of oppositely sexed chimpanzees during the sexual cycle and after ovariectomy. *Journal of Comparative and Physiological Psychology* 37:107–43.

15 · Olfaction and the reproductive behavior of nonhuman primates

E. B. KEVERNE

The processing of sensory information most likely takes place at two distinct levels within the central nervous system. First relayed to various specific nuclei of the thalamus, sensory information is then forwarded, almost certainly, to the neocortex, where sensory processing, which we often surmise and occasionally validate, allows an animal a cognitive awareness of its sensory environment. Sensory input can also produce physiological and behavioral changes of which the animal has no primary cognitive awareness but for which the animal may become secondarily aware. The neural pathways subserving these noncognitive aspects of sensory communication project primarily to the brainstem and diencephalon and produce their effects in even decorticate animals.

Thus, in rodents, tactile cues applied to the perineal region and vagina produce a diencephalic reflex of lordosis, followed by neuroendocrine changes inducing pseudopregnancy (Hansen, Stanfield, and Everitt 1980). These changes occur in estrous females, usually in response to nonfertile mating but also by tactile stimulation with a glass rod, with a high degree of predictability.

Light and the changing of light regimes have been shown to alter patterns of feeding, motor activity, and diurnal rhythms of hormonal secretion in a number of mammalian species (Moore and Zichler 1972; Stephan and Zucker 1972). This is a function of the retinohypothalamic projection to the suprachiasmatic nucleus (Morin, Fitzgerald, Rusak, and Zucker 1977), and in the context of reproduction it is an important pathway whereby changes in daylength affect the onset of the breeding season (Stetson and Watson-Whitmyre 1976).

Finally, olfactory stimuli in mice have been shown to suppress and induce estrus and to block pregnancy, a function subserved by the vomeronasal organ and accessory olfactory pathway projecting directly to the limbic brain, first to the amygdala and then to the hypothalamus (Reynolds and Keverne 1979; Bellringer, Pratt, and Keverne 1980).

Such sensory effects as these, mainly described for rodents, all occur

My thanks to Susan Dilley for assaying the plasma hormones and to Susan Currie for typing the manuscript. The work described involving talapoin monkeys was supported by the Medical Research Council.

with a high degree of predictability in response to specific stimuli and appear to be beyond the animal's volitional control. These sensory effects also have specific neural pathways not directly related to the neocortex. In higher primates the general trend has been toward neocortical dominance of sensory systems, and this may be viewed as part of the evolutionary mechanism emancipating their behavior from stimulus–response activation. With this in mind, I would like to consider the role of olfactory communication in primates, especially in the context of reproductive behavior. Since most primates do not possess the subcortical olfactory projection (vomeronasal, accessory olfactory system), one might predict that their behavioral response to odor cues will be complex, not a simple stimulus–response activity.

In a majority of primate species, no experimental work has investigated the significance of olfactory cues in behavior. However, many field and laboratory observations of prosimians and New and Old World primates suggest that odor cues signal the reproductive state of the female and simultaneously arouse male sexual activity. Scent-marking behavior, the sniffing and mouthing of female genitalia by males, and even the tasting of female urine and vaginal discharge is a prelude to sexual behavior in many species. In the sexual communication of marmosets, where olfactory signals are of importance in most species (see Epple, Alveario, and Katz, chap. 12, this volume), scent marking regularly precedes and follows copulation and is accompanied by intensive sniffing and the licking of the partner's marks and genitals (Epple 1967). Field studies have revealed that the smelling of the females' anogenital region by males is a common, almost routine, occurrence in the vervet, *Cercopithecus aethiops* (Booth 1962), the howler monkey, *Alouatta palliata* (Altmann 1959), the dusky titi, *Callicebus moloch* (Moynihan 1966), the night monkey, *Aotus trivirigatus* (Moynihan 1967), and the patas monkey, *Erythrocebus patas* (Hall, Boelkins, and Goswell 1965). In the talapoin monkey, *Miopithecus talapoin,* males look at or smell the females' perineal area more frequently and for longer periods around the mid-cycle than during the early follicular or late luteal phases of the menstrual cycle (Scruton and Herbert 1970). It was suggested that this greater attention to the females' external genitalia by the males could be related to coincidental changes in the females' vaginal odor. Carpenter (1942) first observed that the vaginal overflow of the rhesus monkey, *Macaca mulatta,* possesses a characteristic odor that might provide an additional sexual stimulus. This may be related to the persistent close following of females by males observed in field studies (Lindberg 1967), from which males might receive some sort of cue as to the state of female receptivity. Jay (1965) noticed a strong-smelling vaginal discharge in the toque macaque, *Macaca sinica,* and observed males to examine the genitalia of most females in the group each day. Female bonnet macaques, *Macaca radiata,* are rarely seen to present for copulation unless solicited by the male, which involves flipping the tail aside for olfactory examination of the genitalia and on occasions the insertion of a finger into the vagina and the subsequent smelling and tasting of secretions (Rahaman and Parthasarathy 1969).

In the stump-tailed macaque, *Macaca arctoides,* the sniffing, fingering, and licking of the perineal area occurs prior to copulation (Blurton-Jones and Trollope 1968). Although olfactory cues from the urine of receptive females are not ruled out, it would appear from these observations that in the macaques, communication of sexual state is by way of vaginal secretions. In the chimpanzee *Pan troglodytes,* van Lawick-Goodall (1968) drew attention to the increased frequency of male olfactory inspection of the female vagina at the approach of estrus. This sometimes simply involved the sniffing of the vaginal area, but males also have been observed to insert a finger into the vagina and then sniff the finger. Occasionally the males use their hands to part the lips of the vulva before they look and sniff. That this behavior is related to the estrous cycle was suggested by the increase in genital inspections at the first signs of sexual swelling. Similar genital inspection involving sniffing has been described for chimpanzees of the Mahali mountains (Nishida 1970). Captive lowland gorillas in the Basel Zoo have been reported to have a distinctive odor during their periods of estrus (Hess 1973). Males have been observed to touch the female genitalia and sniff their hands or fingers, particularly during the proestrous period, and oral–genital contact is common during estrus as a prelude to copulation in both gorillas and orangutans. This wealth of behavioral data clearly indicates that olfactory cues play a significant role in the reproductive life of many primate species regardless of their habitat or social organization.

COMMUNICATION AND SEXUAL BEHAVIOR IN MACAQUES

A striking feature of the sexual interaction of heterosexual pairs of macaque monkeys in a laboratory situation is the cyclicity in male sexual behavior with respect to the females' menstrual cycles (Ball and Hartman 1935; Michael, Herbert, and Welegalla 1967; Bullock, Paris, and Goy 1972; Goy 1979), an observation also reported for free-living monkeys (Southwick, Beg, and Siddiqi 1965). Hence, the endocrine changes occurring in the female at this time are communicated to the male partner and influence his sexual behavior. Of course, a number of sensory channels are available to the male, not the least of which are the effects that changes in endocrine state might have on a female's motivation to interact sexually with male partners. By and large though, female monkeys are prepared to receive the male at any time during their menstrual cycles and rarely refuse his sexual advances (Keverne 1976; Baum, Everitt, Herbert, and Keverne 1977). Females can negate the male's initiative and enhance sexual interactions by showing changes in soliciting behavior which are also influenced by endocrine changes. However, macaque females do not show obvious increases in invitational postures around their ovulatory period when male sexual behavior is highest (Czaja and Bielert 1975). In fact, paradoxically, they may show decreases (Michael and Welegalla 1968; Eaton and Resko 1974). Such is the interest of the male at this time that females have little opportunity to make sexual invitations; the initiative is firmly with their partner. Moreover, in experimental situations, the sexual invitations of females can be markedly increased by the administration of testosterone (Trimble and Herbert 1968); in the absence of female estrogens, however, such invitations go unheeded by the male (Michael and Keverne 1972). Observations such as these lead to the inevitable conclusion that nonbehavioral cues are of some importance in communicating female attractiveness to the male, at least within the close proximity of a laboratory cage.

Of course, this conclusion does not exclude behavioral changes as a means of communication in rhesus monkeys. A whole range of invitational gestures can enhance a female's attractiveness, just as refusals can make a female less attractive. There are no absolute values to attractiveness; it is a concept defined in terms of the behavioral interaction, and a given female under a fixed endocrine treatment can be attractive to male A but unattractive to male B (Keverne 1976). Nevertheless, in general terms, ovariectomized female macaques are sexually uninteresting to males (Michael, Herbert, and Welegalla 1967); however, estradiol-replacement therapy restores male sexual interest (Goldfoot 1971; Eaton and Resko 1974; Johnson and Phoenix 1976) and, by definition, female attractiveness.

In his recent reevaluation of behavioral communication in rhesus monkey dyads (chap. 16, this volume), Goldfoot paid specific attention to the behavioral sequences that follow a solicitation from the female and lead to a mount. Although informative, such a detailed approach in no way invalidates the findings reported earlier from British laboratories, which took into account, with certain differences of interpretation, the sequences outlined by Goldfoot (see Herbert 1974). Thus Goldfoot states that "when males initiate sexual sequences in the absence of prior solicitations from the female, only a third of such approaches progress to an attempted contact." If no contact is made, how, might I ask, does one know that males are initiating a sexual sequence? They could be approaching the female to groom her, to be groomed, to aggress her, to sniff her, or without any motivation whatsoever. Our studies have scored such behavior as an approach and do not include it as a definitive sexual behavior.

The second point that Goldfoot makes is that his female rhesus monkeys were negative to 37% of male contacts and failed to present to an additional 24% of the contacts. This behavior has never been ignored by us; rather, we have scored it as female refusals of male mounting attempts (male success ratio). A major difference, however, is noted in the number of female macaque refusals: in Wisconsin, 61% of male contacts are refused, whereas laboratories in London, Birmingham, and Cambridge consistently report low levels of female refusals (less than 10% of contacts) even in ovariectomized and adrenalectomized females (Baum et al. 1977).

Another major difference between laboratories is seen in the large percentage (33%) of female-initiated sequences that are terminated by the female. We would score this as a female solicit followed by a refusal of the male's mounting attempt. However, no special category was ever designated for this behavioral sequence because of its rare occurrence. Such behavior characterized unfamiliar pairs where the female was fearful of potential male aggression. The ambivalence of the presentation posture in this situation makes sexual identification of the gesture very questionable.

Taken together, the very high number of refusals of male contact (61%) and the extremely high incidence of ambivalent presentations (33%) would suggest a great deal of conflict and stress among the dyads tested in Goldfoot's study. Although I agree with the conclusion that an experienced male can learn to expect that sexual success will follow female solicitation, I am less convinced when such a high percentage of her solicitations are ambivalent and ultimately result in refusal when contact is made. The information that the female's success ratio increases when she receives estrogen must be available to the male at the time the sexual invitation occurs, and it is precisely in this context that nonbehavioral cues are important.

Of the nonbehavioral cues available to the male rhesus macaque, vocalizations are rare and inconsistent in promoting sexual behavior and are certainly not governed by endocrine state.[1] Although tactile cues may be important for the mounting and thrusting performance, they clearly do not initiate male interest, as tactile feedback occurs only after the mounting sequence has started. As for visual cues, rhesus monkeys do not have a sex-skin swelling; rather, they have a red sex-skin coloration, the intensity of which changes with the menstrual cycle and is estrogen dependent (Hisaw and Hisaw 1966). However, topical application of estrogen cream to the sex-skin area promotes an intense red coloration with little or no effect on male sexual interest and behavior (Herbert 1966). Of course, this does not mean that such color changes have no communicatory significance; they may well act as signals over distance. Within the laboratory, however, color changes in themselves are not sufficient to stimulate male mounting behavior.

OLFACTION AND SEXUAL BEHAVIOR IN THE RHESUS MONKEY

With this kind of information in mind, it seemed appropriate to examine the role of olfactory cues in the initiation of male sexual interest and behavior with female partners. The questions asked were: Could female monkeys be made attractive to males by hormone treatments that did not influence either the sex-skin color or proceptive behavior (sexual solicitations), and was this attractiveness communicated olfactorily? Since everyone is agreed that female attractiveness increases in the follicular phase of the cycle when estrogens predominate and that giving estrogen to ovariectomized females restores male sexual interest (Herbert 1970; Phoenix 1973), this was the hormonal manipulation selected. However, to avoid any actions of the hormone in the brain and on soliciting behavior, a very low concentration was administered directly into the female's va-

[1] *Ed. note.* But see Green 1981 for data on sex-related vocalizations in other macaque species.

gina. In this way, that part of the female's anatomy most interesting to the male sexually could be estrogenized without any effect on either sex-skin coloration or sexual solicitations. Moreover, if a mesh barrier was imposed between the male and female, a clearer understanding of the male's interest in the female could be achieved by asking the male to perform a work task in order to gain access to the female. This was achieved by providing the male with a lever that he had to press some 250 times to open a door providing access to the female (Michael and Keverne 1968). At the time it seemed logical to assume that a male working with such dedication for access to a female was a male with some interest in that female. Five males were trained on this schedule and each male had three ovariectomized partners, one of which received injections of estrogen throughout the study to make her attractive. This third female served as the control, and all males pressed the lever to gain access to this female partner; however, four showed little interest and rarely pressed for access to their ovariectomized partners who were not receiving hormone replacement (Figure 15.1). Nor did they show any sexual interest in these females when freely placed together in an open cage at a later time in the day. The male who pressed for access to and mounted all his female partners was dropped from the study; there seemed little point asking this male if estrogen made his female partners attractive when he already found them sufficiently attractive to initiate both his pressing for access and sexual behavior. However, of the remaining eight pairs, the males of which showed no sexual interest in their ovariectomized, estrogenless partners, this idea could be put to the test. The females were given intravaginal estradiol, and their male partners were made anosmic by the insertion of nasal plugs, which anesthetized the olfactory mucosa. In this condition, the males failed to detect any change in the attractiveness of their female partners, although sexual behavior tests with normal control males clearly showed them to be attractive. On reversal of the anosmia, that is, following removal of the nasal plugs, the experimental males pressed for access to the females for the first time and also showed mounting and ejaculations (Figure 15.1). The only novel sensory information available was olfactory; it seemed reasonable to conclude, therefore, that these were the cues the males used. Hence, anosmia prevented the male from realizing that the females had become attractive. However, anosmia did not reduce male sexual behavior in the four control pairs in which the female received injections of estrogen to influence proceptive behavior and with whom a great deal of sexual experience had already been obtained. The failure of anosmia to reduce sexual behavior in the control pairs and the continuance of the males' normal feeding regimen suggest that anosmic procedure is not traumatic, at least not sufficiently traumatic to impair sexual behavior once it has been established.

Subsequent studies show that the female odor cues attractive to males stem from estrogen-influenced vaginal secretions, and chemical analysis of these secretions identify a complex mixture of fatty acids and a number of other substances (Curtis, Ballantine, Keverne, Bonsall, and Michael 1971). It has also been stressed that male response to applied olfactory

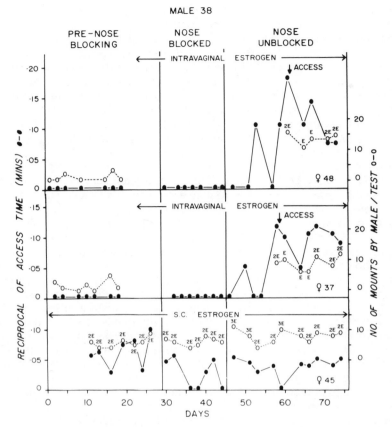

Figure 15.1. Effects of reversible anosmia on the sexual behavior of a male rhesus monkey and on the time taken to obtain access to three female partners in a lever-pressing situation. Lower graph (♀ 45) represents control female. When females received intravaginal estrogen, the male started lever pressing only after olfactory activity had been restored. (○) mounts by male, (●) lever-pressing performance, (E) ejaculations. (From Keverne 1976. Reprinted with permission.)

attractants varies between individuals and is also, in part, dependent on the female partner with whom they are paired (Keverne 1974, 1976). Whereas some estrogen-treated females readily evoke sexual response in the male, others treated in the same manner fail to do so. Hence, the response to olfactory cues in these highly evolved social primates is not stereotyped. In addition, the variability of male behavioral response to the same odor cue with different female partners is a further complication (Figure 15.2). With pairs 67/76, 41/71, and 68/71, estrogen-primed vaginal secretions markedly stimulated male sexual behavior. With the pairs 67/78, 41/79, and 68/78, the social responsiveness of the male increased and the time spent on grooming his female partner lengthened, but no

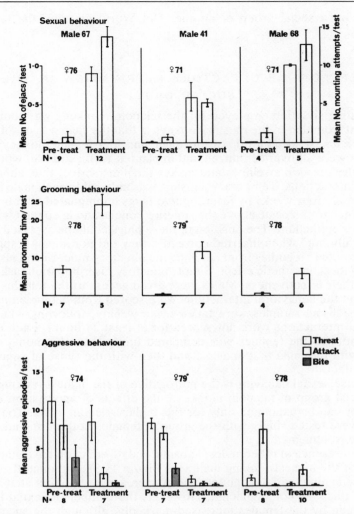

Figure 15.2. Increases in sexual behavior and grooming behavior and the reduction of aggression in rhesus monkeys resulting from applications of pheromone but depending on the female partner. Those pairs marked with an asterisk received applications of ether extracts of vaginal secretions. (From E. B. Keverne 1976. *Journal of the Society of Cosmetic Chemistry* 27:11–23. Reprinted with permission.)

stimulation of sexual activity occurred. With the pairs 67/74, 41/79, 68/78, although no stimulation of sexual activity occurred, a marked reduction was observed in each male's aggressive behavior toward his partner. It is most important to emphasize this lack of stereotyped response to olfactory signals, a result, perhaps, of the normally complex social environment of the primate. This point is illustrated by recent experiments conducted on a social group of another Old World primate, the talapoin monkey.

EFFECTS OF OLFACTORY DEPRIVATION IN THE SOCIAL GROUP

The testing situation employed for the talapoin monkeys was somewhat different from that of the rhesus monkeys in that the cage was of sufficient size to house a social group of four male and four female monkeys. The females were all ovariectomized and implanted with estradiol, which enlarges the sex-skin swellings and makes them attractive, thus enhancing sexual interactions. Temporary anosmia was induced in each male for approximately four weeks by inserting two plugs impregnated with bismuth rodoform paraffin paste above the superior concha and in contact with the olfactory epithelium. The position of these plugs could be assessed radiographically, and when inserted correctly, they did not impair respiration but prevented air eddies from reaching the olfactory mucosa and also produced a local anesthetic effect. Simple olfactory discrimination tasks confirmed their effectiveness. Males were given access to the females twice daily and the behavioral interactions were scored from behind a one-way screen. Plasma samples were taken twice weekly, following which behavioral interactions were not scored for at least 24 hours. Each male's behavior with the females was compared under these hormonal conditions, with the male first anosmic and then with the sense of small normally intact.

Because sexual behavior is the prerogative of the highest-ranking male in a social group of talapoin monkeys, the effects of anosmia on sexual behavior can be examined only for this highest-ranking male. The other males were tested singly with the group of females both during and after anosmic treatment.

In the presence of other males, anosmia had no effect on the ejaculatory scores of the highest-ranking male (see Figure 15.3). Although there is a decrease in male-initiated mounts and an increase in those initiated by the female, this was not significant. However, the number of sexual invitations made by the females increased markedly, although the success of these invitations in initiating a mount declined markedly (Figure 15.4). Thus we see an increase in the females' contributions to sexual interaction but a decrease in their stimulus value when the male was anosmic. Although no overall change in the sexual behavior of the highest-ranking male occurred when he was anosmic, it would be misleading to suggest that his behavior remained the same. His ejaculatory scores remained the same, but the manner in which he distributed them among the females changed even during this short period of anosmia. Under normal conditions, most sexual behavior of the highest-ranking male is observed with high-ranking females (Keverne, Meller, and Martinez-Arias 1978). During the period of anosmia, however, the sexual behavior of the highest-ranking male showed a tendency to decline with the high-ranking females and

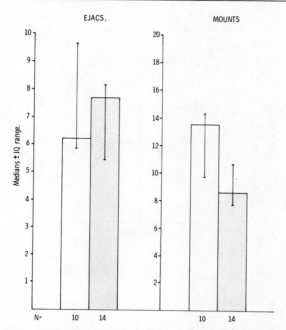

Figure 15.3. Effects of anosmia on the dominant male's sexual behavior in the presence of other males for talapoin monkeys. No decrease in ejaculations or mounting performance is now seen. Anosmic condition is represented by shaded histograms. (From Keverne, 1980. Reprinted with permission.)

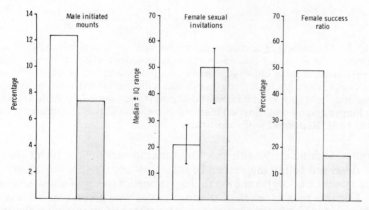

Figure 15.4. Effects of anosmia in the dominant male and female sexual behavior in talapoin monkeys. Although no decreases in male ejaculations were recorded, this male initiated fewer mounts. Female sexual invitations increased, but fewer were successful in initiating a male mount. This suggests that the stimulus value of female sexual invitations is lower for the anosmic male (shaded histogram). (From Keverne, 1980. Reprinted with permission.)

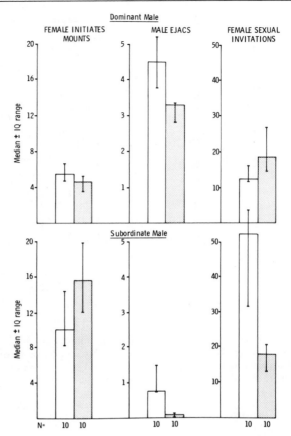

Figure 15.5. The effects of anosmia and social rank on sexual interactions in tala-poin monkeys. Male rank 1 decreases ejaculation scores when anosmic. Although female sexual invitations increase, there is no increase in the number of them initiating mounts. Females initiate more mounts with the lowest-ranking male when he is anosmic, but their overall interest declines (fewer sexual invitations) because he is no longer ejaculating. Shaded histograms indicate anosmic conditions. (From Keverne, 1980. Reprinted with pemission.)

to increase significantly with the low-ranking females, a trend that was rapidly reversed following return of olfactory acuity (Figure 15.4).

In the absence of high-ranking males, subordinates will show sexual behavior. For this reason, each of the four males was given exclusive access to the females in the large group cage. The effect of anosmia on the male's sexual behavior under these conditions varied according to that male's previous sexual experiences. (In the absence of other males, male dominance cannot be assessed and the rank order in Figure 15.5 refers to the status of the male as determined from his earlier interactions in the group and hence his social experience at that time.) With the highest- and lowest-ranking males, ejaculation scores decreased significantly and little change in mounting behavior occurred (Figure 15.5). The second-ranking male showed no change in ejaculations but the number of mounts required to achieve ejaculation increased. The ejaculation scores of the third-ranking male actually increased. This resulted primarily from increase in sexual activity with the most subordinate female in the group, but it was not related to any changes in female-initiated mounts or increased female sexual invitations.

When we examine the effects of anosmia on the sexual interaction of the highest- and lowest-ranking males, we see marked changes in the females' contribution to the interaction, the differences in their behavior depending upon the male (Figure 15.5). With the highest-ranking male, anosmia resulted in the reduction of ejaculations from a daily mean of 6 to 4. Although the females increased sexual invitations to him, the invitations did not initiate more mounts; thus, under anosmic conditions, the stimulus value of female presentation seemed to decline for this male. The sexual behavior of the most subordinate male was more seriously impaired by anosmia, with only 2 ejaculations occurring in the 4-week period, compared with 24 in the same period without anosmia. Interestingly, the females, displaying extremely high levels of sexual invitations, initiated all of this male's sexual behavior without anosmia, but significantly reduced their invitations when he was anosmic. Although the females were less interested in this anosmic male, they nevertheless initiated all his mounting behavior and displayed the same number of sexual invitations to him as they had to the high-ranking male.

Although the effects of anosmia might generally be described as having little significant effect on male ejaculatory behavior, it would be quite misleading to relegate olfactory cues to a redundant role in sexual behavior. Male ejaculation is the end point of an accumulation of interactions in which the part played by the female cannot be overlooked. Thus, although we see little change in the male's ejaculatory scores when he is anosmic, we witness considerable increase in female initiatory behavior. The decrese in the success of that behavior when the male is anosmic can be interpreted only as a loss in the stimulus value, or the attractiveness, of the female. In other words, in attempting to compensate for the loss of nonbehavioral attractiveness (odor), the females increase the behavioral aspects of attractiveness (sexual invitations).

It could be argued that the observed changes in sexual interactions were due not directly to the loss of olfactory sensory information but to the olfactory influences on gonadotrophin release and testosterone secretion. It could also be argued that having plugs up the nose is stressful and that any changes in sexual behavior are secondary to this discomfort. However, our hormone assays found absolutely no anosmic effects on either luteinizing hormone (LH) or testosterone in any male, and of the two stress hormones measured, prolactin and cortisol, no significant increases were seen. This could not, therefore, account for the changes in sexual interaction.

The importance of the social dimension in the influence of olfactory cues on behavior was further illustrated by the decline in ejaculatory performance of the top-ranking male when he was not able to assert his dominance, i.e., in the absence of other males. In the social group, however, in the presence of competition from other males, anosmia had no effect on this male's ejaculation scores, and they were significantly higher than they were when he had been alone with the females. Moreover, although the amount of male sexual behavior did not decrease, the manner in which it was distributed among the females changed markedly with anosmia, the low-ranking females receiving male sexual attention for the first time. Although the significance of this is uncertain, it may be that the unattractiveness of low-ranking females is in some way related to their odor. Certainly, low-ranking females are under some social stress, as evidenced by high levels of prolactin (Keverne 1980), which in turn may affect their food and water intake, their gut cecotroph, and hence their odor characteristics.

CONCLUSIONS

Although these observations do not represent a definitive study on the importance of olfactory cues for sexual interactions in the talapoin monkey, they do illustrate certain points. The experiments show that as a social situation becomes more complex, the response to olfactory cues is correspondingly more complex and less predictable. Thus, although the testing of heterosexual pairs of rhesus monkeys in acoustically and visually isolated booths will show a predictable effect of changing odor cues on behavior, such effects do not necessarily generalize to other situations. Indeed, random selection of female and male rhesus monkeys shows that effects produced by the odor cue in these visually and acoustically isolated booths is by no means reproducible. Both the effect and the kind of behavior produced are dependent upon the pairing. In some pairs agression may decrease, in other grooming may increase, in others behavior may not change, and in some pairs sexual behavior may increase in response to cues from attractive females.

Asking whether olfactory cues are important in the communication of sexual status in social groups of talapoin monkeys introduces yet another dimension of complexity. Since only the highest-ranking male shows sexual behavior in a group situation, the question of status can be asked only of this male. In an anosmic condition, the dominant male appears to continue his usual sexual behavior, but the behavioral interactions that make this behavior possible clearly change, with the females taking more of the initiative. Moreover, when the dominant male was tested with females without the presence of other males, a decline in sexual behavior was observed. Likewise, anosmia produced profound effects on the sexual behavior of the most subordinate male. Although he showed sexual behavior when alone with females prior to being made anosmic, his sexual behavior practically ceases with ejaculations almost completely eliminated when his olfactory sense was denied. There is no doubt, therefore, that olfactory cues communicate something about female attractiveness to the male, but how the male uses this information depends upon the male, his past experiences, and certainly the social context in which he perceives it.

Since primates do not possess a vomeronasal accessory olfactory system with direct projections to the limbic brain, it can be assumed that behavioral interactions as complex as those evoked by changes in olfactory signals are a function of the neocortical projections of the olfactory system. That being so, one might question the appropriateness of describing primates as being microsmatic.

What neuroanatomists usually mean by the term "microsmatic" is that a relatively small area of the brain is given over to the olfactory sense and that the number of olfactory receptors are fewer in number than in a macrosmatic species such as a mouse or rabbit. Now, if receipt and interpretation of the olfactory message relied on these anatomical features, primates certainly could be described as microsmatic. However, two important findings have invalidated this method of classification. The first finding, that single olfactory receptors are not narrowly tuned to particular odors but respond, instead, to a number of different odors, led to the consideration of a pattern of receptor activity as the means of coding odor cues (Gesteland, Lettvin, and Pitts 1965). The second finding is that the neural projections of the olfactory system have access to the neocortex via the thalamus (Powell, Cowan, and Raisman 1965). That olfactory information is coded as a pattern requires a higher order of recognition. Consequently, animals with a greater neural backup are potentially more capable of producing a more sophisticated coding and subsequent analysis of odors. Since the olfactory bulbs seem to act mainly as a filter, and since the decoding of the olfactory message is a more central event, animals with more complex neocortical support systems are able to make more sophisticated use of their olfactory sense. One might argue, therefore, that primates are able to employ their sense of smell in a more sophisticated manner than species such as rabbits or rats. Certainly, simple stimulus–response events are not sufficient to explain the role of olfactory cues in higher primates.

This chapter has not entered into the controversy concerning the chemical composition of estrogen-primed vaginal secretions (Goldfoot et al. 1976; Michael and Bonsall 1977). Since olfactory receptors are not intrinsically specific to single odorants, the olfactory message must be coded in terms of a patterning of neural input. Each odor of a chemical complex will thus generate its own pattern of receptor firing, with considerable receptor overlap among different odorants. Thus, since the amount of vaginal secretions available to males is unknown and since individual fatty acids do not indicate the perceptual qualities of the odor complex, there seems little point in arguing about the amounts of volatile fatty acids present in vaginal secretions.

REFERENCES

Altmann, S. A. 1959. Field observations on a howling monkey society. *Journal of Mammalogy* 40:317–30.

Ball, J., and Hartman, C. G. 1935. Sexual excitability as related to the menstrual cycle in the monkey. *American Journal of Obstetrics and Gynecology* 29: 117–19.

Baum, M. J., Everitt, B. J., Herbert, J., and Keverne, E. B. 1977. Hormonal basis of proceptivity and receptivity in female primates. *Archives of Sexual Behavior* 6:173–91.

Bellringer, J. F., Pratt, H. P. M., and Keverne, E. B. 1980. Involvement of the vomeronasal organ and prolactin in pheromonal induction of delayed implantation in mice. *Journal of Reproduction and Fertility* (U.K.) 59:223–8.

Blurton-Jones, N. G., and Trollope, J. 1968. Social behaviour of stump-tailed macaques in captivity. *Primates* 9:365–94.

Booth, C. 1962. Some observations on behaviour of *Cercopithecus* monkeys. *Annals of the New York Academy of Sciences* 103:477–87.

Bullock, D. W., Paris, C. A., and Goy, R. W. 1972. Sexual behaviour, swelling of the sex skin and plasma progresterone in the pigtail macaque. *Journal of Reproduction and Fertility* (U.K.) 31:225–36.

Carpenter, C. R. 1942. Sexual behaviour of free ranging rhesus monkeys (*Macaca mulatta*). II. Periodicity of oestrous, homosexual autoerotic and non-conformist behaviour. *Journal of Comparative Psychology* 33:143–62.

Curtis, R. F., Ballantine, J. A., Keverne, E. B., Bonsall, R. W., and Michael, R. P. 1971. Identification of primate sexual pheromones and the properties of synthetic attractants. *Nature* 232:396–8.

Czaja, J. A., and Bielert, C. 1975. Female rhesus sexual behaviour and distance to a male partner: relation to stage of the menstrual cycle. *Archives of Sexual Behavior* 4:583–98.

Eaton, G. G., and Resko, J. A. 1974. Ovarian hormones and sexual behaviour in *Macaca nemestrina*. *Journal of Comparative and Physiological Psychology* 86:919–25.

Epple, G. 1967. Vergleichende Untersuchungen über Sexual und Sozialverhalten der Krallenaffen (Hapalidae). *Folia Primatologica* 7:37–65.

Gesteland, R. C., Lettvin, J. Y., and Pitts, W. H. 1965. Chemical transmission in the nose of the frog. *Journal of Physiology* 181:525, 559.

Goldfoot, D. A. 1971. Hormonal and social determinants of sexual behaviour in the pigtail monkey (*Macaca nemestrina*). In *Normal and abnormal developments of brain and behaviour*, ed. G. B. A. Stoelinger and J. J. Van der Werff ten Bosch, pp. 325–42. Leiden: University of Leiden Press.

Goldfoot, D. A., Kravetz, M. A., Goy, R. W., and Freeman, S. K. 1976. Lack of effect of vaginal lavages and aliphatic acids on ejaculatory responses in rhesus monkeys. Behavioural and chemical analyses. *Hormones and Behavior* 7: 1–27.

Goy, R. W. 1979. Sexual compatibility in rhesus monkeys: predicting sexual performance of oppositely sexed pairs of adults. In *Sex, Hormones and Behaviour, Ciba Foundation Symposium* 62:227–49.

Green, S. M. 1981. Sex differences and age gradations in vocalizations of Japanese and lion-tail monkeys (*Macaca fuscata* and *Macaca silenus*). *American Zoologist* 21:165–83.

Hall, K. R. L., Boelkins, R. C., and Goswell, M. J. 1965. Behaviour of patas, *Erythrocebus patas*, in captivity with notes on the natural habitat. *Folia Primatologica* 3:22–49.

Hansen, S., Stanfield, E., and Everitt, B. J. 1980. The role of the ventral bundle noradrenergic neurones in sensory components of sexual behaviour and coitus induced pseudopregnancy. *Nature* 286:152–4.

Herbert, J. 1966. The effects of oestrogen applied directly to the genitalia upon the sexual attractiveness of the female rhesus monkey. *Excerpta Medica International Congress,* ser. 3:212.

1970. Hormones and reproductive behaviour in rhesus and talapoin monkeys. *Journal of Reproduction and Fertility, Supplement* (U.K.) 11:119–40.

1974. Some functions of hormones and the hypothalamus in the sexual activity of primates. *Progress in Brain Research* 41:331–48.

Hess, J. P. 1973. Observations on the sexual behaviour of captive lowland gorillas. In *Comparative ecology and behaviour of primates,* ed. R. P. Michael and J. Crook, pp. 507–81. New York: Academic Press.

Hisaw, F. L., and Hisaw, F. L. 1966. Edema of the sex skin and menstruation in monkeys on repeated oestrogen treatments. *Proceedings of the Society for Experimental Biology and Medicine* 122:66–70.

Jay, P. 1965. Field studies. In *Behaviour of non-human primates,* ed. A. M. Schrier, H. F. Harlow, and F. Stollnitz, pp. 525–92. New York: Academic Press.

Johnson, D. F., and Phoenix, C. H. 1976. Hormonal control of female sexual attractiveness, proceptivity and receptivity in rhesus monkeys. *Journal of Comparative and Physiological Psychology* 90:473–83.

Keverne, E. B. 1974. Sex-attractants in primates. *New Scientist* 63:22–4.

1976. Sexual receptivity and attractiveness in the female rhesus monkey. In *Advances in the study of behaviour,* Vol. 7, ed. J. S. Rosenblatt, R. A. Hinde, E. Shaw, and C. Beer, pp. 155–200. New York: Academic Press.

Keverne, E. B. 1980. Olfaction in the behavior of non-human primates. *Symposium of the Zoological Society of London* 45:313–27.

Keverne, E. B., Meller, R. E., and Martinez-Arias, A. 1978. Dominance, aggression and sexual behaviour in social groups of talapoin monkeys. In *Recent advances in primatology,* Vol. 1, ed. D. J. Chivers and J. Herbert, pp. 53–48. London and New York: Academic.

Lindburg, D. G. 1967. A field study of the reproductive behaviour of the rhesus monkey (*Macaca mulatta*). Ph.D. thesis, University of California, Berkeley.

Michael, R. P., and Bonsall, R. W. 1977. Chemical signals and primate behaviour. In *Chemical signals in vertebrates,* ed. D. Muller-Schwarze and M. M. Mozell, pp. 251–71. New York: Plenum.

Michael, R. P., and Keverne, E. B. 1968. Pheromones: their role in the communication of sexual status in primates. *Nature* 218:746–9.

1972. Differences in the effects of oestrogen and androgen on the sexual motivation of female rhesus monkeys. *Journal of Endocrinology* 55:40–1.

Michael, R. P., and Welegalla, J. 1968. Ovarian hormones and the sexual behaviour of the female rhesus monkey (*Macaca mulatta*) under laboratory conditions. *Journal of Endocrinology* 41:407–20.

Michael, R. P., Herbert, J., and Welegalla, J. 1967. Ovarian hormones and the sexual behaviour of the male rhesus monkey, *Macaca mulatta,* under laboratory conditions. *Journal of Endrocrinology* 39:81–98.

Moore, R. Y., and Zichler, V. B. 1972. Loss of a circadian adrenal corticosteroid rhythm following suprachiasmatic lesions in the rat. *Brain Research* 42:201–6.

Morin, L. P., Fitzgerald, K. M., Rusak, B., and Zucker, I. 1977. Circadian organization and neural mediation of hamster reproductive rhythms. *Psychoneuroendocrinology* 2:73–98.

Moynihan, M. 1966. Communication in the titi monkey (*Callicebus*). *Journal of Zoology* 150:77–127.

1967. Comparative aspects of communication in New World primates. In *Primate ethology,* ed. D. Morris, pp. 236–66. London: Weidenfeld & Nicolson.

Nishida, T. 1970. Social behaviour and relationships among wild chimpanzees of the Mahali mountains. *Primates* 2:47–87.

Phoenix, C. H. 1973. Ejaculation by male rhesus as a function of the female partner. *Hormones and Behavior* 4:365–70.

Powell, T. P. S., Cowan, W. M., and Raisman, G. 1965. The central olfactory connections. *Journal of Anatomy* 99:791–813.

Rahaman, H., and Parthasarathy, M. D. 1969. Studies on the sexual behaviour of bonnet monkeys. *Primates* 10:149–62.

Reynolds, J. M., and Keverne, E. B. 1979. The accessory olfactory system and its role in pheromonally mediated suppression of oestrus. *Journal of Reproduction and Fertility* (U.K.) 57:31–5.

Scruton, D. H., and Herbert, J. 1970. The menstrual cycle and its effect on behaviour in the talapoin monkey (*Miopithecus talapoin*). *Journal of Zoology* 162:419–36.

Southwick, C. H., Beg, M. A., and Siddiqi, M. R. 1965. Rhesus monkeys in north India. In *Primate behavior: Field studies of monkeys and apes*, ed. I. DeVore, pp. 175–96. New York: Holt, Rinehart and Winston.

Stephen, F. K., and Zucker, I. 1972. Rat drinking rhythms: Central visual pathways and endocrine factors mediating responsiveness to environmental illumination. *Physiology and Behavior* 8:315–26.

Stetson, M. H., and Watson-Whitmyre, M. 1976. Nucleus suprachisamatic: the biological clock in the hamster. *Science* 191:197–9.

Trimble, M. R., and Herbert, J. 1968. The effect of testosterone or oestradiol upon the sexual and associated behaviour of the adult female rhesus monkey. *Journal of Endocrinology* 42:171–85.

van Lawick-Goodall, J. 1968. The behaviour of free living chimpanzees in the Gombe Stream Reserve. *Animal Behaviour Monographs* 1, pt. 3:165–311.

16 · Multiple channels of sexual communication in rhesus monkeys: Role of olfactory cues

DAVID A. GOLDFOOT

Nearly 40 years ago F. A. Beach (1942) demonstrated that ablation of any single sensory modality of the male rat did not prevent the animal from copulating. Latencies to ejaculation of sensory-impaired animals were nearly equivalent to intact rats regardless of whether the animals were blinded, subjected to deafferentation of the facial vibrissae, or rendered anosmic.[1] When any two sensory modalities were removed in the same animal, clear deficits in sexual performance occurred, but still copulation could take place. Ablation of all three sensory modalities usually prevented copulation. Similar early studies in rabbits (Brooks 1937) confirmed the multiply coded nature of sensory stimuli that regulate sexual interaction.

The results of those early studies could be interpreted to mean that information regarding the presence of an estrous female was available to the males (1) through several sensory channels, with any single channel sufficient to code the message (redundancy model); (2) through a pattern of sensory channels that together gave the complete message (gestalt or multichannel model); and (3) through a sequence of messages designed to give first distal and then proximal cues to maximize delivery of the information (sequential multichannel model). Various combinations or interactions of 1, 2, and 3 are also possible.

Whereas Beach (1942, 1944) emphasized the potential importance of multichannel messages, most subsequent investigations have approached questions of sexual communication by studying one channel of information at a time, paying minimal attention to the integrated manner by which these messages might be processed. Moreover, until quite recently (Beach 1976), far less attention had been paid to behavioral cues as communicative systems than to the simpler stimulus attributes (color and tur-

[1] The author is aware that this experiment would have failed in some species, e.g., hamster.

This chapter, publication 21-027 of the Wisconsin Regional Primate Research Center, was supported in part from Grants RR00167, MH21312, and BNS-76-14741. I thank R. W. Goy and S. Eisele for allowing me to cite unpublished findings. D. Neff and S. Edwards provided technical support. M. Schatz and J. Kinney typed the manuscript, and Donna McConnell drew the figure.

gescence of sex skin, smell) of sexual partners. In the following discussion, I will attempt to provide an initial hypothesis of an integrated, multichannel system of communication for sexual interaction between rhesus monkeys that emphasizes sequential behavioral signals between the male and the female. The analytical job has just begun, but the evidence is already sufficiently persuasive to indicate that either a gestalt or sequential multichannel model fits the data better than any single-channel sensory model.

The emphasis on multiple channels of communication as necessary for understanding sexual interactions can be understood readily by comparing the behavior of animals in test situations that are designed to optimize reception of a given sensory signal with those that allow more variables to intervene. For example, studies indicate that catarrhine species that have prominent sex-skin swellings at estrus utilize sex-skin appearance as signals during courtship and mate selection (Bielert, chap. 14, this volume). The cue is obvious and can be detected from considerable distances. However, when studies in small group contexts of three females and one male, it is clear that sex-skin condition is not the only signal that is processed before mate selection is accomplished. For example, fully turgescent pig-tailed macaque females of lowest dominance refuse to present to a male when in the presence of aggressive female competition. In such circumstances, the male will not court the turgescent animal but will copulate with the dominant female, even though she might not be in the periovulatory phase of her cycle. If the subordinate female is then isolated from her competition and the male is paired with her alone, she readily copulates and displays full receptivity and he now finds her quite "attractive" (Goldfoot 1971).

Of related significance to this discussion is the role of olfaction, which has been shown to have an influence on the copulatory behavior of laboratory-housed rhesus monkeys. Accordingly, evidence in support of, and contradictory to, a rhesus "releaser" pheromone hypothesis will be considered in detail.

MULTIPLE MODES OF SEXUAL COMMUNICATION IN RHESUS MONKEYS

Copulation typically occurs between adult rhesus monkeys when the female is undergoing an ovulatory cycle and when the male's testosterone levels are above some minimal threshold (see reviews by Michael 1972; Goldfoot 1977). Under the influence of ovarian hormones, the female undergoes several stimulus changes, including a reddening and slight to moderate edema of several areas of skin (face, thighs, perineum; Czaja, Eisele, and Goy 1975), a change in odor quality and odor intensity of vaginal products (Goldfoot, Kravetz, Goy, and Freeman 1976b; Michael and Bonsall 1977; Goldfoot 1981), and a complex series of behavioral changes, which are revealed most easily under naturalistic conditions or in group testing paradigms (Carpenter 1942; Goldfoot 1971; Lindburg 1971; Rowell

1972). Specific vocalizations indicating sexual readiness have not been described for female rhesus monkeys, but the possibility remains that such auditory cues, which have been described for closely related species (*Macaca fuscata:* Green 1975; *M. fascicularis:* Deputte and Goustard 1980), might be given under certain stimulus situations.

The behavior changes that have been described include reductions of interindividual distances (either initiated by the female or tolerated by her when initiated by the male) and the display of several facial and body gestures (presenting, sidling, proximity sitting, head bobbing, shoulder-flexing, hand-reaching, etc.) that have been shown to initiate, maintain, or coordinate sexual interactions. In addition, the female or the male may threaten another individual while simultaneously reducing the distance between herself or himself and a potential consort. The latter behavior sequence apparently serves as a potent mechanism, which recruits the partner to join in the threat and also increases the likelihood of copulation (Zumpe and Michael 1970). Sitting in proximity to the male and following him are other clear indications of sexual proceptivity, provided the testing area is large enough to make such observations reliable. In addition to these proceptive changes, a female is more likely to tolerate mounts or mount attempts during ovulatory cycles than at other times, although both the physical and social conditions of testing strongly influence this variable of receptivity as well (Johnson and Phoenix 1976; Goldfoot 1977).

When studied in the laboratory, and especially when using a pair-test procedure (short duration testing of 15 min to 1 hr, in which two individually housed animals are placed together in an observation cage), most, but not all, of these behavioral tendencies can be observed. In general, sexual interactions in laboratory pair tests are of much higher frequency than those seen in group-living conditions, particularly the mounting behavior of males and presenting postures of the female and her ultimate receptivity, regardless of her ovarian status. Apparently missing from pair tests are the complex social interactions and signals that are usually an integral part of courtship and mating for this species under naturalistic conditions (see Goldfoot 1971). Moreover, the repeated separation and reunion of animals in typical laboratory pair-test studies is thought by several investigators to be an additional source of variance in studies of sexual behavior (Goldfoot, Essock-Vitale, Asa, Thornton, and Leshner 1978; Slob, Wiegand, Goy, and Robinson 1978). Accordingly, considerable caution must be exercised when generalizing from pair-test studies to other situations. For example, Kim Wallen (personal communication) has recently summarized the results of several experiments using pair tests conducted in different laboratories, and he finds that the incidence of copulation and measures of receptivity are inversely related to the size of the cage employed. With very small cages (e.g., 1.5 m diagonal), more copulation and less female refusal behavior is encountered. Thus direct quantitative comparisons of results from one laboratory to the next are fraught with difficulties resulting from such slight procedural differences as the size of cage. Moreover, it is clear from several field and laboratory studies that courtship is modulated by social relations and competition factors that include more animals than the focal pair. A dominant male rhesus monkey can completely inhibit any courtship behavior of a subordinate (Perachio 1978; also, for talapoin, *Miopithecus talapoin,* see Eberhart, Keverne, and Meller 1980), and, likewise, the social dominance of females can moderate proceptive advances of other females, sometimes even overriding ovarian mediation of sexual activity (Goldfoot 1971). Since social factors clearly influence reproductive success in most macaque species, it is obvious from the outset that mechanisms such as pheromonal cues cannot be the exclusive mediator of reproductive probabilities and must be studied in socially complex as well as simple environments to assess their contributory effects as a function of situational complexity.

THE ATTRACTIVENESS HYPOTHESIS

Earlier reports of sexual interactions of rhesus monkeys observed in repeated laboratory pair tests suggested that proceptive and receptive behavior of the female did not change reliably as a function of the ovarian phase of the menstrual cycles, whereas the ejaculatory frequency of the male was very much related to this variable (Michael 1968; Michael, Zumpe, Keverne, and Bonsall 1972). Since behavioral changes of the female were not detected, it was assumed that active sexual solicitation (proceptive behavior) of the female was not the cause of the male's sexual activity at midcycle but that, instead, a nonbehavioral channel of communication was active for reproductive coordination. The male's sexual performance (and motivation) was hypothesized to be influenced by the female's "attractiveness," with emphasis placed on the notion that attractiveness resulted from estrogen-mediated nonbehavioral stimulus changes of the female (Herbert and Trimble 1967; Michael, Saayman, and Zumpe 1967; Trimble and Herbert 1968).

Several criteria have been used to operationally define attractiveness of the female (Keverne 1976a), but all definitions utilize behavior of the male to assess this hypothetical construct and are therefore limited by both the ideographic features of the males used and the situational factors of the assessment conditions. A common measure of attractiveness has been the "female success ratio," which is the percentage of female presentations or other sexual invitations that stimulate mounts or mount attempts by the male in a pair test. It is usually found that this measure increases when females are approaching ovulation or are given exogenous estrogens. No qualitative differences in the detail of these invitational gestures have been described as a function of hormone treatment, so it has been assumed that they "work" in one case and not the other because of nonbehavioral stimulus conditions. An experimental manipulation of sex-skin color has not resulted in an altered success ratio in this species (Trimble and Herbert 1968; but see Bielert, chap. 14, this volume, for positive evidence in a turgescent species), whereas manipulation of vaginal olfactants has increased this measure (Keverne and Michael 1971). Moreover, earlier studies that evaluated sexual performance of females given testosterone reported that invitation rates increased but female success ratios did not (Michael and Zumpe 1977). This finding was taken as additional evidence that the behavioral gestures of the female were not sufficient to alter attractiveness, and that, by implication, a stimulus component constituting attractiveness was missing in testosterone-treated females. More recently, however, studies have indicated that testosterone given to spayed female rhesus monkeys does increase ejaculatory rates and the female success ratio (Wallen and Goy 1977; Goldfoot, Neff, and Edwards, in progress), although in the latter case not necessarily with the same degree of efficacy as estradiol. Moreover, videotaped observations in our laboratory have suggested that a periovulatory or estrogenized female often changes her proceptive strategy from "present-at-distance" to "sit-in-close-proximity" (see Michael and Welegalla 1968; Czaja and Bielert 1975). That is, once the copulatory sequence has commenced, a mid-cycle female might refrain from further presentation postures and, instead, coordinate the series of mounts displayed toward her by sitting very near the male with her back toward him and at appropriate intervals making subtle staccato head, arm, or shoulder movements. Alternatively, if the female is grooming the male between mounts, she might suddenly pivot to a stereotypical sitting posture with her back to the male. All of these postural adjustments seem to induce the male to attempt a mount, and when fully receptive, the female will move to the acceptance posture simultaneously with the male's initial movements toward her. Although this degree of behavioral coordination is fascinating to observe, it is a nightmare to quantify, and differences between laboratories with regard to definitions of male and female sexual initiations, refusals, and so on are related to the details of the various scoring systems and testing parameters in use. Nonetheless, the earlier impression of a lack of behavioral change as a function of hormonal mediation, or of a failure of the female's solicitations to influence a male's sexual response, led many workers back to the suggestion of Carpenter (1942) that vaginal odor might be an important sexual cue for male rhesus monkeys. Interestingly, besides mentioning the possibility of olfactory mediation, Carpenter was one of the first researchers to emphasize that dramatic behavioral changes occurred in females as they approached ovulation.

BEHAVIORAL COMMUNICATION REEVALUATED

Could it be that behavior change can be discounted as a channel of communication in a species as social as the rhesus monkey? Obviously not, and while no laboratory seriously considered that idea, it was believed from success ratio data, and more specifically from the results of the early work with vaginal odorants, that sexual solicitations had far less effect on the male if the proper odor was not available concurrently. A challenge of sorts to the primacy, or at least to the necessity, of vaginal odor as an arousal or attractivity component appeared with the publication of a series of experiments conducted at the Oregon and Wisconsin Regional Primate Research centers (Goldfoot et al. 1976b). These studies reported that both vaginal lavages of estrogenized female rhesus monkeys and aliphatic acid mixtures failed to stimulate copulatory behavior of a large majority of males (19 males and 27 females were evaluated; 3 to 4 males showed marginally positive responses). On the other hand, the experiments presented by Goldfoot and his associates (1976b) were not exact replications of earlier work and therefore did not directly contradict the results of other laboratories. They did raise the question of the generality of the earlier findings, however, and certainly underscored the variability of the phenomenon.

To further examine the role of behavioral changes of the female in the communication of sexual interest, Deborah Neff, Sara Edwards, and I analyzed behavioral sequences between 64 heterosexual pairs of rhesus monkeys (8 males, 8 ovariectomized females, paired in all possible combinations). To better understand the behavioral process that yielded the female success ratio, we specifically examined those sequences in which either the male or the female initiated social interactions. We looked at two basic types of sequences: (1) those in which the initial approach to the female by the male occurred without any observable behavioral solicitation by the female, and (2) those in which the male's approach appeared to be in response to a solicitation gesture of the female. Solicitation gestures could be in the form of presentations, hand reaches, shoulder flexes, head bobs, sidling, or glances. A rapid series of these gestures, for example, a series of hand reaches displayed in quick succession, were analyzed as a single solicitation bout. Some sequences could not be unambiguously labeled as male-initiated or female-initiated, and these were left out of the analyses. Once having detected the onset of a sequence that met our criteria, we looked at the succession of behavioral adjustments that were made by both animals in response to one another. For example, we could determine whether the male attempted to contact the female (touch her with one or both hands on the hips or back quadrant of her body) and whether he mounted her. In turn, we recorded whether the female was negative to the male, that is, whether she ran away, threatened him, and so on, and at which point in the interaction sequence the negative response was first displayed. Similarly, we recorded at which point in the sequence a present was displayed. When neither a positive nor negative response could be observed, we recorded that the female "ignored" the male at that point in the sequence.

Some of these sequential relationships are diagramed in Figure 16.1 for tests in which the ovariectomized females were injected with oil vehicle and for tests in which the females were given daily injections of estradiol benzoate (EB). The diagrams reveal that when males initiate sexual sequences in the absence of prior solicitations from females, only a small percentage of sequences terminate with successful mounts, *regardless of whether the female has been given estrogen.* For example, females given oil injections were overtly negative to the male's unsolicited approach 38.4% of the time. Only about a third of the approaches then progressed to an attempted contact. Females were negative to 37% of those contacts and, in addition, failed to present to an additional 24% of the contacts. The overall result was that males were able to mount females only 13% of

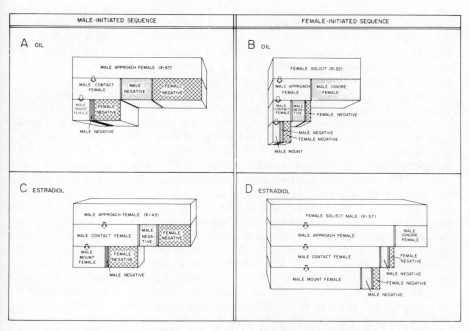

Figure 16.1. Sequences of behavior in rhesus monkey pair tests resulting from the social initiation of males or females: (A) male initiation, female oil-treated; (B) female initiation, female oil-treated; (C) male initiation, female EB-treated; (D) female initiation, female EB-treated. Data are based on 15-min tests of eight males and eight females in all possible pairings for oil condition (64 tests) and EB condition (64 tests). The boxes in the figures are scaled relative to the mean frequency of the first initiation behavior indicated (either male approach or female solicit) and are in proportion to the mean percentages of sequential options that were observed. Sizes of boxes are proportional and in the same scale for all four sections of this figure.

the times that they initiated approach sequences and only 36% of the times that they contacted females. Estradiol treatment tended (nonsignificantly) to reduce negative responses, but males still did not achieve mounts on more then 34% of their initiated approaches. The only circumstance in which the male significantly increased the percentage of his approaches that resulted in mounts was under the estradiol condition, provided that he first waited for a female's solicitation. Estrogenized females very rarely (4%) displayed negative responses to the male's approach or contact when they initiated the interaction. In other words, if the male attended to the female's solicitation *and* to her subsequent responses to his precopulatory behavior, he would be provided with increased information concerning her ultimate receptivity, relying not only on her solicitation behavior but on her lack of negative adjustments following his initial response to her. It would be to his benefit, therefore, to respond more often to her solicitations under such conditions than at times when she showed negativity following his approach or her solicitation. (It should be mentioned that two males were dropped from this study during preliminary pretesting because of their violent attacks upon females and were replaced by gentler animals. Our sample of males is therefore "biased" in this regard, although we believe it is a common and humane practice to eliminate males of this type from laboratory sex tests.)

These data, although restricted to the necessity of artificially dichotomizing sequences into male and female social initiations, provide a realistic extension to previous interpretations of the female success ratio. First, estrogen does indeed increase the likelihood that a higher percentage of solicitations will be responded to by males (from a mean of 42.6% under ovariectomized untreated conditions to 76.8% under estrogen stimulation). However, it is also evident from the data that the female is much more likely to tolerate a male's approach, contact, or mount while under estrogen stimulation, especially so after she displays a solicitation. She is much less likely to attempt to refuse the male by negative gestures when he approaches her and when he contacts her, and she is, in fact, more likely to respond by presenting to these advances or to accommodate and facilitate the mount. Thus, it is not difficult to see that a female success ratio will go up when the female signals cooperation throughout the approach–contact–mount sequence. Given the relatively lower probability of success for males in male-initiated sexual sequences, it seems reasonable to suggest that an experienced male could learn to respond appropriately, that is, to first test a female by initiating a few approach sequences, and then, depending on her initial responses, to let her pace the subsequent sexual interactions. This strategy would obviously result in a higher success ratio score for the female.

To summarize, these sequences clearly reveal that several opportunities for communication via behavioral channels are available during courtship episodes prior to the actual mount or mount attempt. These series of interactions are analogous, perhaps, to a conversation in "body language" composed of a series of signaled intentions and reactions to these intentions. We do not need to accept the argument that some nonbehavioral stimulus aspect of the female must be hypothesized for the differential responses of males to the presentations of estrogen-stimulated females versus nonstimulated females. The initial solicitations of the female might look the same to us in the two conditions, but the "conversations" that proceed from that point are quite different and are under considerable behavioral control of the female.

PHEROMONES, ODOR CUES, AND ODOR NOVELTY

The preceding discussion has served to illustrate that traditional measures of sexual attractivity can be influenced by behavioral characteristics of the female that are mediated in part by ovarian hormones. This demonstration in no way denies the existence of nonbehavioral cues for influencing sexual interactions. It does suggest, however, that reassessments of earlier work and earlier hypotheses that imply a primacy of nonbehavioral cues might be in order. A recent and detailed review of the evidence in support of, and contrary to, the hypothesis that rhesus monkeys possess a sexual pheromonal system of communication has recently been completed (Goldfoot 1981). Briefly, that review concluded that olfactory cues were but one of many communication systems by which the rhesus monkey coordinates its reproductive activity. Moreover, the evidence supports the view that whereas olfactory cues in this species can be involved in sexual mediation, they cannot be considered to be "releaser pheromones" in the classic sense.[2] To meet these criteria, a chemical released to the environment by a conspecific animal should have the properties of (1) compound specificity (only certain chemicals should influence the response), (2) behavior specificity (only certain behaviors should be affected by the chemical[s]), (3) species specificity, and (4) "innateness" of response elicitation (Beauchamp, Doty, Moulton, and Mugford 1976; Goldfoot 1981). None of these properties clearly fits the data collected. Although several articles published in the last decade concerning rhesus monkey sex pheromones persuasively demonstrated a sexually communicatory function for vaginal olfactory substances, problems of interpretation of the findings arose concerning (1) the generality of the phenomenon, (2) the primacy of odor as *the* most important channel conveying attractiveness of a female, (3) evidence of the system being pheromonal as opposed to alternative possibilities, for example, learned associations, and (4) evidence of aliphatic acids being the active components of the vaginal secretions for olfactory mediation of sexual arousal (Goldfoot, Goy, Kravetz, and Freeman 1976a; Rogel 1978). Moreover, several demonstrations illustrate that copulation is not prevented by an inability to detect odors. Anosmic males continue to copulate with females in the laboratory (Michael and Keverne 1968; Goldfoot et al. 1978) and under semi-field conditions (A. Kling, personal communication); males copulate to ejaculation with selected ovariectomized, untreated females or with other males, both of which lack vaginal "pheromones" (Goldfoot 1981); and males often fail to copulate with females who are odor-labeled with mid-cycle vaginal products or with mixtures of short-chain fatty acids, the putative active pheromone (Goldfoot et al. 1976b; Michael, Zumpe, Richter, and Bonsall 1977). Again, this evidence does not imply that odors have no effect, but only that mating and conception can be accomplished without them, and sometimes without evidence that even quantitative behavioral changes occur in their absence (Goldfoot et al. 1978).

Recent experiments have shown that additional interpretive caution is necessary for laboratory experiments that have tested vaginal odorants or the putative active components when no comparable odorous control substance has been utilized. Most odor transfer studies of this nature have employed a pretest, odor test, or post-test design in which only water or diethyl ether has been used as a control substance. Under specific labora-

---

[2] This position does not deny the possibility of "primer pheromones" in this species. This chapter does not consider that issue, which is quite separate from questions of stimuli that directly influence the sexual motivation of the male.

tory conditions, however, sexual arousal was stimulated in six of eight males by the application of odorants entirely unrelated to vaginal products (Goldfoot and Edwards, unpublished, cited in Goldfoot 1981). The increased arousal was equivalent to that obtained with aliphatic acids in the same study and appears to be a phenomenon of disinhibition or a novelty response, perhaps not unlike similar increases in arousal when a new sexual partner is offered (Michael and Zumpe 1978) or when a mirror is placed in the cage (Wilson 1972). This phenomenon of nonspecific olfactory sexual arousal occurs readily in pair-test situations of intact males with ovariectomized females, but only after the same pair has been repeatedly tested and has reached a low asymptote of performance. Arousal effects from olfactants have not occurred in complex social situations where one or more females in a troop of 40 to 45 individuals housed at the Vilas Zoo, Madison, Wisconsin, has been odor-labeled with either novel odorants or aliphatic acids (Goldfoot 1981). Thus, the specific conditions and ultimate function of eliciting a male's increased sexual interest in a female that suddenly smells "different" has not yet been fully resolved.

FEMALES WITHOUT A VAGINA: STUDIES OF PSEUDOHERMAPHRODITES

Studies currently being performed in our laboratory with five prenatally androgenized female rhesus monkeys (pseudohermaphrodites) also serve to highlight the problems of equating vaginal odorants with sexual attractiveness. Females exposed to appropriate amounts of testosterone propionate prenatally are born with no external vagina whatsoever, and yet they are as successful as, and often *more* successful than, normal females at recruiting male mounts and mount attempts. This result has been obtained in situations in which a simultaneous choice of a normal female or a "vaginaless" female was available to the male and in which pseudohermaphrodites and normal females were presented to males sequentially. We believe that behavioral differences, not odor differences, account for the differential "attractiveness" of the pseudohermaphrodites (Goy, Goldfoot, Abbott, and Thornton, study in progress).

Moreover, when tested daily with a male partner throughout an ovulatory cycle, pseudohermaphrodites appear to receive the highest frequencies of mounts at mid-cycle (pseudohermaphrodites have functional ovaries) and the lowest frequencies during the luteal phase (R. W. Goy and S. Eisele, personal communication). Once, again, behavioral tendencies of the pseudohermaphrodite are correlated with these cyclic variations in male performance. At the risk of being overly redundant, it is nonetheless necessary to emphasize that these experiments do not deny a role of olfactory mediation during sexual interactions with normal females, but they clearly show that vaginal odorants need not be present under laboratory conditions to coordinate sexual arousal.

POSSIBLE ROLE OF LEARNED ODOR ASSOCIATIONS

Males reared to adulthood with no opportunities to interact with females (isosexual rearing, Goldfoot and Wallen 1978) showed no special attention to environmental surfaces marked with vaginal products (Goldfoot 1981). Adolescent males that had been reared with females but had never copulated to ejaculation similarly did not attend differentially to vaginal odors over control odors placed on environmental surfaces. In contrast, sexually experienced adolescent and adult males did spend significantly more time sniffing these odorants, therefore suggesting that prior sexual experience is perhaps necessary for these odors to acquire special significance in this species. Thus far, experiments in our laboratory designed to specifically teach experienced males to associate a novel odor to a sexually preferred female have been equivocal. Whereas two males have learned the task, six others have not; consequently, we are unsure of the statistical reliability and validity of these findings (D. A. Goldfoot, unpublished observations).

A MULTIDIMENSIONAL MODEL OF SEXUAL COMMUNICATION

The male and female rhesus monkey coordinate their reproductive behavior within a framework of high social complexity, in which they are attentive to the behavior of each other as well as to others in their social troops. Factors of dominance rank, matrilineage, and social experience have all been suggested to have direct or indirect influence on sexual outcomes. Attraction to, and selection of, a potential mate, in other words, demands much social learning and experience. Although most males achieve physiological puberty at 4 years of age, it might be many years before they successfully attain social adulthood and participate in breeding within the troop.

Analyses of the communication subserving mating, to be comprehensive, must by necessity consider the interplay of a wide variety of social processes and channels of information. Models of olfactory attraction, although contributing to our appreciation of one such channel, cannot be accurately interpreted until efforts are made to show the relationship of that communicative element to several others. In addition, use of the pheromone concept could keep investigators from exploring alternative hypotheses by which olfaction might influence sexual behavior.

It could be the case, for example, that males, rather than possessing an "innate" sexual arousal to specific vaginal odorants, have learned to associate changes in the odor characteristics of their partners with behavioral characteristics that signify a readiness to copulate. Perhaps there is an optimal time for this learning to occur, for example, at the time of the first sexually complete episodes experienced by the male. It is also possible that learned associations to certain odors may be acquired more efficiently with certain kinds of olfactants, thus offering the possibility of a "preparedness" model of learning these associations. Neither of these possibilities has been addressed experimentally in nonhuman primate studies. In addition, the odor novelty phenomenon referred to earlier might be more than a laboratory artifact, it could be related to changes in behavior of the male when the odor character of females changes rather abruptly at the beginning of the seasonal breeding period.

It is clear that a broad and inclusive theoretical model that considers the highly developed associative learning capabilities of this species in addition to the more traditional ethological constructs is needed. In addition, substantial data are needed to determine precisely how and when the various channels of information available to the animal are processed and acted upon. To discount the behavior of the female as an important source of information in these interactions seems not only counterintuitive but also highly unlikely from both evolutionary and experimental viewpoints.

ROLE OF ODOR IN THE MULTIPLE-CHANNEL MODEL

A hypothesized (but still simplified) scenario of how odor might be involved in sexual communication in this species follows: As the breeding season commences, gonadal activity of both adult males and females increases, and visible sex-skin coloration and size alterations take place in both sexes. The details of such sexual coordination, that is, which sex becomes reproductively active first, are not yet known and constitute an active research area for several laboratories (see Vandenbergh and Drickamer 1974). In addition to sex-skin changes, the production of ovarian estrogens induces the synthesis of many volatile vaginal substances, including short-chain fatty acids, which are believed to be products of vaginal flora (Michael and Bonsall 1977). Ovarian hormones also induce changes in the social behavior of females, which serve to reduce male–female distance and to encourage male courtship. The combined behavioral and peripheral changes undoubtedly orient the male to attend to the female, especially to her genitalia. The role of previous social and sexual experience appears to be particularly important at this point. If the male has learned previously to associate copulatory success with the visual and odor qualities of his partner, the peripheral cues would be expected to increase sexual arousal and perhaps even influence transient rises in testosterone of the male. (Preliminary data from talapoins do not support this likelihood, however; Eberhart et al. 1980.) It is also possible that odor cues might begin to be detected by males even prior to the time that the female becomes overtly proceptive, although again direct evidence to this point is either not available or somewhat contradictory (see discussion of Vilas Zoo study, this chapter).

In addition to these stimulus conditions, the male must assess his and his potential partner's social situation, since from previous experience he might have learned that he will not be allowed to copulate if he is within view of a dominant and aggressive male. In turn, the female probably directs her proceptive behavior toward specific males and ignores or excludes others. She might, in fact, aggress one male while simultaneously soliciting another, a strategy that accomplishes for her additional sexual arousal from her chosen partner while reducing the chances of her mating with a nonpreferred male. A consort relationship (short-term pair bond) might result at this time, with repeated copulation occurring for several hours up to several days.

Laboratory studies suggest that a female's proceptive and receptive behavior changes during the menstrual cycle with peak displays indicating readiness to mate during the fertile mid-cycle phase. If that specific mid-cycle condition coincides with a discriminable odor quality, it might be an additional factor that would increase copulatory probabilities to ensure conception. This hypothesized odor quality at mid-cycle might also play a role in determining which particular male copulates with a female during her fertile phase. For example, a specific odor cue that identifies the fertile phase of a female's cycle, and, coincidentally, the time when she is most proceptive and receptive, might be utilized by dominant males to form sexual relationships that would likely result in pregnancy. However, this particular point, that is, whether a unique odor cue develops during the fertile phase of the cycle, is not yet clear. Goldfoot and co-workers (1976b) found that aliphatic acids (the putative ''pheromone'') are highest in vaginal lavages during the luteal phase of the cycle, several days after ovulation, when conception would be unlikely. Although Michael and Bonsall (1977) have reported similar observations, that is, that the highest levels of aliphatic acids appear several days after ovulation, they suggest that viscocity changes of vaginal secretions at mid-cycle might make more of these odorants detectible to males during the fertile period. Keverne (1976b) has suggested that additional chemical components enhance the stimulatory properties of the previously identified aliphatic acids (i.e., phenylpropanoic and parahydroxyphenylpropanoic acids), but the quantitative analyses of these substances throughout the cycle have not been reported. Preliminary results with human judgement of odor qualities of rhesus vaginal secretions (Goldfoot 1981) raised the possibility that other odors, probably not aliphatic acids, might indeed be present only at mid-cycle. In short, more effort and data are needed before this important issue can be clarified.

After ovulation has occurred, copulation frequencies typically decrease and reach fairly low levels during the luteal phase. Both behavioral and nonbehavioral (e.g., odor) hypotheses have been suggested for this decline (Baum, Keverne, Everitt, Herbert, and deGreef 1977; Michael, Bonsall and Zumpe 1978). Michael and Bonsall (1977) and Baum et al. (1977) have raised the interesting possibility that a negative odor cue might be a contributing factor in these cases, but this hypothesis has not been exhaustively tested nor confirmed as yet.

Regardless of the eventual solutions to the several questions raised with regard to odor cues, all researchers apparently agree that the sexual coordination of rhesus monkeys will not be explained exclusively by classical pheromonal models but rather by detailed analyses of the sequences and interactions of all types of information exchange that characterize this complicated, highly social species.

REFERENCES

Baum, M. J., Keverne, E. B., Everitt, B. J., Herbert, J., and de Greef, W. J. 1977. Effects of progesterone and estradiol on sexual attractivity of female rhesus monkeys. *Physiology and Behavior* 18:659–70.

Beach, F. A. 1942. Analysis of the stimuli adequate to elicit mating behavior in the sexually inexperienced male rat. *Journal of Comparative Psychology* 33:163–207.

1944. Experimental studies of sexual behavior in male mammals. *Journal of Clinical Endocrinology and Metabolism* 4:126–34.

1976. Sexual attractivity, proceptivity and receptivity in female mammals. *Hormones and Behavior* 7:105–38.

Beauchamp, G. K., Doty, R. L., Moulton, D. G., and Mugford, R. A. 1976. The pheromone concept in mammalian chemical communication: a critique. In *Mammalian olfaction, reproductive processes and behavior,* ed. R. L. Doty, pp. 143–60. New York: Academic Press.

Brooks, C. McC. 1937. The role of the cerebral cortex and of various sense organs in the excitation and execution of mating activity in the rabbit. *American Journal of Physiology* 120:544–53.

Carpenter, C. R. 1942. Sexual behavior of free-ranging rhesus monkeys (*Macaca mulatta*). II. Periodicity of oestrus, homosexual, autoerotic and nonconformist behavior. *Journal of Comparative Psychology* 33:143–62.

Czaja, J. A., and Bielert, C. F. 1975. Female rhesus sexual behavior and distance to a male partner: relation to stage of the menstrual cycle. *Archives of Sexual Behavior* 4:583–97.

Czaja, J. A., Eisele, S. G., and Goy, R. W. 1975. Cyclic changes in the sexual skin of female rhesus: relationships to mating behavior and successful artificial insemination. *Proceedings of the Federation of American Societies for Experimental Biology* 34:1680–4.

Deputte, B. L., and Goustard, M. 1980. Copulatory vocalizations of female macaques (*M. fascicularis*): variability factors analysis. *Primates* 21:83–99.

Eberhart, J. A., Keverne, E. B., and Meller, R. E. 1980. Social influences on plasma testosterone levels in male talapoin monkeys. *Hormones and Behavior* 14:247–66.

Goldfoot, D. A. 1971. Hormonal and social determinants of sexual behavior in the pigtail monkey (*Macaca nemestrina*). In *Normal and abnormal development of brain and behaviour,* ed. G. B. A. Stoelinga and J. J. van der Werff ten Bosch, pp. 325–42. Leiden: Leiden University Press.

1977. Sociosexual behaviors of nonhuman primates during development and maturity: social and hormonal relationships. In *Behavioral primatology: advances in research and theory,* Vol. 1, ed. A. M. Schrier, pp. 139–84. Hillsdale, N.J.: Erlbaum.

1981. Olfaction, sexual behavior and the pheromone hypothesis in rhesus monkeys: a critique. *American Zoologist* 21:153–64.

Goldfoot, D. A., and Wallen, K. 1978. Development of gender role behaviors in heterosexual and isosexual groups of infant rhesus monkeys. In *Recent advances in primatology* Vol. 1: *Behaviour. Proceedings of the Sixth Congress of the International Primatological Society,* ed. D. J. Chivers and J. Herbert, pp. 155–9. New York: Academic Press.

Goldfoot, D. A., Essock-Vitale, S. M., Asa, C. S., Thornton, J. E., and Leshner, A. I. 1978. Anosmia in male rhesus monkeys does not alter copulatory activity with cycling females. *Science* 199:1095–6.

Goldfoot, D. A., Goy, R. W., Kravetz, M. A., and Freeman, S. K. 1976a. Reply to Michael, Bonsall and Zumpe; reply to Keverne. *Hormones and Behavior* 7:373–8.

Goldfoot, D. A., Kravetz, M. A., Goy, R. W., and Freeman, S. K. 1976b. Lack of effect of vaginal lavages and aliphatic acids on ejaculatory responses in rhesus monkeys: behavioral and chemical analyses. *Hormones and Behavior* 7:1–27.

Goldfoot, D. A., Neff, D. A., and Edwards, S. D. In preparation. Individual differences in proceptive sexual behavior of ovariectomized rhesus monkeys: base rates prior to treatment predict steroid effectiveness.

Goy, R. W., Goldfoot, D. A., Abbott, D., and Thornton, J. In preparation. Sexual proceptivity of adult pseudohermaphroditic rhesus monkeys.

Green, S. 1975. Variation in vocal pattern with social situation in the Japanese monkey (*Macaca fuscata*): a field study. In *Primate behavior,* Vol. 4: *Developments in field and laboratory research,* ed. L. A. Rosenblum, pp. 1–102. New York: Academic Press.

Herbert, J., and Trimble, M. R. 1967. Effect of oestradiol and testosterone on the sexual receptivity and attractiveness of the female rhesus monkey. *Nature* 216:165–6.

Johnson, D. F., and Phoenix, C. H. 1976. The hormonal control of female sexual attractiveness, proceptivity and receptivity in rhesus monkeys. *Journal of Comparative and Physiological Psychology* 90:473–83.

Keverne, E. B. 1976a. Sexual receptivity and attractiveness in the female rhesus monkey. In *Advances in the study of behavior,* Vol. 7, ed. J. S. Rosenblatt, R. A. Hinde, E. Shaw, and C. Beer, pp. 155–96. New York: Academic Press.

1976b. Sexual attractants in primates. *Journal of the Society of Chemists* 27: 257–69.

Keverne, E. B., and Michael, R. P. 1971. Sex attractant properties of ether extracts of vaginal secretions from rhesus monkeys. *Journal of Endocrinology* 51:313–22.

Lindburg, D. G. 1971. The rhesus monkeys in North India: an ecological and behavioral study. *Primate Behavior,* Vol. 2, ed. L. A. Rosenblum, pp. 1–106, New York: Academic Press.

Michael, R. P. 1968. Gonadal hormones and the control of primate behaviour. In *Endocrinology and human behaviour,* ed. R. P. Michael, pp. 69–93. London: Oxford University Press.

1972. Determinants of primate reproductive behavior. *Acta Endocrinologica, Supplement* 166:322–61.

Michael, R. P., and Bonsall, R. W. 1977. Chemical signals and primate behavior. In *Chemical signals in vertebrates,* ed. D. Müller-Schwarze and M. M. Mozell, pp. 251–71. New York: Plenum.

Michael, R. P., and Keverne, E. B. 1968. Pheromones and the communication of sexual status in primates. *Nature* 218:746–9.

Michael, R. P., and Welegalla, J. 1968. Ovarian hormones and the sexual behaviour of the female rhesus monkey (*Macaca mulatta*) under laboratory conditions. *Journal of Endocrinology* 41:407–20.

Michael, R. P., and Zumpe, D. 1977. Effects of androgen administration on sexual invitations by female monkeys (*Macaca mulatta*). *Animal Behaviour* 25: 936–44.

Michael, R. P., and Zumpe, D. 1978. Potency in male rhesus monkeys: effects of continuously receptive females. *Science* 200:451–3.

Michael, R. P., Bonsall, R. W., and Zumpe, D. 1978. Consort bonding and operant behavior by female rhesus monkeys. *Journal of Comparative and Physiological Psychology* 92:837–45.

Michael, R. P., Saayman, G. S., and Zumpe, D. 1967. Sexual attractiveness and receptivity in rhesus monkeys. *Nature* 215:554–6.

Michael, R. P., Zumpe, D., Keverne, E. B., and Bonsall, R. W. 1972. Neuroendocrine factors in the control of primate behavior. *Recent Progress in Hormone Research* 28:665–706.

Michael, R. P., Zumpe, D., Richter, M., and Bonsall, R. W. 1977. Behavioral effects of a synthetic mixture of aliphatic acids in rhesus monkeys (*Macaca mulatta*). *Hormones and Behavior* 9:296–308.

Perachio, A. A. 1978. Hypothalamic regulation of behavioral and hormonal aspects of aggression and sexual performance. In *Recent advances in primatology, Vol. 1: Behaviour. Proceedings of the Sixth Congress of the International Primatological Society,* ed. D. J. Chivers and J. Herbert, pp. 549–64. New York: Academic Press.

Rogel, M. J. 1978. A critical evaluation of the possibility of higher primate reproductive and sexual pheromones. *Psychological Bulletin* 85:810–30.

Rowell, T. E. 1972. Female reproduction cycles and social behavior in primates *Advances in the Study of Behavior,* Vol. 4, ed. D. S. Lehrman, R. A. Hinde, and E. Shaw, pp. 69–105, New York: Academic Press.

Slob, A. K., Wiegand, S. J., Goy, R. W., and Robinson, J. A. 1978. Heterosexual interactions in laboratory-housed stumptail-macaques (*Macaca arctoides*): observations during the menstrual cycle and after ovariectomy. *Hormones and Behavior* 10:193–211.

Trimble, M. R., and Herbert, J. 1968. The effect of testosterone or oestradiol upon the sexual and associated behaviour of the adult female rhesus monkey. *Journal of Endocrinology* 42:171–85.

Vandenbergh, J. G., and Drickamer, L. C. 1974. Reproductive coordination among free-ranging rhesus monkeys. *Physiology and Behavior* 13:373–6.

Wallen, K., and Goy, R. W. 1977. Effects of estradiol benzoate, estrone, and the propionates of testosterone or dihydrotestosterone on sexual and related behaviors of ovariectomized rhesus monkeys. *Horomones and Behavior* 9:228–48.

Wilson, M. I. 1972. The hormonal control of sexual behavior in the rhesus monkey (*Macaca mulatta*). Doctoral dissertation, University of London.

Zumpe, D., and Michael, R. P. 1970. Redirected aggression and gonadal hormones in captive rhesus monkeys (*Macaca mulatta*). *Animal Behaviour* 18:11–19.

Author index

Subject index

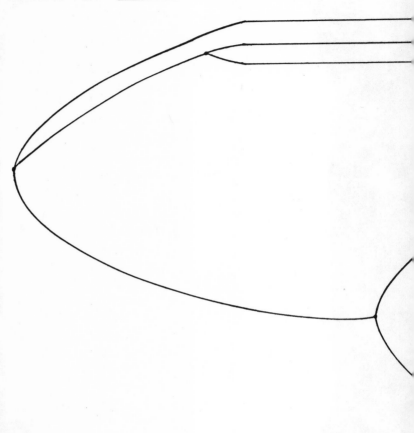

70   60   50   40   30

Millions of Year